D0991342

Francis Bacon

The Advancement of Learning
and
New Atlantis

Francis Bacon

The Advancement of Learning *and* New Atlantis

Edited by **Arthur Johnston**

1974 Clarendon Press · Oxford

Oxford University Press *Ely House, London W.1*

Glasgow New York Toronto Melbourne Wellington
Cape Town Ibadan Nairobi Dar es Salaam Lusaka Addis Ababa
Bombay Calcutta Madras Karachi Lahore Dacca Delhi
Kuala Lumpur Singapore Hong Kong Tokyo

Casebound ISBN 0 19 871035 6
Paperback ISBN 0 19 871069 0

Made and printed in Great Britain by
Richard Clay (The Chaucer Press) Ltd
Bungay, Suffolk

Contents

Introduction

When Bacon looked round the world of learning in his time his comment on it was, 'the last thing anyone would be likely to entertain is an unfamiliar thought'.[1] And Bacon was very conscious that he himself was obsessed by an unfamiliar thought, which gave rise to conclusions that few of his contemporaries seemed willing to listen to. 'I yearn for some secret initiation,' he wrote, 'which, like the coming of April or of Spring, might avail to thaw and loosen your fixed and frozen minds.'[2] All his philosophical writings (mainly in Latin) were attempts to break through 'frozen minds' and explain his radically new idea.

The idea was frighteningly simple, that man was to be 'so firm of mind and purpose as resolutely to compel himself to sweep away all theories and common notions, and to apply the understanding, thus made fair and even, to a fresh examination of particulars'.[3] 'There is no hope,' he wrote, 'except in a new birth of science; that is, in raising it regularly up from experience and building it afresh; which no one (I think) will say has yet been done or thought of.'[4] By a close attention to the particulars of the natural world, Bacon hoped 'to restore and exalt the power and dominion of man himself, of the human race, over the universe'.[5] He hoped to provide men with a method whereby they could make 'better use and management of the human mind'.[6] He confidently expected that the product of his method would be an enlargement of human knowledge of God's works. But knowledge meant power—the power to improve man's lot by means of useful inventions.

If [he wrote], the utility of some one particular invention so impresses men that they exalt to superhuman rank the man who is responsible for it, how much more noble would that discovery be which should contain within itself the potentiality of all particular inventions, and open up to the human spirit a path of direct and easy access to new remoter powers.[7]

To understand the book of God's works, created Nature, was as important for Bacon as to understand the book of his word, the

[1] *Cogitata et Visa* ('Thoughts and Conclusions'), *c.* 1607, trs. B. Farrington in *The Philosophy of Francis Bacon* (Liverpool, 1964), p. 79. Cf. *Novum Organum* i. 90.

[2] *Redargutio Philosophiarum* ('Refutation of Philosophies'), *c.* 1608, in Farrington, p. 128.

[3] *Novum Organum* i. 97. [4] Ibid. [5] Farrington, pp. 92–3.

[6] Ibid. [7] Ibid.

Bible. He did not believe that the two kinds of knowledge would contradict each other. And, he maintained, 'the rule of religion, that a man should show his faith by his works, holds good in natural philosophy too. Science also must be known by works ... The improvement of man's mind and the improvement of his lot are one and the same thing.'[8]

Bacon clearly felt that in entertaining such unfamiliar ideas he was isolated. He saw himself as 'mild and exceedingly placid', as 'one that pitied men', like the Father in the House of Salomon. But, he lamented in 1608, 'I have not even a person with whom I can converse without reserve on such subjects ... why, I met not long ago a certain evil-eyed old fortune-telling woman, who ... prophesied that my offspring should die in the desert.'[9] His radical questioning of Aristotelian science was, he felt, not only unacceptable in itself, but interpreted as 'a disturbing and revolutionary influence' in the state. His 'enthusiasm and non-conformity' were 'a great obstacle' when he came to choose a career.[10] Since Bacon needed a career in law and politics, he had to overcome great prejudice, and was at pains to distinguish between innovations in politics and in science. In politics he was orthodox and authoritarian, 'for political control rests on authority, assent, reputation, opinion, not on demonstration and on truth. But in the arts and sciences, as in mining for minerals, there ought everywhere to be the bustle of new works and further progress.'[11] Thus, when he portrays the political side of his utopia in the *New Atlantis*, he stresses the orderliness of society, characterized by religious and racial toleration, by religious observation, by the incorruptibility of its servants, its honouring of chastity and of family life, its dignity, courtesy, gentleness, and acceptance of hierarchy. He shows that the scientific radicalism of that society has not undermined its political stability, but reinforced it and given it a firm base, as well as providing it with rich consumables.

Science, for Bacon, was innocent knowledge. While he saw himself as an Alexander, daring, in all humility, 'to despise shams',[12] his ideal figure was Solomon. A passage in the *Cogitata et Visa*, probably written two years later than *The Advancement of Learning*, explains the attraction that the figure of Solomon had for him.

[8] Ibid.

[9] *Redargutio Philosophiarum*, in *Works* iii. 559, trs. in *Works* (ed. B. Montagu, 1825–34), xv. 120. Cf. Farrington, p. 104.

[10] Farrington, p. 79. Cf. Preface to *De Interpretatione Naturae*, 1603, in *Works* x. 84 f., and *Novum Organum* i. 62.

[11] Farrington, p. 79.

[12] Farrington, p. 132; *Novum Organum* i. 97.

In the Scriptures, King Solomon, though blessed with empire, gold, splendour of architecture, satellites, servants, ministers and slaves of every kind and degree, with a fleet to boot, and a glorious name and with the flattering admiration of the world, yet prided himself on none of these things. Instead he declared that *It was the glory of God to conceal a thing, the glory of a King to find it out*; as if the divine nature enjoyed the kindly innocence of such hide-and-seek, hiding only in order to be found, and with characteristic indulgence desired the human mind to join Him in this sport. And indeed it is this glory of discovery that is the true ornament of mankind. In contrast with civil business it never harmed any man, never burdened a conscience with remorse. Its blessing and reward is without ruin, wrong or wretchedness to any. For light is in itself pure and innocent; it may be wrongly used, but cannot in its nature defile.[13]

Solomon represented kingly power and a rich, dignified, hedonistic way of life, combined with wisdom, good hierarchic government, legal acumen, and scientific knowledge. His book of Natural History (the ideal of the work Bacon longed to compile) may not have survived, but the Bible text assured Bacon that scientific investigation was innocent and kindly. To pursue it was to accept God's invitation to a game of hide-and-seek, from which would come 'the knowledge of Causes and secret motions of things, and the enlarging of the bounds of Human Empire, to the effecting of all things possible'. As a politician, Bacon was always aware of the danger to the state from economic, social, and religious change. His aim was to direct human energies, 'hard work on an exacting business' into a new channel, that of scientific investigation and invention. 'For the benefits inventors confer extend to the whole human race,' he wrote. He went on, 'And there is this too. Inventions come without force or disturbance to bless the life of mankind, while civil changes rarely proceed without uproar and violence.'[14]

Bacon was aware that in the previous century men had discovered a 'new world' across the Atlantic, and a 'new' learning in the recovered Greek and Latin texts. It would, he thought, 'be a disgrace for mankind if the expanse of the material globe, the land, the seas, the stars, were opened up and brought to light, while, in contrast with this enormous expansion, the bounds of the intellectual globe should be restricted to what was known to the ancients'.[15] Paradoxically, what had been discovered was 'old', 'new' only in the sense

[13] Farrington, p. 92.
[14] Farrington, pp. 91–2. Cf. *Novum Organum* i. 129.
[15] Farrington, p. 94; *Novum Organum* i. 84; *The Advancement of Learning* II. ii. 14.

that it was newly seen by modern men; it had been there a long time. Men were conditioned by these experiences to think of 'discovery' as 'recovery'. In considering how to present his ideas to 'frozen minds' Bacon was willing to use this attitude of mind. In *The Wisdom of the Ancients* (1609) he presented his political and scientific ideas in the form of his interpretations of classical myths, as though the 'new' knowledge were a rediscovery of what had been long lost. The geographical discoveries of a new world presented him with a persuasive image for the intellectual adventurer. So his *New Atlantis* is a story of another geographical discovery in the Pacific. He presents the civilization that is found as being contemporaneous with his own. But its origins are in the far past. Cut off from Hebrew and Greek influences at a distant period, it had first developed its science; later it had been given a special revelation of Christianity and had recognized and incorporated this. It had never suffered the stultifying influence of Plato and Aristotle and the Schoolmen. It is pre-Socratic, Hebrew, scientific, and Christian. Its greatest glory is its House of Salomon or College of the Six Days' Works, a scientific institute established about 300 B.C., in describing which Bacon is presenting a model for the research institute he hoped would be established in England.[16] Although he always looked forward to easing man's lot by producing new inventions, he was not so sanguine as to expect success in this himself. His aim was to encourage men to investigate nature, and his account of Salomon's House is his description of the organization and method he advocated. It is based on a division of scientific labour into collectors of experiments, experimenters, abstracters, and compilers, men who investigate the practical application of discoveries, men who direct new research, those who follow it out, and those who extract fundamental principles from all the work. This is Bacon's dream, and the *raison d'être* of the *New Atlantis*. What such an institute would discover and invent, he does not know—'the knowledge that we now possess will not teach a man even what to *wish*'.[17] But his teams of scientists would work on refrigeration, gardening and agriculture, longevity, brewing and baking, furnaces, light, music, sound, perfumes, engines and motion, mathematics, and would develop microscopes, telescopes, perpetual motions, submarines, and flying

[16] The germ of this institute is in the *Gesta Grayorum* of Christmas 1594 (*Works* viii. 335) and the *Comentarius Solutus* of 26 July 1608 (*Works* xi. 66). Bacon may have written, or revised, the *New Atlantis*, between April and October 1623, when he had hopes that he might become Provost of Eton, and so obtain control of an educational establishment.

[17] Preface to *De Interpretatione Naturae*, in *Works* x. 87.

machines. It was on such topics that he collected his series of one thousand experiments in the *Sylva Sylvarum* ('Forest of Materials') which his chaplain Rawley published with the *New Atlantis* in 1627.

The utopian fable was one way of conveying to his resistant contemporaries the revolutionary idea that Bacon cherished; it does not argue what should be done, but shows a society in which all has been accomplished. He imagined himself as a supreme 'director of studies' at the head of a research institute. But is is clear that the Fellows of the House of Salomon are a select few; Bacon's programme does not require, as Sir Thomas Bodley feared, that every man should 'renounce our common notions, and cancel all our theorems, axioms, rules and tenets, and so to come babes "to the kindom of nature", as we are willed by scriptures to come "to the kingdom of heaven"'. To accept Bacon's ideas, Bodley feared, 'would instantly bring us to barbarism'.[18] Bacon's answer, in the Preface to his *Novum Organum* (1620), was that most men would continue to adhere to the accepted philosophy, which would always be useful 'for supplying matter for disputations or ornaments for discourse, for the professor's lecture and for the business of life' (cf. *Novum Organum* i. 128).

Eighteen years earlier, however, he had not been able to adopt such an urbane attitude. Disappointed in his hopes of obtaining office under the new king, in 1603 he turned seriously to his philosophic work. 'I see signs in the times,' he wrote, 'of the decline and overthrow of that knowledge and erudition which is now in use.' He feared civil wars, religious controversy, the proliferation of speciously easy aids to learning, and did not doubt but that 'that fair-weather learning which is nursed by leisure, blossoms under reward and praise, which cannot withstand the shock of opinion, and is liable to be abused by tricks and quackery, will sink under such impediments as these'.[19] He was convinced that the true learning which he wished to introduce would survive in adverse conditions. But he was not willing that any learning should die. He therefore composed what is now Book I of *The Advancement of Learning*. This is a defence of learning by one who was not a narrow specialist, but who 'had taken all knowledge to be his province'. But it also includes a careful analysis of the internal weaknesses of modern learning which made it vulnerable both to the threats of social upheaval and to the scorn of the active man of affairs and the sus-

[18] Bodley's letter to Bacon, on reading *Cogitata et Visa*, 19 February 1608, in *Works*, ed. Montagu, xii. 85.

[19] *Works* x. 86, the Preface to 'Of the Interpretation of Nature'.

picious man of piety.[20] Bacon set himself the task of defending and praising not merely one branch of learning (as Sidney did with poetry and Alberti with painting) but all learning; and he did it aware that his deepest wish was to persuade men to abandon existing ways of thinking and the so-called knowledge that they thought they possessed. To the humanist doctrine as expressed by Ascham, 'Ye know not what hurt ye do to learning, that care not for words but for matter,' he sharply opposed, 'here therefore is the first distemper of learning, when men study words and not matter.'[21] For the humanist, sound doctrine and morality went hand in hand with eloquence. Bacon was no enemy to eloquence, but he was concerned to expand the bounds of knowledge, which the pursuit of eloquence would never accomplish. Even worse, for Bacon, was the practice of the Schoolmen, who were not renowned for fine writing, but for spinning elaborate systems and speculations out of intense working over of a small number of authors, chiefly Aristotle. 'Vain matter is worse than vain words.' If only their 'thirst for truth and unwearied travail of wit' had been exercised on 'matter, which is the contemplation of the creatures of God', they would have advanced learning and knowledge.

In *The Advancement of Learning* Bacon is trying not to frighten his reader with the startling newness of his ideas about learning, but he cannot resist interspersing his defence of learning in general (which he uses in a vague sense of the best that had been thought and written throughout recorded history) with brilliant attacks on humanists, Schoolmen, and system-builders, which point to his new concept of what is needed. He wants knowledge to be 'in growth', as he felt it would be while it was presented 'in aphorisms and observations'. He felt strongly that for 2,000 years knowledge had been at a stand-still, 'polished, illustrated and accommodated for use and practice' by ingenious minds, but not advanced.

In his more general defence and praise of learning he is moved by the same sense of its dignity and value that inspired Spenser and Samuel Daniel.[22] Though he denies that he is making 'a laudative of learning' or a 'hymn to the Muses', he rises to eloquence at the end of Book I. But his main purpose is to persuade by accumulation of

[20] See G. Bullough, 'Bacon and the Defence of Learning', in *Seventeenth-Century Studies presented to Sir Herbert Grierson* (Oxford, 1938), pp. 1–20.

[21] Roger Ascham, *The Scholemaster* (1570), ed. E. Arber (1897), p. 118; *The Advancement of Learning* I. iv. 3.

[22] Daniel's *Musophilus* (1599) is a verse dialogue in defence of learning. It is dedicated to a friend of Bacon's, Fulke Greville, whose *Treatie of Humane Learning* (probably written *c.* 1620–2) is in part a reply to Bacon's *Advancement*.

arguments and examples in a carefully ordered sequence, and to associate the learned man in his reader's mind with the traditionally accepted types of hero—with the warrior, the statesman, the ruler, the founder of a dynasty, the extender of empire, with the virtuous man and the benefactor, with power, command, empire, and the longing for immortality.[23] There will always be those who 'prefer custom and habit before all excellency', 'the barley-corn before the gem'; but Bacon hoped that his message would find 'fit audience, though few'.

His second book was probably written during the long adjournment of Parliament from January to October 1605 and is more directly addressed to the new king, 'the learnedest king that hath reigned'. At the outset he discusses the Universities, the area in which James might intervene to introduce reforms. No one had absorbed more thoroughly than Bacon the traditional education, which he scornfully described as 'a conducted tour through a potrait-gallery of the ancients'.[24] But the continuance of such an educational programme he felt to be inadequate. 'It would disgrace us, now that the wide spaces of the material globe, the lands and seas, have been broached and explored, if the limits of the intellectual globe should be set by the narrow discoveries of the ancients.'[25] He therefore proposes various practical measures of reform—the provision of books and laboratory equipment; the attraction of the best teachers by higher salaries; links with continental universities; the careful review of the syllabus; the addition of new subjects to the existing professional courses in divinity, law, and medicine, in particular, history, modern languages, politics, and such general philosophical subjects as have no immediate practical value (he is thinking of scientific investigation); the removal of logic and rhetoric from the beginning to the end of the undergraduate course; the appointment of research fellows to investigate those parts of learning so far but thinly explored. These ideas were ignored by James, but were taken up excitedly in the middle of the seventeenth century, though still without practical results.[26]

Bacon devotes the rest of the book to a survey of the whole body of human knowledge. It is a stock-taking before setting out on a voyage, and a charting of the way. In effect Bacon is saying, 'this is our store', and 'in what directions have we so far failed to search for

[23] See J. M. Steadman, 'Beyond Hercules: Bacon and the Scientist as Hero', *Studies in the Literary Imagination*, 4 (Atlanta, 1971), 3–47.

[24] Farrington, p. 120.
[25] Farrington, p. 131. Cf. p. ix above.

[26] See R. F. Jones, *Ancients and Moderns* (St. Louis, 1936), and C. Webster (ed.), *Samuel Hartlib and the Advancement of Learning* (Cambridge, 1970).

new knowledge'? He is not compiling an encyclopaedia of learning, but is asking the question, 'what different kinds of learning is it possible for men to be interested in'? His aim is to omit no feasible and desirable branch of study. With each subject he declares whether it has been adequately pursued by learned men in the past, and indicates whether there are any aspects of it that need further investigation. Often he pauses and gives suggestions how the desired topic ought to be investigated.[27] Most important for his purpose was a method of surveying knowledge which would not commit him to some existing system of thought, and which would ensure that he did not omit any branch—in particular the branches which did not exist but which he thought desirable. Some system of classifying the sciences was essential. There were traditional models available. Medieval and Renaissance encyclopedists organized their accounts of existing knowledge according to some scheme—the six days of Creation,[28] or the seven liberal arts (grammar, rhetoric, and logic, the Trivium of the university curriculum, and arithmetic, geometry, astronomy, and music, the Quadrivium). But these, Bacon would have felt, were the worst possible ways to arrive at sciences not yet in existence. Nor would such classifications have made it possible for Bacon to demonstrate the simple inevitability of the new subjects of study. He is, after all, interested in the *advancement* of learning, and his aim is to persuade his reader.

Bacon is not concerned to compile an encyclopaedia of existing knowledge, but to classify and assess 'what parts thereof lie fresh and waste'. He could have organized his survey on the basis of the nature of knowledge, as, for example, his contemporary Tommaso Campanella (1568–1639) did.[29] For Campanella, knowledge is based on sensation; therefore the basis of all knowledge is history, either divine (from God's word, the Bible) or human (from God's work, nature). In this scheme theology is based on divine history, the natural and moral sciences on human history; the study of the principles that separate the parts of knowledge he calls metaphysics. Bacon would have agreed with this classification in its general terms. But he goes on to ask the question, 'how do we learn?', in addition to Campanella's question, 'from what do we learn?' And his answer is that we learn about things divine through the soul, and about

[27] This was part of his programme. The 'Description of the Intellectual Globe', c. 1612, is an attempt to show how astronomy should be studied.

[28] See V. K. Whitaker, *Francis Bacon's Intellectual Milieu* (Los Angeles, 1962).

[29] For a full survey of attempts to classify the sciences, see Robert Flint, *Philosophy as Scientia Scientiarum and a History of the Classifications of the Sciences* (1904).

natural phenomena through the faculties of the mind. Things divine and the nature of the soul he will leave to theologians to discuss; these are not matters that men have neglected to study—though Bacon cannot resist some account and comments. But natural phenomena and the means whereby men learn about them, i.e. the faculties of the mind, interest him profoundly. The concept of the faculties was of long standing; what is new is Bacon's decision to classify human learning in terms of what he regarded as the three basic faculties, Memory, Imagination, and Reason (see pp. 264–5 below).[30] For this he has been accused of adopting a 'subjective position', of classifying knowledge in terms which suggest that the rational soul exercises its faculties in isolation. But Bacon was aware of the danger to the advancement of learning, and of the distortions and inaccuracies, produced by rigid categories, however useful such categories might be in facilitating understanding and explanation. He warns his reader (II. ix. 1) 'that all divisions of knowledge be accepted rather for lines and veins, than for sections and separations; and that the continuance and entireness of knowledge be preserved'.[31]

Book II, which analyses all the identifiable branches of the tree of knowledge, exists as a unity. Its aim is to persuade the reader that the process of learning is not merely the acquiring of what is already known, but the investigating of areas that are as yet unexplored. It is a programme for mankind, in which Bacon is the guide, the herald, the trumpeter, calling others to the action, to occupy the new territory that he has charted. His function is to identify those areas of which he can say 'that this part of knowledge is wanting, to my judgment standeth plainly confessed'. Because Bacon is himself convinced that his scheme of classification is important, he rigorously pursues it throughout Book II. It has the double attraction for him of being a secular system, and of being based on what he accepted as psychological facts. He frequently explains his dislike of 'systems', the 'over-early and peremptory reduction of knowledge into arts and methods' (pp. 34, 135 below). He objects to the imposition of rational patterns on man's experience of the universe,

[30] Later on, when he discusses ethics, he adds other traditional faculties—understanding, appetite, will. For a full discussion, see K. R. Wallace, *Francis Bacon on the Nature of Man* (Urbana, 1967).

[31] Bacon would have been aware of the divisions of the sciences derived from Aristotle and the Stoics. The former divided into Speculative (physic, mathematic, theology or metaphysics), Practical (ethics, economic, politic), and Artistic (rhetoric, poetic), with logic as the tool necessary for all the sciences. The latter divided into Logic (including grammar and rhetoric), Ethics and Physics (including theology). See the last chapter of Locke's *Essay concerning Human Understanding* (1690).

calling these 'the Idols of the Theatre or Systems' (*Novum Organum* i. 62). The patterns may be Aristotelian or Platonist, scholastic or humanist, alchemical or magical; all are unacceptable to Bacon. 'For God forbid that we should give out a dream of our own imagination for a pattern of the world' (*Works* iv. 32–3). But his own love of system was, in spite of his denials, intense. The Great Instauration which he planned, with its six parts, is a 'system'—though Bacon believed that, because it was built on observed facts, it would be a 'true' system. Most of his contemporaries responded imaginatively to the divers elements that made up a coherent world-picture, an amalgam of medieval and Renaissance images and patterns that held in solution an explanation of the universe and man's place in it. Its essential elements—the theory of correspondences, the parallels of microcosm and macrocosm, the great chain of being—were patterns of images that Bacon could not accept, to which he did not respond, because they were static. His book is called *The Advancement of Learning*, his basic image that of a ship setting out into the open sea on a voyage of discovery. His greatness lies in his attempt to suggest a new way of understanding the world, which would lead to a new 'system', through images of movement. His main images are derived from his reading of the first two chapters of the book of Genesis—images of light, water, plants, seeds, fruit, the tree of knowledge, light and darkness, the river and the fountain, man naming the creatures, the garden. But these traditional images are experienced freshly by Bacon, because they are not for him part of a static system but part of a forward-moving sequence. For example, the first image in Genesis is of 'darkness over the face of the abyss' and a mighty wind sweeping over the surface of the waters, and 'God said, "Let there be light," and there was light; and God saw the light, that it was good'. This, for Bacon, becomes an image of light playing on water. It is represented pictorially on the title-page of the *Sylva Sylvarum* (1627) where, between the pillars of Hercules (representing the limits of geographical and, metaphorically, of intellectual adventure and discovery) is the globe of the intellectual world, resting on the sea. From above there shines Light, with the words 'And God saw the light, that it was good.' Light for Bacon is divine illumination, is knowledge, is the transmission of learning.[32] When he read the opening of Book vii of Virgil's *Aeneid* he responded to and remembered the image there of light shining on the water (p. 85 below), partly because this is one of the images to which he re-

[32] See B. Vickers, *Francis Bacon and Renaissance Prose* (Cambridge, 1968), pp. 189 f., and C. Spurgeon, *Shakespeare's Imagery* (Cambridge, 1965), pp. 17 f.

sponded deeply, partly because it comes in a context of Aeneas setting out in a ship for a voyage in which the winds blow him to his destined home. For Aeneas is the adventurer, the warrior, and the founder of an empire, and so an image for Bacon of his own intellectual ideal.

Another of the significant images from Genesis to which Bacon responded was that of water, in particular the reference in Genesis 2: 5–6 to the two sources of life-giving water, rain and the fountain or spring. It becomes for him an image of knowledge, 'The knowledge of man is as the waters, some descending from above, and some springing from beneath; the one informed by the light of nature, the other inspired by divine revelation' (p. 83 below). On this basis he distinguishes between divinity and philosophy in his analysis. Equally, the image enables him to suggest the preciousness of learning, and the care and labour that must be taken to collect and preserve it, 'so this excellent liquor of knowledge, whether it descend from divine inspiration, or spring from human sense, would soon perish and vanish to oblivion, if it were not preserved in books, traditions, conferences, and places appointed, as universities, colleges, and schools, for the receipt and comforting of the same' (p. 61 below). Knowledge, like water, must not trickle away.

Bacon has a passionate sense not only of the intellectual world but also of the details of the physical world around him. He has been excited by what he has read, so that much of his imagery is drawn from his reading of classical history, of myth, and of the Bible. But he has also been excited by his experience of gardening and of experimenting, of observing colours and listening to music, of serving on committees and pleading in the law-courts. In the *Advancement*, for example, he is concerned with the areas of knowledge that are 'waste' and 'uncultivated'. He sees himself as the gardener, not 'taking [the] seeds and slips, and rearing them first into plants, and so uttering them in pots, when they are in flower, and in their best state', but 'uttering rather seeds than plants' and 'sowing with the basket [rather] than with the hand' (*Works* x. 301). His garden of learning would, like the princely garden he describes in his essay 'Of Gardens', bloom all the year round, and be fruitful, like the original garden of Eden. 'God Almightie first planted a Garden. And indeed it is the purest of human pleasures.' All the creative acts in Genesis were symbols for Bacon of knowledge and its cultivation, dissemination, growth, and value. He would restore that Paradise, and men by their labours would make another Eden, such as Bacon describes in Bensalem.

Another characteristic Baconian quality is his love of ceremony and ritual. The dignified ceremonial behaviour of the inhabitants of Bensalem at all times, and in particular the entrance of the Father of the House of Salomon, and the ritual of the Feast of the Tirsan, are matched by Bacon's sense of ceremony in his writing. At the beginning of the *Advancement* he sees himself as an ancient Hebrew, his work for the King as a legal counsellor and his writing of his book are presented as the fulfilling of the rituals of daily sacrifice and free-will offerings. He is living out the ritualistic and mythic patterns he has absorbed from his reading. And he wants James to live out the patterns he is urging on him, to be a Solomon in wise rule and learning, a Hermes combining priest, king, and philosopher, an Alexander providing a new Aristotle with means to experiment, a Jehovah rewarding the daily sacrifice and free-will offerings with favours.

Bacon felt that the greatest obstacle to the progress of learning was that men despaired of making progress (*Novum Organum* i. 92). To this he opposed his own sense of hope, born of his response to 'that wonderful voyage' of Columbus across the Atlantic, and the prophecy of Daniel, 'many shall go to and fro, and knowledge shall be increased', clearly indicating, he thought, that 'the through passage of the world . . . and the advancement of the sciences are destined by Fate, that is, by Divine Providence, to meet in the same age' (*Novum Organum* i. 93). Bacon's only doubt was how to present what he had to say to an audience that he felt was resistant to his message. He did not doubt the rightness of his contempt for existing systems, but only the wisdom of saying so too openly. He did not doubt his own ability to pile up telling reasons for his contempt and telling arguments in favour of his own methods. For there is nothing that Bacon does so well as to compile a list—lists of flowering plants, lists of desiderata, of *magnalia naturae*, of instances of Heat (Book ii of the *Novum Organum* is a series of lists), of arguments *pro* and *contra*, of natural histories, even, in the 'Promus of Formularies and Elegancies', of proverbs, turns of speech, and forms of morning and evening salutation. His idea that science should begin again at the beginning and attend to particulars, and that scientists should be content to compile innumerable facts and observations of natural history, is an extension of his own love of the list. He found as a member of Parliament that his colleagues always elected him to prepare and deliver their cases, because he alone had compiled a list of all the points made or to be made on any business. The rest were tongue-tied in comparison. And his lists were neat, organized, and carefully presented with a 'First', 'Secondly', 'Thirdly'. Book II of the *Advancement* is based on a list of all the subjects Bacon could think of that man might study, organized in an interesting sequence so that the list sounds inevitable and natural. In Book I the reader feels that there can surely be no more examples in Plutarch or in classical and European history to support the thesis that learning and national glory go hand in hand. The list of university reforms in Book II is surely complete. When the Father of the House of Salomon describes what the Fathers do and what equipment they have, the list comes helter-skelter. Bacon is distrustful of the patterns that lists can be forced into, and the connections that can be made between the separate items (see p. 135 below). So he would prefer to offer nothing but the lists. Our image of Bacon therefore is of a man with pen and paper, thinking, noting, arranging his thoughts under headings, sanely observing how men actually behave, what happens in an experiment. But the urge to compile his lists is not an end in itself. It serves the larger purpose of presenting man with a vision, of the recovery of knowledge and power, of a new start, of hope, of co-operative effort for the relief of man's estate, of the advancement of learning.

The reception of *The Advancement of Learning* must have been a disappointment to Bacon. Its publication coincided with the Gunpowder Plot, which doubtless diverted men's minds from his book. He certainly felt that it had been a mistake to write it in English, and he began to arrange for its translation into Latin, so that it would reach a European audience. His next book was in Latin, *De Sapientia Veterum* [of the Wisdom of the Ancients] (1609), in which he used a familiar Renaissance method of interpreting classical fables as a means of explaining his ideas to an audience that did not seem disposed to listen. Throughout the next decade he was busy as the chief legal officer of the crown, but he steadily worked at his *Novum Organum*, an exposition of the inductive method to be used in scientific investigation. When he published this in Latin in 1620 he told King James that this was but the same argument as the *Advancement* 'sunk deeper', and he was pleased to find that his former book was 'well tasted in the universities here, and the English colleges abroad' (*Works* xiv. 120). The failure of the *Novum Organum* to persuade James to supply funds for the compiling of a natural and experimental history 'which must be the main foundation of a true and active philosophy' (ibid.), followed by Bacon's disgrace and fall from office, convinced him that it was to the future he must look. 'I have lost much time with this age,' he wrote. 'These modern languages will at one time or other play the bank-rowtes with books' (*Works* xiv. 429). In 1623 he therefore published his *Advancement* in a new form and in Latin. The *De Augmentis Scientiarum* expands his Book II to eight books and incorporates a great deal of new material. He thought of this as a necessary first part of his Great Instauration of the Sciences, a systematization of the entire field of learning. The *Novum Organum* was to be the second part, describing the method to be used in investigating the Forms of nature. The remainder of the great Baconian system is a series of fragments. The two works reprinted in this volume represent the beginning of the Instauration and a vision of its completion in an island Paradise that is 'a mirror worthy to hold men's eyes', 'a picture of our salvation in heaven', 'a land of angels'. Bacon's hopes of a regained Paradise are based on the creation of an ideal state achieved through the secular advancement of learning.

Chronological Table

1561	Francis Bacon born (22 Jan.) at York House in the Strand, son of Sir Nicholas Bacon, Lord Keeper of the Seal, and Anne Cooke (his second wife) daughter of Edward VI's tutor.
1573	Bacon enters Trinity College, Cambridge (April).
1576	Admitted to Gray's Inn (June).
	Went to Paris with the Ambassador, Sir Amyas Paulet.
1579	Death of his father (22 Feb.). Returned to England.
1582	Admitted Utter Barrister at Gray's Inn (27 June).
1584	M.P. for Melcombe in Dorsetshire (23 Nov.). (He was a Member for various constituencies until he went to the Lords in 1618.)
1588	Acquaintance with Essex begins.
1592	*A Conference of Pleasure* written (publ. 1870).
1594	*Gesta Grayorum*, a device presented at Gray's Inn.
1595–6	*Formularies and Elegancies* compiled (publ. 1859, 1883).
1597	*Essays, Colours of Good and Evil, Meditationes Sacrae* published.
1601	Trial and execution of Essex.
	Bacon drew up a *Declaration of the Practises and Treasons attempted and committed by Robert late Earl of Essex.*
1603	Death of Elizabeth (24 Mar.), accession of James 1.
	Bacon knighted (23 July).
	Valerius Terminus of the Interpretation of Nature (publ.1734).
	De Interpretatione Naturae Proaemium [Preface to 'of the Interpretation of Nature'] (publ. 1653).
1604	*Apology in certain imputations concerning the late Earl of Essex.*
	Bacon appointed King's Counsel (18 Aug.).
1605	*The Advancement of Learning.*
1606	Marriage to Alice Barnham (10 May).
1607	Bacon appointed Solicitor General (25 June).
	Cogitata et Visa [*Thoughts and Conclusions*] (publ. 1653).
	Filum Labyrinthi [*The Clue to the Labyrinth*] (publ. 1734).
1608	*Temporis Partus Masculus* [*The Masculine Birth of Time*] (publ. 1653).
1609	*De Sapientia Veterum* [*Of The Wisdom of the Ancients*].
1610	Death of Bacon's mother.
1612	*Essays* (enlarged) published.
1612	*Descriptio Globi Intellectualis* [*Description of the Intellectual Globe*] (publ. 1653).
	Thema Coeli [*Theory of the Heaven*] (publ. 1653).
1613	Appointed Attorney General (26 Oct.).
1616	Made a Privy Councillor (9 June).

1617	Made Lord Keeper (7 Mar.).
1618	Made Lord Chancellor (4 Jan.).
	Created Baron Verulam (12 July).
1620	*Novum Organum* published (Oct.).
1621	Created Viscount St. Albans (27 Jan.).
	Meeting of Parliment (30 Jan.).
	Sentenced by House of Lords for corruption (3 May).
	Limited pardon sealed by James I (Nov.).
1622	*History of Henry VII* published (Mar.).
1622–3	Published monthly instalments of his Natural History—*Historia Ventorum* [*History of Winds*], *Historia Vitae et Mortis* [*History of Life and Death*].
1623	*De Augmentis Scientiarum* published (Oct.).
1624	Published *Apophthegms* and *Translation of the Psalms.*
1625	Death of James I (27 Mar.).
	Essays (enlarged) published.
1626	Bacon died at Highgate (9 April).
1627	*Sylva Sylvarum* and *New Atlantis* published by Bacon's chaplain, William Rawley.

Note on the Texts

The Twoo Bookes of Francis Bacon. Of the proficience and advancement of Learning, divine and humane was printed in 1605. The two books were entered separately in the Stationers' Register, on 19 August and 19 September, the entries stating that they were 'to be printed bothe in Englishe and Lattine'. Only the English text was however printed, and this was reprinted in 1629 and in 1633. The text used here is that prepared by William Aldis Wright in 1868, based on that of 1605, 'corrected where necessary by the Errata and by the subsequent editions of 1629 and 1633'. The numbering of the sections and paragraphs is that provided by Wright.

Bacon's marginal notes have been moved to the beginning of the paragraph in which they occur. Their function in the original was to draw attention to important topics, but Bacon did not provide them consistently throughout the book. In this edition the table of contents provides a brief analysis of *The Advancement of Learning*, and this analysis has been used in the text itself, the headings being supplied in square brackets.

The *New Atlantis* was first printed with the *Sylva Sylvarum* in 1627. It is printed here from Spedding's edition in the *Works* iii. 129–68.

References to Greek and Latin authors are given in a form to facilitate reference to the English translations in the Loeb editions. References to Aristotle's *Ethics* are to the *Nicomachean Ethics.*

In the glossing of individual words a * at the end of a footnote indicates that this is the first use of the word, or the first use in the sense given, recorded in the *OED*. It must be remembered that this does not necessarily mean that Bacon invented the word, or was the first to use it with that particular meaning. A † at the end of a footnote indicates that this is a usage that antedates the first usage given in the *OED*.

Reading List

Unless otherwise stated the place of publication is London.

1. *Editions:*

Works. Ed. J. Spedding, R. L. Ellis, and D. D. Heath. 1857–74. Vols. 1–7 contain the works, with English translations; vols. 8–14 contain the life, including letters, speeches, and occasional papers.

Philosophical Works. Ed. J. M. Robertson. 1905. Prints the philosophical works from Ellis and Spedding, giving their English versions only of the Latin works.

Selections. Ed. C. J. Dixon (1963); ed. S. Warhaft (Toronto and London, 1965); ed. A. Johnston (1965).

The Advancement of Learning. Ed. T. Markby (1852); ed. G. W. Kitchin (1861, repr. Everyman's Library, 1915, 1973); ed. W. A. Wright (Oxford, 1868 and frequently reprinted); ed. T. Case (with *New Atlantis*, Oxford, 1906 and frequently reprinted).

New Atlantis. Ed. G. C. Moore Smith (Cambridge, 1900); ed. A. B. Gough (Oxford, 1924).

2. *Commentaries:*

(*a*) *Books:*

ANDERSON, F. M. *The Philosophy of Francis Bacon*. Chicago, 1948.

BROAD, C. D. *The Philosophy of Francis Bacon*. Cambridge, 1926, repr. in *Ethics and the History of Philosophy*. 1952.

FARRINGTON, B. *Bacon, Philosopher of Industrial Science*. 1951.

——*The Philosophy of Francis Bacon*. Liverpool, 1964. This includes translations of three Latin works not translated in Spedding—*Temporis Partus Masculus, Redargutio Philosophiarum*, and *Cogitata et Visa*.

JAMES, D. G. *The Dream of Learning: an Essay on the Advancement of Learning, Hamlet and King Lear*. Oxford, 1951.

LEMMI, C. W. *The Classical Deities in Bacon*. Baltimore, 1933 (on *The Wisdom of the Ancients*).

PADHI, SHANTI. *Serpent and Columbine: four chapters on Bacon and The Advancement of Learning*. New Delhi, 1969.

ROSSI, PAOLO. *Francis Bacon: from Magic to Science*. 1968.

VICKERS, B. *Francis Bacon and Renaissance Prose*. Cambridge, 1968.

——(ed.) *Essential Articles for the Study of Francis Bacon*. Hamden, Connecticut, 1968.

WALLACE, K. R. *Francis Bacon on Communication and Rhetoric*. Chapel Hill, 1943.

——*Francis Bacon on the Nature of Man: the Faculties of the Soul*. Urbana, 1967.

WHITE, H. B. *Peace Among the Willows: the Political Philosophy of Francis Bacon*. The Hague, 1968 (on the *New Atlantis*).

WHITTAKER, V. K. *Francis Bacon's Intellectual Milieu*. Los Angeles, 1962.

(*b*) *Articles:*

ADAMS, R. 'The Social Responsibilities of Science in *Utopia, New Atlantis* and after', *Journal of the History of Ideas*, 10 (1949), 374–98.

BIERMAN, J. 'Science and Society in the *New Atlantis* and other Renaissance Utopias', *PMLA*, 78 (1963), 492–500.

BIERMAN, J. '*New Atlantis* revisited', *Studies in the Literary Imagination* (Atlanta, Georgia), 4 (1971), 121–41.

BLODGETT, E. 'Campanella and the *New Atlantis*', *PMLA*, 46 (1931), 763–80.

BULLOUGH, G. 'Bacon and the Defence of Learning', in *Seventeenth Century Studies presented to Sir Herbert Grierson* (Oxford, 1938), pp. 1–20.

COCHRANE, R. 'Francis Bacon and the Architect of Fortune', *Studies in the Renaissance*, 5 (1958), 176–95.

COLIE, R. L. 'Cornelis Drebbel and Salomon de Caus: Two Jacobean Models for Salomon's House', *Huntington Library Quarterly*, 18 (1955), 245–60.

CRANE, R. S. 'The Relation of Bacon's *Essays* to his Programme for the Advancement of Learning', in *Schelling Anniversary Papers* (New York, 1923), pp. 87–105.

HARRISON, J. L. 'Bacon's View of Rhetoric, Poetry and the Imagination', *Huntington Library Quarterly*, 20 (1957), 107–25.

KNIGHTS, L. C. 'Bacon and the Dissociation of Sensibility', in *Explorations* (1946), pp. 92–111.

METZ, R. 'Bacon's Part in the Intellectual Movement of his Time', in *Seventeenth Century Studies presented to Sir Herbert Grierson* (Oxford, 1938), pp. 21–32.

PRIOR, M. E. 'Bacon's Man of Science', *Journal of the History of Ideas*, 15 (1954), 348–70.

REDPATH, R. T. H. 'Bacon and the Advancement of Learning', in *Pelican Guide to English Literature*, ed. B. Ford, 2 (1956), pp. 369–85.

RIGHTER, A. 'Francis Bacon', in *The English Mind*, eds. H. S. Davies and G. Watson (Cambridge, 1964), pp. 7–29.

VICKERS, B. 'Swift and the Baconian Idol', in *The World of Jonathan Swift*, ed. Vickers (Oxford, 1968).

3. *Important background books:*

BOAS, MARIE. *The Scientific Renaissance, 1540–1630*. 1962, repr. 1970.

BRIGGS, R. *The Scientific Revolution in the Seventeenth Century*. 1969.

CAWLEY, R. R. *Unpathed Waters: Studies in the Influence of the Voyages on Elizabethan Literature*. Princeton, 1940.

CRAIG, H. *The Enchanted Glass: the Elizabethan Mind in Literature.* Oxford, 1936, repr. 1950.

FISCH, H. *Jerusalem and Albion: the Hebraic Factor in Seventeenth-Century Literature.* 1964.

GILBERT, N. W. *Renaissance Concepts of Method.* New York, 1960, repr. 1963.

HADYN, H. *The Counter Renaissance.* New York, 1950.

HILL, C. *Intellectual Origins of the English Revolution.* 1965.

HOWELL, W. S. *Logic and Rhetoric in England, 1500–1700.* Princeton, 1956.

JONES, R. F. *Ancients and Moderns: a Study of the Rise of the Scientific Movement in Seventeenth Century England.* St. Louis, 1936, repr. 1961.

KEARNEY, H. F. *Origins of the Scientific Revolution.* 1964, repr. 1966.

WILLEY, B. *The Seventeenth Century Background.* 1934, 1949.

WILSON, F. P. *Elizabethan and Jacobean.* Oxford, 1945.

4. *Biographies:*

CHURCH, R. W. *Bacon* (English Men of Letters). 1884.

FARRINGTON, B. *Francis Bacon, Philosopher of Science,* 1951.

GROWTHER, J. G. *Francis Bacon: the First Statesman of Science.* 1960.

ANDERSON, F. H. *Francis Bacon: His Career and Thought.* Los Angeles, 1962.

The Advancement of Learning

The First Book of Francis Bacon of the Proficience and Advancement of Learning Divine and Human

TO THE KING

1. There were under the law, excellent King, both daily sacrifices and freewill offerings;[n] the one proceeding upon ordinary observance, the other upon a devout cheerfulness: in like manner there belongeth to kings from their servants both tribute of duty and presents of affection. In the former of these I hope I shall not live to be wanting, according to my most humble duty, and the good pleasure of your Majesty's employments: for the latter, I thought it more respective to make choice of some oblation, which might rather refer to the propriety and excellency of your individual person, than to the business of your crown and state.

2. Wherefore, representing your Majesty many times unto my mind, and beholding you not with the inquisitive eye of presumption, to discover that which the Scripture telleth me is inscrutable, but with the observant eye of duty and admiration; leaving aside the other parts of your virtue and fortune, I have been touched, yea, and possessed with an extreme wonder at those your virtues and faculties, which the Philosophers call intellectual; the largeness of your capacity,[n] the faithfulness of your memory, the swiftness of your apprehension, the penetration of your judgement, and the facility and order of your elocution: and I have often thought, that of all the persons living that I have known, your Majesty were the best instance to make a man of Plato's opinion, that all knowledge is but remembrance, and that the mind of man by nature knoweth all things, and hath but her own native and original notions (which by the strangeness and darkness of this tabernacle of the body are sequestered) again revived and restored: such a light of nature I have observed in your Majesty, and such a readiness to take flame and blaze from the least occasion presented, or the least spark of

offerings: the superior character 'n' is used throughout to direct the reader to the Notes, pp. 251–91.
proceeding upon: resulting from.
respective: appropriate.
oblation: offering.
propriety: particularity.
inscrutable: Proverbs 25:3.
remembrance: Plato in *Meno* 81 d and *Phaedo* 72 e.

another's knowledge delivered. And as the Scripture saith of the wisest king, 'That his heart was as the sands of the sea'; which though it be one of the largest bodies, yet it consisteth of the smallest and finest portions; so hath God given your Majesty a composition of understanding admirable, being able to compass and comprehend the greatest matters, and nevertheless to touch and apprehend the least; whereas it should seem an impossibility in nature, for the same instrument to make itself fit for great and small works. And for your gift of speech, I call to mind what Cornelius Tacitus saith of Augustus Caesar: *Augusto profluens, et quae principem deceret, eloquentia fuit.* For if we note it well, speech that is uttered with labour and difficulty, or speech that savoureth of the affectation of art and precepts, or speech that is framed after the imitation of some pattern of eloquence, though never so excellent; all this hath somewhat servile, and holding of the subject.[n] But your Majesty's manner of speech is indeed prince-like, flowing as from a fountain, and yet streaming and branching itself into nature's order, full of facility and felicity, imitating none, and inimitable by any. And as in your civil estate there appeareth to be an emulation and contention of your Majesty's virtue with your fortune; a virtuous disposition with a fortunate regiment; a virtuous expectation (when time was) of your greater fortune, with a prosperous possession thereof in the due time; a virtuous observation of the laws of marriage, with most blessed and happy fruit of marriage; a virtuous and most Christian desire of peace, with a fortunate inclination in your neighbour princes thereunto: so likewise in these intellectual matters, there seemeth to be no less contention between the excellency of your Majesty's gifts of nature and the universality and perfection of your learning. For I am well assured that this which I shall say is no amplification at all, but a positive and measured truth; which is, that there hath not been since Christ's time any king or temporal monarch, which hath been so learned in all literature and erudition, divine and human. For let a man seriously and diligently revolve and peruse the succession of the emperors of Rome, of which

sands of the sea: said of Solomon in I Kings 4:29.
a composition of understanding admirable: i.e. a combination of the powers of comprehending which is wonderful.
Augusto profluens: Quoted from memory from Tacitus, *Annals* xiii. 3: 'Augustus had an easy and fluent way of speaking, such as became a sovereign.'
felicity: a happy faculty in speech, an admirable appropriateness or grace of invention or expression.
regiment: government or control of oneself.
amplification: exaggeration.

Caesar the Dictator, who lived some years before Christ, and Marcus Antoninus were the best learned; and so descend to the emperors of Grecia, or of the West, and then to the lines of France, Spain, England, Scotland, and the rest, and he shall find this judgement is truly made. For it seemeth much in a king, if, by the compendious extractions of other men's wits and labours, he can take hold of any superficial ornaments and shows of learning; or if he countenance and prefer learning and learned men: but to drink indeed of the true fountains of learning, nay, to have such a fountain of learning in himself, in a king, and in a king born, is almost a miracle. And the more, because there is met in your Majesty a rare conjunction, as well of divine and sacred literature, as of profane and human; so as your Majesty standeth invested of that triplicity, which in great veneration was ascribed to the ancient Hermes;[n] the power and fortune of a king, the knowledge and illumination of a priest, and the learning and universality of a philosopher. This propriety inherent and individual attribute in your Majesty deserveth to be expressed not only in the fame and admiration of the present time, nor in the history or tradition of the ages succeeding, but also in some solid work, fixed memorial, and immortal monument, bearing a character or signature both of the power of a king and the difference and perfection of such a king.

3. Therefore I did conclude with myself, that I could not make unto your Majesty a better oblation than of some treatise tending to that end, whereof the sum will consist of these two parts; the former concerning the excellency of learning and knowledge, and the excellency of the merit and true glory in the augmentation and propagation thereof: the latter, what the particular acts and works are, which have been embraced and undertaken for the advancement of learning; and again, what defects and undervalues I find in such particular acts: to the end that though I cannot positively or affirmatively advise your Majesty, or propound unto you framed particulars, yet I may excite your princely cogitations to visit the excellent treasure of your own mind, and thence to extract particulars for this purpose, agreeable to your magnanimity and wisdom.

[DEFENCE OF LEARNING AGAINST DIVINES]

I. 1. In the entrance to the former of these, to clear the way, and as it were to make silence, to have the true testimonies concerning the

propriety: particular character. *difference*: distinguishing mark.
undervalues: insufficiency in worth.

dignity of learning to be better heard, without the interruption of tacit objections; I think good to deliver it from the discredits and disgraces which it hath received, all from ignorance; but ignorance severally disguised; appearing sometimes in the zeal and jealousy of divines; sometimes in the severity and arrogancy of politiques; and sometimes in the errors and imperfections of learned men themselves.

2. I hear the former sort say, that knowledge is of those things which are to be accepted of with great limitation and caution: that the aspiring to overmuch knowledge was the original temptation and sin whereupon ensued the fall of man: that knowledge hath in it somewhat of the serpent, and therefore where it entereth into a man it makes him swell; *Scientia inflat*: that Salomon gives a censure, 'That there is no end of making books, and that much reading is weariness of the flesh'; and again in another place, 'That in spacious knowledge there is much contristation, and that he that increaseth knowledge increaseth anxiety': that Saint Paul gives a caveat, 'That we be not spoiled through vain philosophy': that experience demonstrates how learned men have been arch-heretics, how learned times have been inclined to atheism, and how the contemplation of second causes doth derogate from our dependence upon God, who is the first cause.[n]

3. To discover then the ignorance and error of this opinion, and the misunderstanding in the grounds thereof, it may well appear these men do not observe or consider that it was not the pure knowledge of nature and universality, a knowledge by the light whereof man did give names unto other creatures in Paradise,[n] as they were brought before him, according unto their proprieties, which gave the occasion to the fall: but it was the proud knowledge of good and evil, with an intent in man to give law unto himself, and to depend no more upon God's commandments, which was the form of the temptation. Neither is it any quantity of knowledge, how great soever, that can make the mind of man to swell; for nothing can fill, much less extend the soul of man, but God and the contemplation of God; and therefore Salomon, speaking of the two principal senses of inquisition, the eye and the ear, affirmeth that 'the eye is never satisfied with seeing, nor the ear with hearing'; and if there be

severally: differently, variously.
politiques: politicians.
Scientia inflat: 1 Corinthians 8:1 'Knowledge puffeth up.'
Salomon: The spelling is that of the Vulgate: Ecclesiastes 12:12 and 1:18.
St. Paul: In Colossians 2:8.
grounds: fundamental principles.
universality: the whole world.
proprieties: particular natures.
Salomon: In Ecclesiastes 1:8.

no fulness, then is the continent greater than the content: so of knowledge itself, and the mind of man, whereto the senses are but reporters, he defineth likewise in these words, placed after that Kalendar or Ephemerides which he maketh of the diversities of times and seasons for all actions and purposes; and concludeth thus: 'God hath made all things beautiful, or decent, in the true return of their seasons: Also he hath placed the world in man's heart, yet cannot man find out the work which God worketh from the beginning to the end': declaring not obscurely, that God hath framed the mind of man as a mirror or glass, capable of the image of the universal world, and joyful to receive the impression thereof, as the eye joyeth to receive light; and not only delighted in beholding the variety of things and vicissitude of times, but raised also to find out and discern the ordinances and decrees, which throughout all those changes are infallibly observed. And although he doth insinuate that the supreme or summary law of nature, which he calleth, 'The work which God worketh from the beginning to the end', is not possible to be found out by man; yet that doth not derogate from the capacity of the mind, but may be referred to the impediments,[n] as of shortness of life, ill conjunction of labours, ill tradition of knowledge over from hand to hand, and many other inconveniences, whereunto the condition of man is subject. For that nothing parcel of the world is denied to man's inquiry and invention, he doth in another place rule over, when he saith, 'The spirit of man is as the lamp of God, wherewith he searcheth the inwardness of all secrets.' If then such be the capacity and receipt of the mind of man, it is manifest that there is no danger at all in the proportion or quantity of knowledge, how large soever, lest it should make it swell or outcompass itself; no, but it is merely the quality of knowledge, which, be it in quantity more or less, if it be taken without the true corrective thereof, hath in it some nature of venom or malignity, and some effects of that venom, which is ventosity or swelling. This corrective spice, the mixture whereof maketh knowledge so sovereign, is charity, which the Apostle immediately addeth to the former clause: for so he saith, 'Knowledge bloweth up, but charity buildeth up'; not unlike unto that which he delivereth in another place: 'If

continent: then is that which contains greater than that which is contained.
Ephemerides: Ecclesiastes 3:1–11.
capable of: able to receive.
nothing parcel: no part.
The spirit: Proverbs 20:27.
derogate from: detract from, lessen.
rule over: lay down authoritatively.
ventosity: inflated, puffed up.
sovereign: potent in a superlative degree.
Apostle: St. Paul in I Corinthians 8:1 and 13:1.

I spake', saith he, 'with the tongues of men and angels, and had not charity, it were but as a tinkling cymbal'; not but that it is an excellent thing to speak with the tongues of men and angels, but because, if it be severed from charity, and not referred to the good of men and mankind, it hath rather a sounding and unworthy glory, than a meriting and substantial virtue. And as for that censure of Salomon, concerning the excess of writing and reading books, and the anxiety of spirit which redoundeth from knowledge; and that admonition of Saint Paul, 'That we be not seduced by vain philosophy'; let those places be rightly understood, and they do indeed excellently set forth the true bounds and limitations, whereby human knowledge is confined and circumscribed; and yet without any such contracting or coarctation, but that it may comprehend all the universal nature of things; for these limitations are three: the first, 'That we do not so place our felicity in knowledge, as we forget our mortality': the second, 'That we make application of our knowledge, to give ourselves repose and contentment, and not distaste or repining': the third, 'That we do not presume by the contemplation of nature to attain to the mysteries of God.' For as touching the first of these, Salomon doth excellently expound himself in another place of the same book, where he saith: 'I saw well that knowledge recedeth as far from ignorance as light doth from darkness; and that the wise man's eyes keep watch in his head, whereas the fool roundeth about in darkness: but withal I learned, that the same mortality involveth them both.' And for the second, certain it is, there is no vexation or anxiety of mind which resulteth from knowledge otherwise than merely by accident; for all knowledge and wonder (which is the seed of knowledge) is an impression of pleasure in itself: but when men fall to framing conclusions out of their knowledge, applying it to their particular, and ministering to themselves thereby weak fears or vast desires, there groweth that carefulness and trouble of mind which is spoken of: for then knowledge is no more *Lumen siccum*,[n] whereof Heraclitus the profound said, *Lumen siccum optima anima*; but it becometh *Lumen madidum*, or *maceratum*, being steeped and infused in the humours of the affections. And as for the third point, it deserveth to be a little stood upon, and not to be lightly passed over: for if any man shall think by view

meriting: deserving. *Salomon*: In Ecclesiastes 12:12.
St. Paul: Colossians 2:8. *coarctation*: restriction.*
distaste: unpleasantness.* *Salomon*: In Ecclesiastes 2:13–14.
roundeth about: roams about.
wonder: Plato says that wonder is the incentive to discovery in *Theaetetus* 155d.
particular: own individual case. *carefulness*: anxiety.

and inquiry into these sensible and material things to attain that light, whereby he may reveal unto himself the nature or will of God, then indeed is he spoiled by vain philosophy: for the contemplation of God's creatures and works produceth (having regard to the works and creatures themselves) knowledge, but having regard to God, no perfect knowledge, but wonder, which is broken knowledge. And therefore it was most aptly said by one of Plato's school, 'That the sense of man carrieth a resemblance with the sun, which (as we see) openeth and revealeth all the terrestrial globe; but then again it obscureth and concealeth the stars and celestial globe: so doth the sense discover natural things, but it darkeneth and shutteth up divine.' And hence it is true that it hath proceeded, that divers great learned men have been heretical, whilst they have sought to fly up to the secrets of the Deity by the waxen wings of the senses.[n] And as for the conceit that too much knowledge should incline a man to atheism,[n] and that the ignorance of second causes should make a more devout dependence upon God, which is the first cause; first, it is good to ask the question which Job asked of his friends: 'Will you lie for God, as one man will do for another, to gratify him?' For certain it is that God worketh nothing in nature but by second causes: and if they would have it otherwise believed, it is mere imposture, as it were in favour towards God; and nothing else but to offer to the author of truth the unclean sacrifice of a lie. But further, it is an assured truth, and a conclusion of experience, that a little or superficial knowledge of philosophy may incline the mind of man to atheism, but a further proceeding therein doth bring the mind back again to religion. For in the entrance of philosophy, when the second causes, which are next unto the senses, do offer themselves to the mind of man, if it dwell and stay there it may induce some oblivion of the highest cause; but when a man passeth on further, and seeth the dependence of causes, and the works of Providence, then, according to the allegory of the poets, he will easily believe that the highest link of nature's chain must needs be tied to the foot of Jupiter's chair.[n] To conclude therefore, let no man upon a weak conceit of sobriety or an ill-applied moderation think or maintain, that a man can search too far, or be too well studied in the book of God's word, or in the book of God's works, divinity or philosophy; but rather

broken knowledge: fragmentary or disjointed knowledge.
one of Plato's school: Philo of Alexandria, *On Dreams* i. 83–4.
discover: uncover, lay bare. *proceeded*: resulted.
conceit: conception. *Job*: In Job 13:7–9.
in favour towards God: i.e. for God's sake.
sobriety: moderation in its widest sense, sobermindedness.

let men endeavour an endless progress or proficience in both; only let men beware that they apply both to charity, and not to swelling; to use, and not to ostentation; and again, that they do not unwisely mingle or confound these learnings together.[n]

[DEFENCE OF LEARNING AGAINST POLITICIANS]

II. 1. And as for the disgraces which learning receiveth from politiques, they be of this nature; that learning doth soften men's minds, and makes them more unapt for the honour and exercise of arms; that it doth mar and pervert men's dispositions for matter of government and policy, in making them too curious and irresolute by variety of reading, or too peremptory or positive by strictness of rules and axioms, or too immoderate and overweening by reason of the greatness of examples, or too incompatible and differing from the times by reason of the dissimilitude of examples; or at least, that it doth divert men's travails from action and business, and bringeth them to a love of leisure and privateness; and that it doth bring into states a relaxation of discipline, whilst every man is more ready to argue than to obey and execute. Out of this conceit, Cato, surnamed the Censor, one of the wisest men indeed that ever lived, when Carneades the philosopher came in embassage to Rome, and that the young men of Rome began to flock about him, being allured with the sweetness and majesty of his eloquence and learning, gave counsel in open senate that they should give him his dispatch with all speed, lest he should infect and enchant the minds and affections of the youth, and at unawares bring in an alteration of the manners and customs of the state. Out of the same conceit or humour did Virgil, turning his pen to the advantage of his country, and the disadvantage of his own profession, make a kind of separation between policy and government, and between arts and sciences, in the verses so much renowned, attributing and challenging the one to the Romans, and leaving and yielding the other to the Grecians: *Tu regere imperio populos, Romane, memento, Hae tibi erunt artes, &c.* So likewise we see that Anytus, the accuser of Socrates, laid it as an article of

proficience: progress.* *swelling*: inflation by pride, vanity.
politiques: politicians. *curious*: difficult to satisfy, cautious.
travails: labours.
Cato: The story is told in Plutarch's 'Life' of Marcus Cato, 22.
in embassage: on an embassy.* *at unawares*: unperceived.
challenging: laying claim to.
Tu regere: Virgil, *Aeneid* vi. 852 f. 'Be it thy task, O Roman, to rule over subject peoples; these shall be thy arts, etc.'
accuser of Socrates: In Plato's *Apology* 23 d, and Xenophon's *Memorabilia* I. i. 1.

charge and accusation against him, that he did, with the variety and power of his discourses and disputations, withdraw young men from due reverence to the laws and customs of their country, and that he did profess a dangerous and pernicious science, which was, to make the worse matter seem the better, and to suppress truth by force of eloquence and speech.

2. But these and the like imputations have rather a countenance of gravity than any ground of justice: for experience doth warrant, that both in persons and in times there hath been a meeting and concurrence in learning and arms, flourishing and excelling in the same men and the same ages. For as for men, there cannot be a better nor the like instance, as of that pair, Alexander the Great[n] and Julius Caesar the Dictator; whereof the one was Aristotle's scholar in philosophy, and the other was Cicero's rival in eloquence: or if any man had rather call for scholars that were great generals, than generals that were great scholars, let him take Epaminondas the Theban,[n] or Xenophon the Athenian;[n] whereof the one was the first that abated the power of Sparta, and the other was the first that made way to the overthrow of the monarchy of Persia. And this concurrence is yet more visible in times than in persons, by how much an age is [a] greater object than a man. For both in Egypt, Assyria, Persia, Grecia, and Rome, the same times that are most renowned for arms, are likewise most admired for learning,[n] so that the greatest authors and philosophers and the greatest captains and governors have lived in the same ages. Neither can it otherwise be: for as in man the ripeness of strength of the body and mind cometh much about an age, save that the strength of the body cometh somewhat the more early, so in states, arms and learning, whereof the one correspondeth to the body, the other to the soul of man, have a concurrence or near sequence in times.

3. And for matter of policy and government, that learning should rather hurt, than enable thereunto, is a thing very improbable: we see it is accounted an error to commit a natural body to empiric physicians, which commonly have a few pleasing receipts whereupon they are confident and adventurous, but know neither the causes

science: knowledge, learning.
countenance of gravity: appearance of importance.
near sequence in times: According to Aristotle (*Rhetoric* II. xiv) the body is in its prime from 30 to 35, the mind at 49.
enable: qualify.
empiric: empirical, not based on scientific knowledge but used because believed to have proved its efficacy in practice.*
adventurous: rashly daring.

of diseases, nor the complexions[n] of patients, nor peril of accidents, nor the true method of cures: we see it is a like error to rely upon advocates or lawyers, which are only men of practice and not grounded in their books, who are many times easily surprised when matter falleth out besides their experience, to the prejudice of the causes they handle: so by like reason it cannot be but a matter of doubtful consequence if states be managed by empiric statesmen, not well mingled with men grounded in learning. But contrariwise, it is almost without instance contradictory that ever any government was disastrous that was in the hands of learned governors. For howsoever it hath been ordinary with politique men to extenuate and disable learned men by the names of *pedantes*; yet in the records of time it appeareth in many particulars that the governments of princes in minority (notwithstanding the infinite disadvantage of that kind of state) have nevertheless excelled the government of princes of mature age, even for that reason which they seek to traduce, which is, that by that occasion the state hath been in the hands of *pedantes*: for so was the state of Rome for the first five years, which are so much magnified, during the minority of Nero, in the hands of Seneca a *pedanti*: so it was again, for ten years' space or more, during the minority of Gordianus the younger, with great applause and contentation in the hands of Misitheus a *pedanti*; so was it before that, in the minority of Alexander Severus, in like happiness, in hands not much unlike, by reason of the rule of the women, who were aided by the teachers and preceptors.[n] Nay, let a man look into the government of the bishops of Rome, as by name, into the government of Pius Quintus and Sextus Quintus in our times, who were both at their entrance esteemed but as pedantical friars, and he shall find that such popes do greater things, and proceed upon truer principles of estate, than those which have ascended to the papacy from an education and breeding in affairs of estate and courts of princes; for although men bred in learning are perhaps to seek in

peril of accidents: dangerousness of symptoms.
contradictory: The earliest example of the word (OED).*
extenuate: disparage.
disable: to pronounce incapable, hence depreciate, belittle.
traduce: expose to contempt.
magnified: made much of, extolled.
Pius Quintus: Pope Pius V (1565–72), who was Pope when the battle of Lepanto was won.
Sextus Quintus: Pope Sixtus V (1585–90), who restored the papal finances and built extensively in Rome.
estate: In its first usage here Bacon means administration, in the second he means simply 'state'.

points of convenience and accommodating for the present, which the Italians call *ragioni di stato*, whereof the same Pius Quintus could not hear spoken with patience, terming them inventions against religion and the moral virtues; yet on the other side, to recompense that, they are perfect in those same plain grounds of religion, justice, honour, and moral virtue, which if they be well and watchfully pursued, there will be seldom use of those other, no more than of physic in a sound or well-dieted body. Neither can the experience of one man's life furnish examples and precedents for the events of one man's life. For as it happeneth sometimes that the grandchild, or other descendant, resembleth the ancestor more than the son; so many times occurrences of present times may sort better with ancient examples than with those of the later or immediate times: and lastly, the wit of one man can no more countervail learning than one man's means can hold way with a common purse.

4. And as for those particular seducements or indispositions of the mind for policy and government, which learning is pretended to insinuate; if it be granted that any such thing be, it must be remembered withal, that learning ministereth in every of them greater strength of medicine or remedy than it offereth cause of indisposition or infirmity. For if by a secret operation it make men perplexed and irresolute, on the other side by plain precept it teacheth them when and upon what ground to resolve; yea, and how to carry things in suspense without prejudice, till they resolve. If it make men positive and regular, it teacheth them what things are in their nature demonstrative, and what are conjectural, and as well the use of distinctions and exceptions, as the latitude of principles and rules. If it mislead by disproportion or dissimilitude of examples, it teacheth men the force of circumstances, the errors of comparisons, and all the cautions of application; so that in all these it doth rectify more

points of convenience . . .: i.e. are perhaps deficient in matters of expediency and temporizing.
ragioni di stato: reasons of state.
sort: fit, agree with.
immediate times: those times which are not separated by any interval from the present.*
wit: mental capacity, intellect.
countervail: equal, match.
hold: i.e. than one man's income can compete with the Exchequer.
ministereth: supplies.
every: each.
positive: dogmatic, confident in opinion or assertion.†
regular: rigid in adhering to rule.
demonstrative: capable of proof by demonstration.†
latitude: range, scope.*

effectually that it can pervert. And these medicines it conveyeth into men's minds much more forcibly by the quickness and penetration of examples. For let a man look into the errors of Clement the seventh, so lively described by Guicciardine,[n] who served under him, or into the errors of Cicero,[n] painted out by his own pencil in his Epistles to Atticus, and he will fly apace from being irresolute. Let him look into the errors of Phocion,[n] and he will beware how he be obstinate or inflexible. Let him but read the fable of Ixion,[n] and it will hold him from being vaporous or imaginative. Let him look into the errors of Cato the second,[n] and he will never be one of the Antipodes, to tread opposite to the present world.

5. And for the conceit that learning should dispose men to leisure and privateness, and make men slothful; it were a strange thing if that which accustometh the mind to a perpetual motion and agitation should induce slothfulness: whereas contrariwise it may be truly affirmed, that no kind of men love business for itself but those that are learned; for other persons love it for profit, as an hireling, that loves the work for the wages; or for honour, as because it beareth them up in the eyes of men, and refresheth their reputation, which otherwise would wear; or because it putteth them in mind of their fortune, and giveth them occasion to pleasure and displeasure; or because it exerciseth some faculty wherein they take pride, and so entertaineth them in good humour and pleasing conceits toward themselves; or because it advanceth any other their ends. So that as it is said of untrue valours, that some men's valours are in the eyes of them that look on; so such men's industries are in the eyes of others, or at least in regard of their own designments: only learned men love business as an action according to nature, as agreeable to health of mind as exercise is to health of body, taking pleasure in the action itself, and not in the purchase: so that of all men they are the most indefatigable, if it be towards any business which can hold or detain their mind.

6. And if any man be laborious in reading and study and yet idle in business and action, it groweth from some weakness of body or softness of spirit; such as Seneca speaketh of: *Quidam tam sunt umbratiles, ut putent in turbido esse quicquid in luce est*; and not

quickness: liveliness. *lively*: vividly.
vaporous: inclined to be fanciful.* *wear*: diminish, fade.
designments: designs, that which is planned.
not in the purchase: i.e. not in that which is purchased by it.
softness: effeminacy.
Quidam tam: Seneca, *Epistles to Lucilius* i. 3. 'Some men live so much in the shade, that whenever they are in the light they seem to be in trouble.'

of learning: well may it be that such a point of a man's nature may make him give himself to learning, but it is not learning that breedeth any such point in his nature.

7. And that learning should take up too much time or leisure; I answer, the most active or busy man that hath been or can be, hath (no question) many vacant times of leisure, while he expecteth the tides and returns of business (except he be either tedious and of no dispatch, or lightly and unworthily ambitious to meddle in things that may be better done by others), and then the question is but how those spaces and times of leisure shall be filled and spent; whether in pleasures or in studies; as was well answered by Demosthenes to his adversary Aeschines, that was a man given to pleasure and told him 'That his orations did smell of the lamp': 'Indeed (said Demosthenes) there is a great difference between the things that you and I do by lamp-light.' So as no man need doubt that learning will expulse business, but rather it will keep and defend the possession of the mind against idleness and pleasure, which otherwise at unawares may enter to the prejudice of both.

8. Again, for that other conceit that learning should undermine the reverence of laws and government, it is assuredly a mere depravation and calumny, without all shadow of truth. For to say that a blind custom of obedience should be a surer obligation than duty taught and understood, it is to affirm, that a blind man may tread surer by a guide than a seeing man can by a light. And it is without all controversy, that learning doth make the minds of men gentle, generous, maniable, and pliant to government; whereas ignorance makes them churlish, thwart, and mutinous: and the evidence of time doth clear this assertion, considering that the most barbarous, rude, and unlearned times have been most subject to tumults, seditions, and changes.

9. And as to the judgement of Cato the Censor,[n] he was well punished for his blasphemy against learning, in the same kind wherein he offended; for when he was past threescore years old, he was

expecteth: awaits.
tides: proper times; also with the sense of the ebbing and flowing of business.
tedious: dilatory, slow. *dispatch*: promptitude in dealing with affairs.*
Demosthenes: In Plutarch's 'Life' of Demosthenes, 8:12, this story is told of Pytheas, not Aeschines.
doubt: fear. *expulse*: expel, drive out.
depravation: defamation.
gentle, generous: mild in disposition, free from meanness or prejudice.
maniable: manageable, tractable.*
thwart: disposed to resist, stubborn.
clear: prove, make manifest.*

taken with an extreme desire to go to school again, and to learn the Greek tongue, to the end to peruse the Greek authors; which doth well demonstrate that his former censure of the Grecian learning was rather an affected gravity, than according to the inward sense of his own opinion. And as for Virgil's verses, though it pleased him to brave the world in taking to the Romans the art of empire, and leaving to others the arts of subjects; yet so much is manifest that the Romans never ascended to that height of empire, till the time they had ascended to the height of other arts. For in the time of the two first Caesars, which had the art of government in greatest perfection, there lived the best poet, Virgilius Maro; the best historiographer, Titus Livius; the best antiquary, Marcus Varro; and the best, or second orator, Marcus Cicero, that to the memory of man are known. As for the accusation of Socrates,[n] the time must be remembered when it was prosecuted; which was under the Thirty Tyrants, the most base, bloody, and envious persons that have governed; which revolution of state was no sooner over, but Socrates, whom they had made a person criminal, was made a person heroical, and his memory accumulate with honours divine and human; and those discourses of his which were then termed corrupting of manners, were after acknowledged for sovereign medicines of the mind and manners, and so have been received ever since till this day. Let this therefore serve for answer to politiques, which in their humorous severity, or in their feigned gravity, have presumed to throw imputations upon learning; which redargution nevertheless (save that we know not whether our labours may extend to other ages) were not needful for the present, in regard of the love and reverence towards learning, which the example and countenance of two so learned princes, Queen Elizabeth and your Majesty, being as Castor and Pollux, *lucida sidera*, stars of excellent light and most benign influence, hath wrought in all men of place and authority in our nation.

Virgil's verses: see p. 10 and n. above.
two first Caesars: i.e. Julius and his adopted son Octavius, called Augustus.
envious: i.e. malicious.
accumulate: i.e. heaped with.
humorous: Here in the obsolete sense 'fanciful'.
redargution: refutation.
lucida sidera: Horace, *Odes* I. iii. 2. 'Shining stars.'

[DEFENCE OF LEARNING AGAINST THE DISCREDITS ARISING FROM THE FORTUNES AND MANNERS OF LEARNED MEN]

III. 1. Now therefore we come to that third sort of discredit or diminution of credit that groweth unto learning from learned men themselves, which commonly cleaveth fastest: it is either from their fortune, or from their manners, or from the nature of their studies. For the first, it is not in their power; and the second is accidental; the third only is proper to be handled: but because we are not in hand with true measure, but with popular estimation and conceit, it is not amiss to speak somewhat of the two former. The derogations therefore which grow to learning from the fortune or condition of learned men, are either in respect of scarcity of means, or in respect of privateness of life and meanness of employments.

2. Concerning want, and that it is the case of learned men usually to begin with little, and not to grow rich so fast as other men, by reason they convert not their labours chiefly to lucre and increase, it were good to leave the common place in commendation of poverty to some friar to handle, to whom much was attributed by Machiavel in this point; when he said, 'That the kingdom of the clergy had been long before at an end, if the reputation and reverence towards the poverty of friars had not borne out the scandal of the superfluities and excesses of bishops and prelates.' So a man might say that the felicity and delicacy of princes and great persons had long since turned to rudeness and barbarism, if the poverty of learning had not kept up civility and honour of life: but without any such advantages, it is worthy the observation what a reverent and honoured thing poverty of fortune was for some ages in the Roman state, which nevertheless was a state without paradoxes. For we see what Titus Livius saith in his introduction: *Caeterum aut me amor negotii suscepti fallit, aut nulla unquam respublica nec major, nec sanctior, nec bonis exemplis ditior fuit; nec in quam tam serae avaritia luxuriaque immigraverint; nec ubi tantus ac tam diu paupertati ac parsimoniae honos fuerit.* We see likewise, after that the state of

not in hand with: not dealing with. *by reason*: because.
common place: theme, topic. *Machiavel*: In his *Discourses on Livy* III. i.
borne out: sustained. *delicacy*: luxuriousness.
rudeness: want of culture, of refinement, of learning.
civility: refinement, civilization.
Caeterum aut: From the Preface to Livy's History of Rome. 'If I am not led away by the task which I have undertaken, there never was a republic greater, more religious, and more abounding in good examples than the Roman, nor one that so long withstood avarice and luxury, or so much honoured poverty and thrift.'

Rome was not itself, but did degenerate, how that person that took upon him to be counsellor to Julius Caesar after his victory where to begin his restoration of the state, maketh it of all points the most summary to take away the estimation of wealth: *Verum haec et omnia mala pariter cum honore pecuniae desinent; si neque magistratus, neque alia vulgo cupienda, venalia erunt.* To conclude this point, as it was truly said, that *Rubor est virtutis color,* though sometimes it come from vice; so it may be fitly said that *Paupertas est virtutis fortuna,* though sometimes it may proceed from misgovernment and accident. Surely Salomon hath pronounced it both in censure, *Qui festinat ad divitias non erit insons*; and in precept; *Buy the truth, and sell it not; and so of wisdom and knowledge*; judging that means were to be spent upon learning, and not learning to be applied to means. And as for the privateness or obscureness (as it may be in vulgar estimation accounted) of life of contemplative men; it is a theme so common to extol a private life, not taxed with sensuality and sloth, in comparison and to the disadvantage of a civil life, for safety, liberty, pleasure, and dignity, or at least freedom from indignity, as no man handleth it but handleth it well; such a consonancy it hath to men's conceits in the expressing, and to men's consents in the allowing. This only I will add, that learned men forgotten in states and not living in the eyes of men, are like the images of Cassius and Brutus in the funeral of Junia; of which not being represented, as many others were, Tacitus saith, *Eo ipso praefulgebant, quod non visebantur.*[n]

3. And for meanness of employment, that which is most traduced to contempt is that the government of youth is commonly allotted to them; which age, because it is the age of least authority, it is transferred to the disesteeming of those employments wherein youth is

summary: important.
Verum haec: From the 'Speech to Caesar on the Administration of the State', viii, attributed to Sallust. 'But these, and all evils, will disappear when wealth is no longer honoured, and when the magistracies and other objects of general ambition are not procurable by money.'
Rubor est:] Attributed to Diogenes the Cynic in Diogenes Laertius, *Lives of Eminent Philosophers* vi. 54. 'A blush is virtue's colour.'
Paupertas est: 'Poverty is virtue's fortune.'
Qui festinat: Proverbs 28:20. 'He that makes haste to be rich shall not be innocent.'
Buy the truth: Proverbs 23:23.
extol a private life: to praise a life withdrawn from participation in active affairs of state.
consonancy: agreement, harmony.
allowing: approving.
traduced: exposed to contempt.
disesteeming: despising.*

conversant, and which are conversant about youth. But how unjust this traducement is (if you will reduce things from popularity of opinion to measure of reason) may appear in that we see men are more curious what they put into a new vessel than into a vessel seasoned; and what mould they lay about a young plant than about a plant corroborate; so as the weakest terms and times of all things use to have the best applications and helps. And will you hearken to the Hebrew rabbins?[n] 'Your young men shall see visions, and your old men shall dream dreams'; say they youth is the worthier age, for that visions are nearer apparitions of God than dreams? And let it be noted, that howsoever the condition of life of *pedantes*[n] hath been scorned upon theatres, as the ape of tyranny; and that the modern looseness or negligence hath taken no due regard to the choice of schoolmasters and tutors; yet the ancient wisdom of the best times did always make a just complaint, that states were too busy with their laws and too negligent in point of education: which excellent part of ancient discipline hath been in some sort revived of late times by the colleges of the Jesuits;[n] of whom, although in regard of their superstition I may say, *Quo meliores, eo deteriores*; yet in regard to this, and some other points concerning human learning and moral matters, I may say, as Agesilaus said to his enemy Pharnabazus, *Talis quum sis, utinam noster esses.* And thus much touching the discredits drawn from the fortunes of learned men.

4. As touching the manners of learned men, it is a thing personal and individual: and no doubt there be amongst them, as in other professions, of all temperatures: but yet so as it is not without truth which is said, that *Abeunt studia in mores*, studies have an influence and operation upon the manners of those that are conversant in them.

5. But upon an attentive and indifferent review, I for my part cannot find any disgrace to learning can proceed from the manners of learned men[n] not inherent to them as they are learned; except it be a fault (which was the supposed fault of Demosthenes, Cicero, Cato the second, Seneca, and many more) that because the times they read of are commonly better than the times they live in, and the

corroborate: grown strong. *use*: are accustomed.
Your young men: Joel 2:28. Cf. Essay 42.
Quo meliores: Diogenes Laertius, *Lives of Eminent Philosophers* vi. 46. 'The better they are, the worse they are.'
Talis quum sis: In Plutarch's 'Life' of Agesilaus, 12. 'You are so good that I wish you were on our side.'
temperatures: dispositions.
Abeunt studia: Ovid, *Heroides* xv. 83. 'One's studies pass into one's character.'
indifferent: impartial.

duties taught better than the duties practised, they contend sometimes too far to bring things to perfection, and to reduce the corruption of manners to honesty of precepts or examples of too great height. And yet hereof they have caveats enough in their own walks. For Solon, when he was asked whether he had given his citizens the best laws, answered wisely, 'Yea of such as they would receive': and Plato, finding that his own heart could not agree with the corrupt manners of his country, refused to bear place or office; saying, 'That a man's country was to be used as his parents were, that is, with humble persuasions, and not with contestations.' And Caesar's counsellor put in the same caveat, *Non ad vetera instituta revocans quae jampridem corruptis moribus ludibrio sunt*; and Cicero noteth this error directly in Cato the second, when he writes to his friend Atticus; *Cato optime sentit, sed nocet interdum reipublicae; loquitur enim tanquam in republica Platonis, non tanquam in faece Romuli*. And the same Cicero doth excuse and expound the philosophers for going too far and being too exact in their prescripts, when he saith, *Isti ipsi praeceptores virtutis et magistri videntur fines officiorum paulo longius quam natura vellet protulisse, ut cum ad ultimum animo contendissemus, ibi tamen, ubi oportet, consisteremus*: and yet himself might have said, *Monitis sum minor ipse meis*; for it was his own fault, though not in so extreme a degree.

6. Another fault likewise much of this kind hath been incident to learned men; which is, that they have esteemed the preservation, good, and honour of their countries or masters before their own fortunes or safeties. For so saith Demosthenes unto the Athenians;

contend: strive earnestly. *Solon*: In Plutarch's 'Life' of Solon 15.
Plato: *Crito* 51 c. Bacon probably found this in Cicero's *Letters to his Friends* I. ix. 18.
contestations: controversies.
Non ad vetera: From the 'Speech to Caesar on the Administration of the State', v, attributed to Sallust. 'Do not attempt to restore things to their original institutions which, because of the long corruption of morals, are held in contempt.'
Cato optime: Quoted from memory by Bacon, from Cicero's *Letters to Atticus* II. i. 3. 'Cato's opinions are excellent, but occasionally he does harm to the state: for he speaks as though he were living in Plato's Republic, and not in the filth of Rome.'
expound: interpret the motives or reasons.*
Isti ipsi: Cicero, *Pro Murena* xxi. 65. 'Those very teachers of virtue themselves seem to have fixed the standard of duty somewhat higher than nature can bear, in order that when we strive our utmost to attain it, we may nevertheless reach the level we ought to.'
Monitis sum: Ovid, *The Art of Love* ii. 548. 'I do not act up to my own precepts.'
Demosthenes: *On the Chersonese*, 71.

'If it please you to note it, my counsels unto you are not such whereby I should grow great amongst you, and you become little amongst the Grecians; but they be of that nature, as they are sometimes not good for me to give, but are always good for you to follow.' And so Seneca, after he had consecrated that *Quinquennium Neronis* to the eternal glory of learned governors, held on his honest and loyal course of good and free counsel, after his master grew extremely corrupt in his government. Neither can this point otherwise be; for learning endueth men's minds with a true sense of the frailty of their persons, the casualty of their fortunes, and the dignity of their soul and vocation: so that it is impossible for them to esteem that any greatness of their own fortune can be a true or worthy end of their being and ordainment; and therefore are desirous to give their account to God, and so likewise to their masters under God (as kings and the states that they serve) in these words; *Ecce tibi lucrefeci*, and not *Ecce mihi lucrefeci*: whereas the corrupter sort of mere politiques, that have not their thoughts established by learning in the love and apprehension of duty, nor never look abroad into universality, do refer all things to themselves, and thrust themselves into the centre of the world, as if all lines should meet in them and their fortunes; never caring in all tempests what becomes of the ship of estates, so they may save themselves in the cockboat of their own fortune: whereas men that feel the weight of duty and know the limits of self-love, use to make good their places and duties, though with peril; and if they stand in seditious and violent alterations, it is rather the reverence which many times both adverse parts do give to honesty, than any versatile advantage of their own carriage. But for this point of tender sense and fast obligation of duty which learning doth endue the mind withal, howsoever fortune may tax it, and many in the depth of their corrupt principles may despise it, yet it will receive an open allowance, and therefore needs the less disproof or excusation.

7. Another fault incident commonly to learned men, which may be more probably defended than truly denied, is, that they fail some-

Quinquennium Neronis: the first five years of Nero's reign. Seneca was Nero's tutor, and later one of his chief advisers from A.D. 54 to 63.
casualty: uncertainty.
ordainment: divine appointment.*
Ecce tibi: 'Lo, I have made profit for you.' Cf. Matthew 25:20.
Ecce mihi: 'Lo, I have made profit for myself.'
cockboat: a small ship's boat. *use to*: are accustomed to.
stand in: stand firm in, keep their place. *versatile*: inconstant, changeable.*
allowance: approval.
excusation: excuse.

times in applying themselves to particular persons: which want of exact application ariseth from two causes; the one, because the largeness of their mind can hardly confine itself to dwell in the exquisite observation or examination of the nature and customs of one person: for it is a speech for a lover, and not for a wise man, *Satis magnum alter alteri theatrum sumus.* Nevertheless I shall yield, that he that cannot contract the sight of his mind as well as disperse and dilate it, wanteth a great faculty. But there is a second cause, which is no inability, but a rejection upon choice and judgement. For the honest and just bounds of observation by one person upon another, extend no further but to understand him sufficiently, whereby not to give him offence, or whereby to be able to give him faithful counsel, or whereby to stand upon reasonable guard and caution in respect of a man's self. But to be speculative into another man to the end to know how to work him, or wind him, or govern him, proceedeth from a heart that is double and cloven and not entire and ingenuous; which as in friendship it is want of integrity, so towards princes or superiors is want of duty. For the custom of the Levant, which is that subjects do forbear to gaze or fix their eyes upon princes, is in the outward ceremony barbarous, but the moral is good: for men ought not by cunning and bent observations to pierce and penetrate into the hearts of kings, which the scripture hath declared to be inscrutable.

8. There is yet another fault (with which I will conclude this part) which is often noted in learned men, that they do many times fail to observe decency and discretion in their behaviour and carriage, and commit errors in small and ordinary points of action, so as the vulgar sort of capacities do make a judgement of them in greater matters by that which they find wanting in them in smaller. But this consequence doth oft deceive men, for which I do refer them over to that which was said by Themistocles, arrogantly and uncivilly being applied to himself out of his own mouth, but, being applied to the general state of this question, pertinently and justly; when being invited to touch a lute he said 'He could not fiddle, but he could make a small town a great state.'

applying: adapting. *application*: deference, compliance.*
exquisite: careful, minute.
Satis magnum: Seneca, *Letters to Lucilius* 7, where it is quoted from Epicurus, who applies it to himself and his friend. 'We are a large enough theatre for one another.'
speculative: given to pry into, inquisitive.*
Levant: See Herodotus, *History* i. 99.
bent: crooked, twisted, and so sinister. *inscrutable*: Proverbs 25:3.
Themistocles: In Plutarch's 'Life' of Themistocles, 2.

So no doubt many may be well seen in the passages of government and policy, which are to seek in little and punctual occasions. I refer them also to that which Plato said of his master Socrates,[n] whom he compared to the gallipots of apothecaries, which on the outside had apes and owls and antiques but contained within sovereign and precious liquors and confections; acknowledging that to an external report he was not without superficial levities and deformities, but was inwardly replenished with excellent virtues and powers. And so much touching the point of manners of learned men.

9. But in the mean time I have no purpose to give allowance to some conditions and courses base and unworthy, wherein divers professors of learning have wronged themselves and gone too far; such as were those trencher philosophers which in the later age of the Roman state were usually in the houses of great persons, being little better than solemn parasites; of which kind, Lucian[n] maketh a merry description of the philosopher that the great lady took to ride with her in her coach, and would needs have him carry her little dog, which he doing officiously and yet uncomely, the page scoffed and said, 'That he doubted the philosopher of a Stoic would turn to be a Cynic.' But above all the rest, the gross and palpable flattery, whereunto many not unlearned have abased and abused their wits and pens, turning (as Du Bartas[n] saith) Hecuba into Helena, and Faustina into Lucretia, hath most diminished the price and estimation of learning. Neither is the modern dedication of books and writings, as to patrons, to be commended: for that books (such as are worthy the name of books) ought to have no patrons but truth and reason. And the ancient custom was to dedicate them only to private and equal friends, or to entitle the books with their names: or if to kings and great persons, it was to some such as the argument of the book was fit and proper for: but these and the like courses may deserve rather reprehension than defence.

10. Not that I can tax or condemn the morigeration or application of learned men to men in fortune. For the answer was good that Diogenes made to one that asked him in mockery, 'How it came to

seen: skilled. *passages*: transactions.
are to seek: are at a loss. *punctual*: small.*
gallipots: glazed earthen pots used for medicines.
antiques: grotesque figures. *levities*: smoothnesses.
divers: several. *trencher*: a plate.
uncomely: in an uncomely manner.
doubted: feared. *price*: preciousness, worth.
morigeration: obsequiousness.* *application*: deference, compliance.*
Diogenes: Diogenes Laertius, *Lives of Eminent Philosophers* ii. 69. The answer was given by Aristippus, the disciple of Socrates.

pass that philosophers were the followers of rich men, and not rich men of philosophers?' He answered soberly, and yet sharply, 'Because the one sort knew what they had need of, and the other did not.' And of the like nature was the answer which Aristippus made, when having a petition to Dionysius, and no ear given to him, he fell down at his feet; whereupon Dionysius stayed and gave him the hearing, and granted it; and afterward some person, tender on the behalf of philosophy, reproved Aristippus that he would offer the profession of philosophy such an indignity as for a private suit to fall at a tyrant's feet: but he answered, 'It was not his fault, but it was the fault of Dionysius, that had his ears in his feet.' Neither was it accounted weakness but discretion in him that would not dispute his best with Adrianus Caesar; excusing himself, 'That it was reason to yield to him that commanded thirty legions.' These and the like applications and stooping to points of necessity and convenience cannot be disallowed; for though they may have some outward baseness, yet in a judgement truly made they are to be accounted submissions to the occasion and not to the person.

[DISCREDITS TO LEARNING FROM THE FOLLIES OF THE LEARNED—CICERONIANISM, SCHOLASTICISM, ALCHEMY, AND NATURAL MAGIC]

IV. 1. Now I proceed to those errors and vanities which have intervened amongst the studies themselves of the learned, which is that which is principal and proper to the present argument; wherein my purpose is not to make a justification of the errors, but by a censure and separation of the errors to make a justification of that which is good and sound, and to deliver that from the aspersion of the other. For we see that it is the manner of men to scandalize and deprave that which retaineth the state and virtue, by taking advantage upon that which is corrupt and degenerate: as the heathens in the primitive church used to blemish and taint the Christians with the faults and corruptions of heretics. But nevertheless I have no meaning at this time to make an exact animadversion of the errors and impediments in matters of learning, which are more secret and remote from vulgar opinion, but only to speak unto such as do fall under or near unto a popular observation.

Aristippus: ibid. ii. 70.
Adrianus Caesar: Spartianus in his life of Hadrian 15, tells this story of the philosopher Favorinus.
it was reason: it was reasonable. *deprave*: defame.
the state: original, sound condition.
animadversion: observation, consideration.*

2. There be therefore chiefly three vanities in studies, whereby learning hath been most traduced. For those things we do esteem vain, which are either false or frivolous, those which either have no truth or no use: and those persons we esteem vain, which are either credulous or curious; and curiosity is either in matter or words: so that in reason as well as in experience there fall out to be these three distempers (as I may term them) of learning: the first, fantastical learning; the second, contentious learning; and the last, delicate learning; vain imaginations, vain altercations, and vain affectations; and with the last I will begin. Martin Luther,[n] conducted (no doubt) by an higher providence, but in discourse of reason, finding what a province he had undertaken against the bishop of Rome and the degenerate traditions of the church, and finding his own solitude, being no ways aided by the opinions of his own time, was enforced to awake all antiquity, and to call former times to his succours to make a party against the present time: so that the ancient authors,[n] both in divinity and in humanity, which had long time slept in libraries, began generally to be read and revolved. This by consequence did draw on a necessity of a more exquisite travail in the languages original, wherein those authors did write, for the better understanding of those authors, and the better advantage of pressing and applying their words. And thereof grew again a delight in their manner of style and phrase, and an admiration of that kind of writing; which was much furthered and precipitated by the enmity and opposition that the propounders of those primitive but seeming new opinions had against the schoolmen; who were generally of the contrary part, and whose writings were altogether in a different style and form; taking liberty to coin and frame new terms of art to express their own sense, and to avoid circuit of speech, without regard to the pureness, pleasantness, and (as I may call it) lawfulness of the phrase or word. And again, because the great labour then was with the people (of whom the Pharisees were wont to say, *Execrabilis ista turba, quae non novit legem*), for the winning and persuading of them there grew of necessity in chief price and request eloquence and variety of discourse, as the fittest and forciblest access into the capacity of the vulgar sort: so that these four causes concurring,

curious: difficult to satisfy.
fantastical: existing only in the imagination. *delicate*: affected.
discourse of reason: an obsolete phrase meaning the process of reasoning.
province: sphere of action. *revolved*: considered.
exquisite travail: careful labour.
pressing: emphasizing. *circuit*: circumlocution.
Execrabilis: John 7:49. 'This multitude which knoweth not the law are accursed.'

the admiration of ancient authors, the hate of the schoolmen, the exact study of languages, and the efficacy of preaching, did bring in an affectionate study of eloquence and copie of speech,[n] which then began to flourish. This grew speedily to an excess; for men began to hunt more after words than matter; more after the choiceness of the phrase, and the round and clean composition of the sentence, and the sweet falling of the clauses, and the varying and illustration of their works with tropes and figures, than after the weight of matter, worth of subject, soundness of argument, life of invention, or depth of judgement. Then grew the flowing and watery vein of Osorius,[n] the Portugal bishop, to be in price. Then did Sturmius spend such infinite and curious pains upon Cicero the Orator, and Hermogenes the Rhetorician, besides his own books of Periods and Imitation, and the like. Then did Car of Cambridge and Ascham with their lectures and writings almost deify Cicero and Demosthenes, and allure all young men that were studious unto that delicate and polished kind of learning. Then did Erasmus take occasion to make the scoffing echo, *Decem annos consumpsi in legendo Cicerone*; and the echo answered in Greek *One, Asine*.[n] Then grew the learning of the schoolmen to be utterly despised as barbarous.[n] In sum, the whole inclination and bent of those times was rather towards copie than weight.

3. Here therefore is the first distemper of learning, when men study words and not matter; whereof, though I have represented an example of late times, yet it hath been and will be *secundum majus et minus* in all time. And how is it possible but this should have an operation to discredit learning, even with vulgar capacities, when they see learned men's works like the first letter of a patent, or limned book;[n] which though it hath large flourishes, yet it is but a letter? It seems to me that Pygmalion's frenzy is a good emblem or portraiture of this vanity: for words are but the images of matter; and except they have life of reason and invention, to fall in love with them is all one as to fall in love with a picture.

4. But yet notwithstanding it is a thing not hastily to be condemned, to clothe and adorn the obscurity even of philosophy itself with sensible and plausible elocution. For hereof we have great examples in Xenophon, Cicero, Seneca, Plutarch, and of Plato also

affectionate: ambitious.
distemper: disorder, 'disease'.
secundum: 'to a greater or less degree'.
Pygmalion's frenzy: Pygmalion fell in love with a statue. Ovid, *Metamorphoses* x. 243.
sensible and plausible: such as makes a strong impression and wins approval.

in some degree; and hereof likewise there is great use: for surely, to the severe inquisition of truth and the deep progress into philosophy, it is some hindrance; because it is too early satisfactory to the mind of man, and quencheth the desire of further search, before we come to a just period.[n] But then if a man be to have any use of such knowledge in civil occasions, of conference, counsel, persuasion, discourse, or the like, then shall he find it prepared to his hands in those authors which write in that manner. But the excess of this is so justly contemptible, that as Hercules, when he saw the image of Adonis, Venus' minion, in a temple, said in disdain, *Nil sacri es*;[n] so there is none of Hercules' followers in learning, that is, the more severe and laborious sort of inquirers into truth, but will despise those delicacies and affectations, as indeed capable of no divineness.[n] And thus much of the first disease or distemper of learning.

5. The second which followeth is in nature worse than the former: for as substance of matter is better than beauty of words, so contrariwise vain matter is worse than vain words: wherein it seemeth the reprehension of Saint Paul was not only proper for those times, but prophetical for the times following; and not only respective to divinity, but extensive to all knowledge: *Devita profanas vocum novitates, et oppositiones falsi nominis scientiae.* For he assigneth two marks and badges of suspected and falsified science: the one, the novelty and strangeness of terms; the other, the strictness of positions, which of necessity doth induce oppositions, and so questions and altercations. Surely, like as many substances in nature which are solid do putrify and corrupt into worms; so it is the property of good and sound knowledge to putrify and dissolve into a number of subtle, idle, unwholesome, and (as I may term them) vermiculate questions, which have indeed a kind of quickness and life of spirit, but no soundness of matter or goodness of quality. This kind of degenerate learning did chiefly reign amongst the schoolmen:[n] who having sharp and strong wits, and abundance of leisure, and small variety of reading, but their wits being shut up in the cells of a few authors (chiefly Aristotle their dictator) as their persons were shut up in the cells of monasteries and colleges, and

period: end, conclusion. *minion*: darling.
extensive: far-reaching, capable of being extended.*
Devita: I Timothy 6:20. 'Turn a deaf ear to empty and worldly chatter and the contradictions of so-called "knowledge".'
strictness of positions: i.e. rigidity of logical propositions; see §6 below.
vermiculate: intricate. Bacon, who is the first to use this word (OED), is thinking of the putrifying body 'dissolving' into worms.*
quickness: animation, lifelikeness.

knowing little history, either of nature or time, did out of no great quantity of matter and infinite agitation of wit spin out unto us those laborious webs of learning which are extant in their books. For the wit and mind of man, if it work upon matter, which is the contemplation of the creatures of God, worketh according to the stuff and is limited thereby; but if it work upon itself, as the spider worketh his web, then it is endless, and brings forth indeed cobwebs of learning, admirable for the fineness of thread and work, but of no substance or profit.

6. This same unprofitable subtility or curiosity is of two sorts; either in the subject itself that they handle, when it is a fruitless speculation or controversy (whereof there are no small number both in divinity and philosophy), or in the manner or method of handling of a knowledge, which amongst them was this; upon every particular position or assertion to frame objections, and to those objections, solutions; which solutions were for the most part not confutations, but distinctions: whereas indeed the strength of all sciences is, as the strength of the old man's faggot, in the bond. For the harmony of a science, supporting each part the other, is and ought to be the true and brief confutation and suppression of all the smaller sort of objections. But, on the other side, if you take out every axiom, as the sticks of the faggot, one by one, you may quarrel with them and bend them and break them at your pleasure: so that as was said of Seneca, *Verborum minutiis rerum frangit pondera*, so a man may truly say of the schoolmen, *Quaestionum minutiis scientiarum frangunt soliditatem*. For were it not better for a man in a fair room to set up one great light, or branching candlestick of lights, than to go about with a small watch candle into every corner? And such is their method, that rests not so much upon evidence of truth proved by arguments, authorities, similitudes, examples, as upon particular confutations and solutions of every scruple, cavillation, and objection; breeding for the most part one question as fast as it solveth another; even as in the former resemblance, when you carry the light into one corner, you darken the rest; so that the fable and fiction of Scylla seemeth to be a lively image of this kind of philosophy or knowledge; which was trans-

agitation: disturbance, commotion.*

Verborum minutiis: Quoted from memory from Quintilian, *Institutions* X. i. 130. 'That he broke up the weight of the matter by his verbal subtleties.'

Quaestionum: 'They break up the solidity of the sciences by hair-splitting questions.'

cavillation: the making of quibbling objections.

formed into a comely virgin for the upper parts; but then *Candida succinctam latrantibus inguina monstris*: so the generalities of the schoolmen are for a while good and proportionable; but then when you descend into their distinctions and decisions, instead of a fruitful womb for the use and benefit of man's life, they end in monstrous altercations and barking questions. So as it is not possible but this quality of knowledge must fall under popular contempt, the people being apt to contemn truth upon occasion of controversies and altercations, and to think they are all out of their way which never meet; and when they see such digladiation about subtilties, and matter of no use or moment, they easily fall upon that judgement of Dionysius of Syracusa. *Verba ista sunt senum otiosorum.*

7. Notwithstanding, certain it is that if those schoolmen to their great thirst of truth and unwearied travail of wit had joined variety and universality of reading and contemplation, they had proved excellent lights, to the great advancement of all learning and knowledge; but as they are, they are great undertakers indeed, and fierce with dark keeping.[n] But as in the inquiry of the divine truth, their pride inclined to leave the oracle of God's word, and to vanish in the mixture of their own inventions; so in the inquisition of nature, they ever left the oracle of God's works, and adored the deceiving and deformed images which the unequal mirror of their own minds, or a few received authors or principles, did represent unto them. And thus much for the second disease of learning.

8. For the third vice or disease of learning, which concerneth deceit or untruth, it is of all the rest the foulest; as that which doth destroy the essential form of knowledge, which is nothing but a representation of truth: for the truth of being and the truth of knowing are one, differing no more than the direct beam and the beam reflected. This vice therefore brancheth itself into two sorts; delight in deceiving and aptness to be deceived; imposture and credulity; which, although they appear to be of a diverse nature, the one seeming to proceed of cunning and the other of simplicity, yet certainly they do for the most part concur: for, as the verse noteth,

Candida: Virgil, *Eclogues* vi. 75. 'There were barking monsters all about her loins.'
proportionable: well-proportioned.†
digladiation: literally, fighting with swords; hence, contention, wrangling.
Verba ista: Diogenes Laertius, *Lives of Eminent Philosophers* iii. 18. 'It is old men's idle talk.' Cf. *Novum Organum* i. 71.
undertakers: students, those who undertake serious study.*
unequal: inadequate because uneven.

Percontatorem fugito, nam garrulus idem est,

an inquisitive man is a prattler; so upon the like reason a credulous man is a deceiver: as we see it in fame, that he that will easily believe rumours, will as easily augment rumours and add somewhat to them of his own; which Tacitus wisely noteth, when he saith, *Fingunt simul creduntque*: so great an affinity hath fiction and belief.

9. This facility of credit and accepting or admitting things weakly authorized or warranted, is of two kinds according to the subject: for it is either a belief of history, or, as the lawyers speak, matter of fact; or else of matter of art and opinion. As to the former, we see the experience and inconvenience of this error in ecclesiastical history; which hath too easily received and registered reports and narrations of miracles wrought by martyrs, hermits, or monks of the desert, and other holy men, and their relics, shrines, chapels, and images: which though they had a passage for a time by the ignorance of the people, the superstitious simplicity of some, and the politic toleration of others, holding them but as divine poesies; yet after a period of time, when the mist began to clear up, they grew to be esteemed but as old wives' fables, impostures of the clergy, illusions of spirits, and badges of Antichrist, to the great scandal and detriment of religion.

10. So in natural history, we see there hath not been that choice and judgement used as ought to have been; as may appear in the writings of Plinius,[n] Cardanus, Albertus,[n] and divers of the Arabians,[n] being fraught with much fabulous matter, a great part not only untried, but notoriously untrue, to the great derogation of the credit of natural philosophy with the grave and sober kind of wits: wherein the wisdom and integrity of Aristotle is worthy to be observed; that, having made so diligent and exquisite a history of living creatures,[n] hath mingled it sparingly with any vain or feigned matter: and yet on the other side hath cast all prodigious narrations, which he thought worthy the recording, into one book: excellently discerning that matter of manifest truth, such whereupon observation and rule was to be built, was not to be mingled or weakened with matter of doubtful credit; and yet again, that rarities and reports that seem

Percontatorem: Horace, *Epistles* I. xviii. 69. 'Shun the inquisitive man, for he is a babbler.'
fame: here means report, rumour.
Fingunt simul: Adaptation of a phrase from Tacitus, *Annals* v. 10. 'Those who are prone to invent are also prone to believe.'
had a passage: were generally accepted.

uncredible are not to be suppressed or denied to the memory of men.

11. And as for the facility of credit which is yielded to arts and opinions, it is likewise of two kinds; either when too much belief is attributed to the arts themselves, or to certain authors in any art. The sciences themselves, which have had better intelligence and confederacy with the imagination of man than with his reason, are three in number; astrology, natural magic, and alchemy:[n] of which sciences, nevertheless, the ends or pretences are noble. For astrology pretendeth to discover that correspondence or concatenation which is between the superior globe and the inferior: natural magic pretendeth to call and reduce natural philosophy from variety of speculations to the magnitude of works: and alchemy pretendeth to make separation of all the unlike parts of bodies which in mixtures of nature are incorporate. But the derivations and prosecutions to these ends,[n] both in the theories and in the practices, are full of error and vanity; which the great professors themselves have sought to veil over and conceal by enigmatical writings, and referring themselves to auricular traditions and such other devices, to save the credit of impostures. And yet surely to alchemy this right is due, that it may be compared to the husbandman whereof Aesop makes the fable; that, when he died, told his sons that he had left unto them gold buried under ground in his vineyard; and they digged over all the ground, and gold they found none; but by reason of their stirring and digging the mould about the roots of their vines, they had a great vintage the year following: so assuredly the search and stir to make gold hath brought to light a great number of good and fruitful inventions and experiments, as well for the disclosing of nature as for the use of man's life.[n]

12. And as for the overmuch credit that hath been given unto authors in sciences, in making them dictators, that their words should stand, and not consuls to give advice; the damage is infinite that sciences have received thereby, as the principal cause that hath kept them low at a stay without growth or advancement. For hence it hath comen, that in arts mechanical the first deviser comes shortest, and time addeth and perfecteth; but in sciences the first author goeth furthest, and time leeseth and corrupteth. So we see, artillery, sailing, printing,[n] and the like, were grossly managed at the first, and by time accommodated and refined: but contrariwise, the philosophies and sciences of Aristotle, Plato, Democritus, Hippocrates,

arts: branches of learning.
pretences: expressed aim.
grossly: clumsily, unskilfully.
intelligence: a relation of intercourse.
auricular: hearsay, oral.*

Euclides, Archimedes, of most vigour at the first and by time degenerate and imbased; whereof the reason is no other, but that in the former many wits and industries have contributed in one; and in the latter many wits and industries have been spent about the wit of some one, whom many times they have rather depraved than illustrated. For as water will not ascend higher than the level of the first springhead from whence it descendeth, so knowledge derived from Aristotle, and exempted from liberty of examination, will not rise again higher than the knowledge of Aristotle. And therefore although the position be good, *Oportet discentem credere*, yet it must be coupled with this, *Oportet edoctum judicare*; for disciples do owe unto masters only a temporary belief and a suspension of their own judgement till they be fully instructed, and not an absolute resignation or perpetual captivity: and therefore, to conclude this point, I will say no more, but so let great authors have their due, as time, which is the author of authors, be not deprived of his due, which is, further and further to discover truth. Thus have I gone over these three diseases of learning; besides the which there are some other rather peccant[n] humours than formed diseases, which nevertheless are not so secret and intrinsic but that they fall under a popular observation and traducement, and therefore are not to be passed over.

[OTHER ERRORS OF LEARNED MEN WHICH HINDER THE PROGRESS AND CREDIT OF LEARNING]

V. 1. The first of these is the extreme affecting of two extremities: the one antiquity, the other novelty; wherein it seemeth the children of time do take after the nature and malice of the father. For as he devoureth his children, so one of them seeketh to devour and suppress the other; while antiquity envieth there should be new additions, and novelty cannot be content to add but it must deface: surely the advice of the prophet is the true direction in this matter, *State super vias antiquas, et videte quaenam sit via recta et bona et ambulate in ea.* Antiquity deserveth that reverence, that men should

imbased: depraved.
Oportet discentem: Aristotle, *Sophistici Elenchi* 165b. 3. 'The learner should take things on trust.'
Oportet edoctum: 'The learned man should exercise his judgement.'
intrinsic: internal, and so hidden.
father: i.e. Chronos (which is the Greek word for time) who devoured his children at birth.
envieth: begrudges.
State super: Jeremiah 6:16. 'Stand in the ancient paths and look about, and see which is the straight and right way, and so walk in it.'

make a stand thereupon and discover what is the best way; but when the discovery is well taken, then to make progression. And to speak truly, *Antiquitas saeculi juventus mundi*.[n] These times are the ancient times, when the world is ancient, and not those which we account ancient *ordine retrogrado*, by a computation backward from ourselves.

2. Another error induced by the former is a distrust that anything should be now to be found out, which the world should have missed and passed over so long time; as if the same objection were to be made to time, that Lucian maketh to Jupiter and other the heathen gods; of which he wondereth that they begot so many children in old time, and begot none in his time; and asketh whether they were become septuagenary, or whether the law *Papia*,[n] made against old men's marriages, had restrained them. So it seemeth men doubt lest time is become past children and generation; wherein contrariwise we see commonly the levity and unconstancy of men's judgements, which till a matter be done, wonder that it can be done; and as soon as it is done, wonder again that it was no sooner done: as we see in the expedition of Alexander into Asia, which at first was prejudged as a vast and impossible enterprise; and yet afterwards it pleaseth Livy to make no more of it than this, *Nil aliud quam bene ausus vana contemnere*. And the same happened to Columbus[n] in the western navigation. But in intellectual matters it is much more common; as may be seen in most of the propositions of Euclid;[n] which till they be demonstrate, they seem strange to our assent; but being demonstrate, our mind accepteth of them by a kind of relation[n] (as the lawyers speak) as if we had known them before.

3. Another error, that hath also some affinity with the former, is a conceit that of former opinions or sects after variety and examination[n] the best hath still prevailed and suppressed the rest; so as if a man should begin the labour of a new search, he were but like to light upon somewhat formerly rejected, and by rejection brought into oblivion: as if the multitude, or the wisest for the multitude's sake, were not ready to give passage rather to that which is popular and superficial, than to that which is substantial and profound; for the truth is, that time seemeth to be of the nature of a river or stream, which carrieth down to us that which is light and blown up, and sinketh and drowneth that which is weighty and solid.

Lucian: not Lucian but Seneca, quoted by Lactantius *De Falsa Religione*, i. 16.
septuagenary: seventy years old.*
Nil aliud: Livy ix. 17. 'It was nothing but daring to despise groundless fears.'
still: invariably.

4. Another error, of a diverse nature from all the former, is the over-early and peremptory reduction of knowledge into arts and methods;[n] from which time commonly sciences receive small or no augmentation. But as young men, when they knit and shape perfectly, do seldom grow to a further stature; so knowledge, while it is in aphorisms and observations, it is in growth: but when it once is comprehended in exact methods, it may perchance be further polished and illustrate and accommodated for use and practice; but it increaseth no more in bulk and substance.

5. Another error which doth succeed that which we last mentioned, is, that after the distribution of particular arts and sciences, men have abandoned universality, or *philosophia prima*:[n] which cannot but cease and stop all progression. For no perfect discovery can be made upon a flat or a level: neither is it possible to discover the more remote and deeper parts of any science, if you stand but upon the level of the same science, and ascend not to a higher science.

6. Another error hath proceeded from too great a reverence, and a kind of adoration of the mind and understanding of man; by means whereof, men have withdrawn themselves too much from the contemplation of nature, and the observations of experience, and have tumbled up and down in their own reason and conceits. Upon these intellectualists,[n] which are notwithstanding commonly taken for the most sublime and divine philosophers, Heraclitus gave a just censure, saying, 'Men sought truth in their own little worlds, and not in the great and common world'; for they disdain to spell, and so by degrees to read in the volume of God's works: and contrariwise by continual meditation and agitation of wit do urge and as it were invocate their own spirits to divine and give oracles unto them, whereby they are deservedly deluded.

7. Another error that hath some connexion with this latter is, that men have used to infect their meditations, opinions, and doctrines, with some conceits which they have most admired, or some sciences which they have most applied; and given all things else a tincture according to them, utterly untrue and unproper. So hath Plato intermingled his philosophy with theology,[n] and Aristotle with logic; and the second school of Plato, Proclus and the rest, with the mathematics. For these were the arts which had a kind of primogeniture with them severally. So have the alchemists[n] made a philosophy out of a few experiments of the furnace; and Gilbertus[n] our countryman hath made a philosophy out of the observations of a

knit: become compact or firm.* *illustrate*: i.e. rendered illustrious.
Heraclitus: in Sextus Empiricus, *Against the Logicians* i. 133.

loadstone. So Cicero, when, reciting the several opinions of the nature of the soul, he found a musician that held the soul was but a harmony, saith pleasantly, *Hic ab arte sua non recessit, &c.* But of these conceits Aristotle speaketh seriously and wisely when he saith, *Qui respiciunt ad pauca de facili pronunciant.*

8. Another error is an impatience of doubt, and haste to assertion without due and mature suspension of judgement. For the two ways of contemplation are not unlike the two ways of action[n] commonly spoken of by the ancients: the one plain and smooth in the beginning, and in the end impassable; the other rough and troublesome in the entrance, but after a while fair and even: so it is in contemplation; if a man will begin with certainties, he shall end in doubts; but if he will be content to begin with doubts, he shall end in certainties.

9. Another error is in the manner of the tradition[n] and delivery of knowledge, which is for the most part magistral and peremptory, and not ingenuous and faithful; in a sort as may be soonest believed, and not easiliest examined. It is true that in compendious treatises for practice that form is not to be disallowed: but in the true handling of knowledge, men ought not to fall either on the one side into the vein of Velleius the Epicurean, *Nil tam metuens, quam ne dubitare aliqua de re videretur*; nor on the other side into Socrates his ironical doubting of all things; but to propound things sincerely with more or less asseveration, as they stand in a man's own judgement proved more or less.

10. Other errors there are in the scope that men propound to themselves, whereunto they bend their endeavours; for whereas the more constant and devote kind of professors of any science ought to propound to themselves to make some additions to their science, they convert their labours to aspire to certain second prizes: as to be a profound interpreter or commenter, to be a sharp champion or defender, to be a methodical compounder or abridger, and so the patrimony of knowledge cometh to be sometimes improved, but seldom augmented.

Hic ab arte: Cicero, *Tusculan Disputations* I. x. 20. 'This man has not gone outside the limits of his own art.'
Qui respiciunt: Aristotle, *On Coming-to-be and Passing-away* I. 2 (316a). 'Those who take into account only a few considerations easily make dogmatic pronouncements.'
magistral: dogmatic.*
Nil tam: Cicero, *On the Nature of the Gods* I. viii. 18. 'Fearing nothing so much as to appear doubtful about anything.' Cf. *Novum Organum* i. 67.
Socrates: cf. II. xiii. 4 below.
scope: mark to aim at, and so aim, object.

11. But the greatest error of all the rest is the mistaking or misplacing of the last or furthest end of knowledge.[n] For men have entered into a desire of learning and knowledge, sometimes upon a natural curiosity and inquisitive appetite; sometimes to entertain their minds with variety and delight; sometimes for ornament and reputation; and sometimes to enable them to victory of wit and contradiction; and most times for lucre and profession; and seldom sincerely to give a true account of their gift of reason, to the benefit and use of men: as if there were sought in knowledge a couch whereupon to rest a searching and restless spirit; or a terrace for a wandering and variable mind to walk up and down with a fair prospect; or a tower of state for a proud mind to raise itself upon; or a fort or commanding ground for strife and contention; or a shop for profit or sale; and not a rich storehouse for the glory of the Creator and the relief of man's estate. But this is that which will indeed dignify and exalt knowledge, if contemplation and action may be more nearly and straitly conjoined and united together than they have been; a conjunction like unto that of the two highest planets, Saturn, the planet of rest and contemplation, and Jupiter, the planet of civil society and action.[n] Howbeit, I do not mean, when I speak of use and action, that end before-mentioned of the applying of knowledge to lucre and profession; for I am not ignorant how much that diverteth and interrupteth the prosecution and advancement of knowledge, like unto the golden ball thrown before Atalanta, which while she goeth aside and stoopeth to take up, the race is hindered,

Declinate cursus, aurumque voulbile tollit.[n]

Neither is my meaning, as was spoken of Socrates, to call philosophy down from heaven to converse upon the earth; that is, to leave natural philosophy aside, and to apply knowledge only to manners and policy. But as both heaven and earth do conspire and contribute to the use and benefit of man; so the end ought to be, from both philosophies to separate and reject vain speculations, and whatsoever is empty and void, and to preserve and augment whatsoever is solid and fruitful: that knowledge may not be as a courtesan, for pleasure and vanity only, or as a bond-woman, to acquire and gain to her master's use; but as a spouse, for generation, fruit, and comfort.

12. Thus have I described and opened, as by a kind of dissection,

enable: qualify.
Socrates: In Cicero, *Tusculan Disputations* V. iv. 10.
converse: dwell.

those peccant humours (the principal of them) which have not only given impediment to the proficience of learning, but have given also occasion to the traducement thereof: wherein if I have been too plain, it must be remembered, *fidelia vulnera amantis, sed dolosa oscula malignantis.* This I think I have gained, that I ought to be the better believed in that which I shall say pertaining to commendation; because I have proceeded so freely in that which concerneth censure. And yet I have no purpose to enter into a laudative of learning, or to make a hymn to the Muses (though I am of opinion that it is long since their rites were duly celebrated), but my intent is, without varnish or amplification justly to weigh the dignity of knowledge in the balance with other things, and to take the true value thereof by testimonies and arguments divine and human.

[DIVINE PROOFS OF THE DIGNITY OF LEARNING]

VI. 1. First therefore let us seek the dignity of knowledge in the arch-type[n] or first platform, which is in the attributes and acts of God, as far as they are revealed to man and may be observed with sobriety; wherein we may not seek it by the name of learning; for all learning is knowledge acquired, and all knowledge in God is original: and therefore we must look for it by another name, that of wisdom or sapience, as the scriptures call it.

2. It is so then, that in the work of the creation[n] we see a double emanation of virtue from God; the one referring more properly to power, the other to wisdom; the one expressed in making the subsistence of the matter, and the other in disposing the beauty of the form. This being supposed, it is to be observed that for anything which appeareth in the history of the creation, the confused mass and matter of heaven and earth was made in a moment; and the order and disposition of that chaos or mass was the work of six days; such a note of difference it pleased God to put upon the works of power, and the works of wisdom; wherewith concurreth, that in the former it is not set down that God said, 'Let there be heaven and earth,' as it is set down of the works following; but actually, that God made heaven and earth: the one carrying the style of a manufacture, and the other of a law, decree, or counsel.

fidelia vulnera: Proverbs 27:6. 'Faithful are the wounds of a friend, but the kisses of an enemy are deceitful.'
laudative: eulogy.* *amplification*: exaggeration.
arch-type: the original pattern or model from which copies are made.
platform: model, pattern. *scriptures call it*: see Proverbs 8:22–31.
subsistence: substratum.

3. To proceed to that which is next in order from God to spirits; we find, as far as credit is to be given to the celestial hierarchy of that supposed Dionysius[n] the senator of Athens, the first place or degree is given to the angels of love, which are termed seraphim; the second to the angels of light, which are termed cherubim; and the third, and so following places, to thrones, principalities, and the rest, which are all angels of power and ministry; so as the angels of knowledge and illumination are placed before the angels of office and domination.

4. To descend from spirits and intellectual forms to sensible and material forms, we read the first form that was created was light,[n] which hath a relation and correspondence in nature and corporal things to knowledge in spirits and incorporal things.

5. So in the distribution of days we see the day wherein God did rest and contemplate his own works, was blessed above all the days wherein he did effect and accomplish them.

6. After the creation was finished, it is set down unto us that man was placed in the garden to work therein; which work, so appointed to him, could be no other than work of contemplation; that is, when the end of work is but for exercise and experiment, not for necessity; for there being then no reluctation of the creature, nor sweat of the brow, man's employment must of consequence have been matter of delight in the experiment, and not matter of labour for the use. Again, the first acts which man performed in Paradise consisted of the two summary parts of knowledge; the view of creatures, and the imposition of names.[n] As for the knowledge which induced the fall, it was, as was touched before, not the natural knowledge of creatures, but the moral knowledge of good and evil; wherein the supposition was, that God's commandments or prohibitions were not the originals of good and evil, but that they had other beginnings, which man aspired to know; to the end to make a total defection from God and to depend wholly upon himself.

7. To pass on: in the first event or occurrence after the fall of man, we see (as the scriptures have infinite mysteries, not violating at all the truth of the story or letter) an image of the two estates, the contemplative state and the active state, figured in the two persons of Abel and Cain,[n] and in the two simplest and most primitive trades of life; that of the shepherd (who, by reason of his leisure, rest in a place, and living in view of heaven, is a lively image of a contemplative life), and that of the husbandman: where we see again the

reluctation: unwillingness.* *summary*: most important.

favour and election of God went to the shepherd, and not to the tiller of the ground.

8. So in the age before the flood, the holy records within those few memorials which are there entered and registered, have vouchsafed to mention and honour the name of the inventors and authors of music and works in metal.[n] In the age after the flood, the first great judgement of God upon the ambition of man was the confusion of tongues; whereby the open trade and intercourse of learning and knowledge was chiefly imbarred.

9. To descend to Moyses[n] the lawgiver, and God's first pen: he is adorned by the scriptures with this addition and commendation, 'That he was seen in all the learning of the Egyptians'; which nation we know was one of the most ancient schools of the world: for so Plato brings in the Egyptian priest saying unto Solon, 'You Grecians are ever children; you have no knowledge of antiquity, nor antiquity of knowledge.' Take a view of the ceremonial law of Moyses;[n] you shall find, besides the prefiguration of Christ, the badge or difference of the people of God, the exercise and impression of obedience, and other divine uses and fruits thereof, that some of the most learned Rabbins have travailed profitably and profoundly to observe, some of them a natural, some of them a moral, sense or reduction of many of the ceremonies and ordinances. As in the law of the leprosy, where it is said, 'If the whiteness have overspread the flesh, the patient may pass abroad for clean; but if there be any whole flesh remaining, he is to be shut up for unclean'; one of them noteth a principle of nature, that putrefaction is more contagious before maturity than after: and another noteth a position of moral philosophy, that men abandoned to vice do not so much corrupt manners, as those that are half good and half evil. So in this and very many other places in that law, there is to be found, besides the theological sense, much aspersion of philosophy.

10. So likewise in that excellent book of Job,[n] if it be revolved with diligence, it will be found pregnant and swelling with natural philosophy; as for example, cosmography, and the roundness of the world, *Qui extendit aquilonem super vacuum, et appendit terram super nihilum*; wherein the pensileness of the earth, the pole of the north, and the finiteness or convexity of heaven are manifestly

within: among. *That he was seen*: Acts 7:22.
You Grecians: Plato, *Timaeus* 22 b. *difference*: a distinguishing mark.
If the whiteness: Leviticus 13:12–13. *aspersion*: intermixture.*
Qui extendit: Job 26:7 '[He] stretcheth out the north over the empty space, and hangeth the earth upon nothing.'
pensileness: suspended condition.*

touched. So again, matter of astronomy; *Spiritus ejus ornavit caelos, et obstetricante manu ejus eductus est Coluber tortuosus.* And in another place, *Nunquid conjungere valebis micantes stellas Pleiadas, aut gyrum Arcturi poteris dissipare?* Where the fixing of the stars, ever standing at equal distance, is with great elegancy noted. And in another place, *Qui facit Arcturum, et Oriona, et Hyadas, et interiora Austri*; where again he takes knowledge of the depression of the southern pole, calling it the secrets of the south, because the southern stars were in that climate unseen. Matter of generation; *Annon sicut lac mulsisti me, et sicut caseum coagulasti me?* &c. Matter of minerals; *Habet argentum venarum suarum principia: et auro locus est in quo conflatur, ferrum de terra tollitur, et lapis solutus calore in aes vertitur*: and so forwards in that chapter.

11. So likewise in the person of Salomon the king, we see the gift or endowment of wisdom and learning, both in Salomon's petition[n] and in God's assent thereunto, preferred before all other terrene and temporal felicity. By virtue of which grant or donative of God Salomon became enabled not only to write those excellent parables or aphorisms concerning divine and moral philosophy; but also to compile a natural history[n] of all verdure, from the cedar upon the mountain to the moss upon the wall (which is but a rudiment between putrefaction and an herb), and also of all things that breathe or move. Nay, the same Salomon the king, although he excelled in the glory of treasure and magnificent buildings, of shipping and navigation, of service and attendance, of fame and renown, and the like, yet he maketh no claim to any of those glories, but only to the glory of inquisition of truth; for so he saith expressly, 'The glory of God is to conceal a thing, but the glory of the king is to find it out';

Spiritus ejus: Job 26:13. 'By his spirit he hath garnished the heavens; his hand hath formed the crooked serpent.'
Nunquid conjungere: Job 38:31. 'Canst thou bind the glittering stars of the Pleiades, or scatter the array of Arcturus?'
fixing: fixed position.
Qui facit: Job 9:9. 'Which maketh Arcturus, Orion, and Pleiades, and the secrets of the south.'
depression: A term in astronomy, signifying the angular distance of the pole beneath the horizon.
climate: region of the earth.
Annon sicut: Job 10:10. 'Hast thou not poured me out as milk and curdled me like cheese?'
Habet argentum: Job 28:1–2. 'Surely there is a vein for the silver, and a place for gold, where they fine it. Iron is taken out of the earth, and brass is molten out of the stone.'
parables or aphorisms: i.e. the book of Proverbs and The Wisdom of Solomon (in the Apocrypha).
The glory of God: In Proverbs 25:2.

as if, according to the innocent play of children, the Divine Majesty took delight to hide his works, to the end to have them found out; and as if kings could not obtain a greater honour than to be God's playfellows[n] in that game; considering the great commandment of wits and means, whereby nothing needeth to be hidden from them.

12. Neither did the dispensation of God vary in the times after our Saviour came into the world; for our Saviour himself did first show his power to subdue ignorance, by his conference with the priests and doctors of the law, before he showed his power to subdue nature by his miracles. And the coming of the Holy Spirit was chiefly figured and expressed in the similitude and gift of tongues, which are but *vehicula scientiae*.

13. So in the election of those instruments, which it pleased God to use for the plantation of the faith, notwithstanding that at the first he did employ persons altogether unlearned, otherwise than by inspiration, more evidently to declare his immediate working, and to abase all human wisdom or knowledge; yet nevertheless that counsel of his was no sooner performed, but in the next vicissitude and succession he did send his divine truth into the world, waited on with other learnings, as with servants or handmaids: for so we see Saint Paul, who was only learned amongst the Apostles, had his pen most used in the scriptures of the New Testament.

14. So again we find that many of the ancient bishops and fathers of the Church were excellently read and studied in all the learning of the heathen; insomuch that the edict of the Emperor Julianus[n] (whereby it was interdicted unto Christians to be admitted into schools, lectures, or exercises of learning) was esteemed and accounted a more pernicious engine and machination against the Christian Faith, than were all the sanguinary prosecutions of his predecessors; neither could the emulation and jealousy of Gregory[n] the first of that name, bishop of Rome, ever obtain the opinion of piety or devotion; but contrariwise received the censure of humour, malignity and pusillanimity, even amongst holy men; in that he designed to obliterate and extinguish the memory of heathen antiquity and authors. But contrariwise it was the Christian church, which, amidst the inundations of the Scythians on the one side from

conference: Luke 2:46.
vehicula: Acts 2:1 f. 'Carriers of knowledge.'
learnings: branches of knowledge. *only learned*: i.e. the only learned.
engine: contrivance requiring skill, hence a plot.
emulation: ambitious rivalry, in a bad sense.
jealousy: vehemence of feeling against something.
opinion: reputation. *humour*: caprice.

the north-west, and the Saracens from the east, did preserve in the sacred lap and bosom thereof the precious relics even of heathen learning, which otherwise had been extinguished as if no such thing had ever been.

15. And we see before our eyes, that in the age of ourselves and our fathers, when it pleased God to call the Church of Rome to account for their degenerate manners and ceremonies, and sundry doctrines obnoxious and framed to uphold the same abuses; at one and the same time it was ordained by the Divine Providence, that there should attend withal a renovation and new spring of all other knowledges. And, on the other side we see the Jesuits, who partly in themselves and partly by the emulation and provocation of their example, have much quickened and strengthened the state of learning, we see (I say) what notable service and reparation they have done to the Roman see.[n]

16. Wherefore to conclude this part, let it be observed, that there be two principal duties and services, besides ornament and illustration, which philosophy and human learning do perform to faith and religion. The one, because they are an effectual inducement to the exaltation of the glory of God. For as the Psalms and other scriptures do often invite us to consider and magnify the great and wonderful works of God, so if we should rest only in the contemplation of the exterior of them as they first offer themselves to our senses, we should do a like injury unto the majesty of God, as if we should judge or construe of the store of some excellent jeweller, by that only which is set out toward the street in his shop. The other, because they minister a singular help and preservative against unbelief and error. For our Saviour saith, 'You err, not knowing the scriptures, nor the power of God'; laying before us two books[n] or volumes to study, if we will be secured from error; first the scriptures, revealing the will of God, and then the creatures expressing his power; whereof the latter is a key unto the former: not only opening our understanding to conceive the true sense of the scriptures, by the general notions of reason and rules of speech; but chiefly opening our belief, in drawing us into a due meditation of the omnipotency of God, which is chiefly signed and engraven upon his works. Thus much therefore for divine testimony and evidence concerning the true dignity and value of learning.

emulation: here the word is used in its modern sense.
Psalms: e.g. Psalms, 19, 66, 104.
You err: Matthew 22:29. Cf. *Novum Organum* i. 89.
signed: stamped as a sign.*

[HUMAN PROOFS OF THE DIGNITY OF LEARNING—HEATHEN APOTHEOSIS OF INVENTORS, CIVILIZING QUALITY OF LEARNING, THE LEARNED RULER IN PEACE AND WAR]

VII. 1. As for human proofs, it is so large a field, as in a discourse of this nature and brevity it is fit rather to use choice of those things which we shall produce, than to embrace the variety of them. First therefore, in the degrees of human honour amongst the heathen, it was the highest to obtain to a veneration and adoration as a God. This unto the Christians is as the forbidden fruit. But we speak now separately of human testimony: according to which, that which the Grecians call *apotheosis*, and the Latins *relatio inter divos*, was the supreme honour which man could attribute unto man: specially when it was given, not by a formal decree or act of state, as it was used among the Roman Emperors, but by an inward assent and belief. Which honour, being so high, had also a degree or middle term: for there were reckoned above human honours heroical and divine: in the attribution and distribution of which honours we see antiquity made this difference: that whereas founders and uniters of states and cities, lawgivers, extirpers of tyrants, fathers of the people, and other eminent persons in civil merit, were honoured but with the titles of worthies or demi-gods; such as were Hercules, Theseus, Minos, Romulus, and the like:[n] on the other side, such as were inventors and authors of new arts, endowments, and commodities towards man's life, were ever consecrated amongst the gods themselves; as was Ceres, Bacchus, Mercurius, Apollo, and others;[n] and justly; for the merit of the former is confined within the circle of an age or a nation; and is like fruitful showers, which though they be profitable and good, yet serve but for that season, and for a latitude of ground where they fall; but the other is indeed like the benefits of heaven, which are permanent and universal. The former again is mixed with strife and perturbation; but the latter hath the true character of Divine Presence, coming in *aura leni*, without noise or agitation.

2. Neither is certainly that other merit of learning, in repressing the inconveniences which grow from man to man, much inferior to the former, of relieving the necessities which arise from nature; which merit was lively set forth by the ancients in that feigned relation of Orpheus' theatre,[n] where all beasts and birds assembled; and

obtain: attain. *apotheosis*: deification.
extirpers: those who root out or destroy.
aura leni: 'with a gentle breeze'. *agitation*: disturbance.*

forgetting their several appetites, some of prey, some of game, some of quarrel, stood all sociably together listening unto the airs and accords of the harp; the sound whereof no sooner ceased, or was drowned by some louder noise, but every beast returned to his own nature: wherein is aptly described the nature and condition of men, who are full of savage and unreclaimed desires, of profit, of lust, of revenge; which as long as they give ear to precepts, to laws, to religion, sweetly touched with eloquence and persuasion of books, of sermons, of harangues, so long is society and peace maintained; but if these instruments be silent, or that sedition and tumult make them not audible, all things dissolve into anarchy and confusion.

3. But this appeareth more manifestly, when kings themselves, or persons of authority under them, or other governors in commonwealths and popular estates, are endued with learning. For although he might be thought partial to his own profession, that said 'Then should people and estates be happy, when either kings were philosophers, or philosophers kings'; yet so much is verified by experience, that under learned princes and governors there have been ever the best times: for howsoever kings may have their imperfections in their passions and customs; yet if they be illuminate by learning, they have those notions of religion, policy, and morality, which do preserve them and refrain them from all ruinous and peremptory errors and excesses; whispering evermore in their ears, when counsellors and servants stand mute and silent. And senators or counsellors likewise, which be learned, do proceed upon more safe and substantial principles, than counsellors which are only men of experience: the one sort keeping dangers afar off, whereas the other discover them not till they come near hand, and then trust to the agility of their wit to ward or avoid them.

4. Which felicity of times under learned princes (to keep still the law of brevity, by using the most eminent and selected examples) doth best appear in the age which passed from the death of Domitianus the emperor until the reign of Commodus; comprehending a succession of six princes,[n] all learned, or singular favourers and advancers of learning, which age for temporal respects was the most happy and flourishing that ever the Roman empire (which then was a model of the world) enjoyed; a matter revealed and prefigured unto Domitian in a dream the night before he was slain; for he thought there was grown behind upon his shoulders a neck and a

Then should people: Plato, *Republic* Book v, 473.
refrain: restrain. *model*: epitome.
Domitian: Suetonius, *The Lives of the Caesars* VIII. xxiii.

head of gold: which came accordingly to pass in those golden times which succeeded: of which princes we will make some commemoration; wherein although the matter will be vulgar, and may be thought fitter for a declamation than agreeable to a treatise infolded[n] as this is, yet because it is pertinent to the point in hand, *Neque semper arcum tendit Apollo*, and to name them only were too naked and cursory, I will not omit it altogether. The first was Nerva; the excellent temper of whose government is by a glance in Cornelius Tacitus touched to the life: *Postquam divus Nerva res olim insociabiles miscuisset, imperium et libertatem.* And in token of his learning, the last act of his short reign left to memory was a missive to his adopted son Trajan, proceeding upon some inward discontent at the ingratitude of the times, comprehended in a verse of Homer's:

Telis, Phoebe, tuis lacrymas ulciscere nostras.[n]

5. Trajan,[n] who succeeded, was for his person not learned: but if we will hearken to the speech of our Saviour, that saith, 'He that receiveth a prophet in the name of a prophet shall have a prophet's reward,' he deserveth to be placed amongst the most learned princes: for there was not a greater admirer of learning or benefactor of learning; a founder of famous libraries, a perpetual advancer of learned men to office, and a familiar converser with learned professors and preceptors, who were noted to have then most credit in court. On the other side, how much Trajan's virtue and government was admired and renowned, surely no testimony of grave and faithful history doth more lively set forth, than that legend tale of Gregorius Magnus,[n] bishop of Rome, who was noted for the extreme envy he bare towards all heathen excellency: and yet he is reported, out of the love and estimation of Trajan's moral virtues, to have made unto God passionate and fervent prayers for the delivery of his soul out of hell: and to have obtained it, with a caveat that he should make no more such petitions. In this prince's time also the persecutions against the Christians received intermission, upon the certificate of Plinius Secundus, a man of excellent learning and by Trajan advanced.

infolded: involved, intricate.
Neque semper: Horace, *Odes* II. x. 19. 'Apollo does not always bend his bow.'
Nerva: Emperor, A.D. 96–8.
glance: reference.
Postquam divus: Quoted from memory from Tacitus, *Agricola* 3. 'The divine Nerva reconciled things which did not go together before, namely, sovereignty and freedom.'
He that: Matthew 10:21.
legend: legendary. A nonce usage.
envy: hostility.
excellency: excellence.
certificate: attestation.
Plinius: Pliny the Younger in his *Epistles* x. 96.

6. Adrian,[n] his successor, was the most curious man that lived, and the most universal inquirer; insomuch as it was noted for an error in his mind, that he desired to comprehend all things, and not to reserve himself for the worthiest things: falling into the like humour that was long before noted in Philip of Macedon; who, when he would needs over-rule and put down an excellent musician in an argument touching music, was well answered by him again, 'God forbid, sir (saith he), that your fortune should be so bad, as to know these things better than I.' It pleased God likewise to use the curiosity of this emperor as an inducement to the peace of his Church in those days. For having Christ in veneration, not as a God or Saviour but as a wonder or novelty, and having his picture in his gallery, matched with Apollonius (with whom in his vain imagination he thought he had some conformity), yet it served the turn to allay the bitter hatred of those times against the Christian name, so as the Church had peace during his time. And for his government civil, although he did not attain to that of Trajan's in glory of arms or perfection of justice, yet in deserving of the weal of the subject he did exceed him. For Trajan erected many famous monuments and buildings; insomuch as Constantine the Great in emulation was wont to call him *Parietaria*, wall-flower, because his name was upon so many walls: but his buildings and works were more of glory and triumph than use and necessity. But Adrian spent his whole reign, which was peaceable, in a perambulation or survey of the Roman empire; giving order and making assignation where he went, for re-edifying of cities, towns, and forts decayed; and for cutting of rivers and streams, and for making bridges and passages, and for policing of cities and commonalties with new ordinances and constitutions, and granting new franchises and incorporations; so that his whole time was a very restoration of all the lapses and decays of former times.

7. Antoninus Pius, who succeeded him, was a prince excellently learned, and had the patient and subtle wit of a schoolman; insomuch as in common speech (which leaves no virtue untaxed) he was called *Cymini Sector*, a carver or a divider of cummin seed, which is one of the least seeds; such a patience he had and settled spirit,

curious: minute in inquiry.
God forbid: In Plutarch, *How a Flatterer may be distinguished from a friend* 27.
weal: welfare. *assignation*: authoritative orders.
policing: regulating. *commonalties*: corporations.
Antoninus Pius: Emperor, A.D. 138–61.
schoolman: a medieval scholastic, who treated of logic, metaphysics, and theology.
Cymini: In Dio's *Roman History* lxx. 3.

to enter into the least and most exact differences of causes; a fruit no doubt of the exceeding tranquillity and serenity of his mind; which being no ways charged or incumbered, either with fears, remorses, or scruples, but having been noted for a man of the purest goodness, without all fiction or affectation, that hath reigned or lived, made his mind continually present and entire. He likewise approached a degree nearer unto Christianity, and became, as Agrippa said unto St. Paul, 'half a Christian'; holding their religion and law in good opinion, and not only ceasing persecution, but giving way to the advancement of Christians.

8. There succeeded him the first *Divi fratres*,[n] the two adoptive brethren, Lucius Commodus Verus, son to Aelius Verus, who delighted much in the softer kind of learning, and was wont to call the poet Martial his Virgil; and Marcus Aurelius Antoninus; whereof the latter, who obscured his colleague and survived him long, was named the Philosopher: who, as he excelled all the rest in learning, so he excelled them likewise in perfection of all royal virtues; insomuch as Julianus the emperor, in his book intituled *Caesares*, being as a pasquil or satire to deride all his predecessors, feigned that they were all invited to a banquet of the gods, and Silenus the jester sat at the nether end of the table, and bestowed a scoff on every one as they came in; but when Marcus Philosophus came in, Silenus was gravelled and out of countenance, not knowing where to carp at him; save at the last he gave a glance at his patience towards his wife. And the virtue of this prince, continued with that of his predecessor, made the name of Antoninus so sacred in the world, that though it were extremely dishonoured in Commodus, Caracalla, and Heliogabalus, who all bare the name, yet when Alexander Severus refused the name because he was a stranger to the family, the senate with one acclamation said, *Quomodo Augustus, sic et Antoninus.* In such renown and veneration was the name of these two princes in those days, that they would have had it as a perpetual addition in all the emperors' style. In this emperor's time also the Church for the most part was in peace; so as in this sequence of six princes we do see the blessed effects of learning in sovereignty, painted forth in the greatest table of the world.

St. Paul: cf. Acts 26:28.
divi fratres: divine brothers. 'Divus' was the epithet of the deified emperors.
Caesares: Julian, *Caesares* xviii. *pasquil*: lampoon.
gravelled: puzzled. *glance*: a satirical allusion.†
Quomodo: Lampridius, *Life of Severus* 5–10. 'Let the name of Antoninus be as the name of Augustus.'
painted: depicted. *table*: picture.

9. But for a tablet or picture of smaller volume (not presuming to speak of your Majesty that liveth), in my judgement the most excellent is that of Queen Elizabeth,[n] your immediate predecessor in this part of Britain; a prince that, if Plutarch were now alive to write lives by parallels, would trouble him I think to find for her a parallel amongst women. This lady was endued with learning in her sex singular, and rare even amongst masculine princes; whether we speak of learning, of language, or of science, modern or ancient, divinity or humanity: and unto the very last year of her life she accustomed to appoint set hours for reading, scarcely any young student in an university more daily or more duly. As for her government, I assure myself, I shall not exceed, if I do affirm that this part of the island never had forty-five years of better times; and yet not through the calmness of the season, but through the wisdom of her regiment. For if there be considered of the one side, the truth of religion established, the constant peace and security, the good administration of justice, the temperate use of the prerogative, not slackened, nor much strained, the flourishing state of learning, sortable to so excellent a patroness, the convenient estate of wealth and means, both of crown and subject, the habit of obedience, and the moderation of discontents; and there be considered on the other side the differences of religion, the troubles of neighbour countries, the ambition of Spain, and opposition of Rome; and then that she was solitary and of herself: these things I say considered, as I could not have chosen an instance so recent and so proper, so I suppose I could not have chosen one more remarkable or eminent to the purpose now in hand, which is concerning the conjunction of learning in the prince with felicity in the people.

10. Neither hath learning an influence and operation only upon civil merit and moral virtue, and the arts or temperature of peace and peaceable government; but likewise it hath no less power and efficacy in enablement towards martial and military virtue and prowess; as may be notably represented in the examples of Alexander the Great and Caesar the dictator, mentioned before, but now in fit place to be resumed: of whose virtues and acts in war there needs no note or recital, having been the wonders of time in that kind: but of their affections towards learning, and perfections in learning, it is pertinent to say somewhat.

regiment: government.
sortable: befitting.
convenient: proper.
discontents: grievances.*
temperature: disposition.
enablement: a qualification.*
Alexander: The following stories are told in Plutarch's 'Life' of Alexander.

11. Alexander[n] was bred and taught under Aristotle the great philosopher, who dedicated divers of his books of philosophy unto him: he was attended with Callisthenes[n] and divers other learned persons, that followed him in camp, throughout his journeys and conquests. What price and estimation he had learning in doth notably appear in these three particulars: first, in the envy he used to express that he bare towards Achilles, in this, that he had so good a trumpet of his praises as Homer's verses: secondly, in the judgement or solution he gave touching that precious cabinet of Darius, which was found among his jewels; whereof question was made what thing was worthy to be put into it; and he gave his opinion for Homer's works: thirdly, in his letter to Aristotle, after he had set forth his books of nature, wherein he expostulateth with him for publishing the secrets or mysteries of philosophy; and gave him to understand that himself esteemed it more to excel other men in learning and knowledge than in power and empire. And what use he had of learning doth appear, or rather shine, in all his speeches and answers, being full of science and use of science, and that in all variety.

12. And herein again it may seem a thing scholastical, and somewhat idle, to recite things that every man knoweth; but yet, since the argument I handle leadeth me thereunto, I am glad that men shall perceive I am as willing to flatter (if they will so call it) an Alexander, or a Caesar, or an Antoninus, that are dead many hundred years since, as any that now liveth: for it is the displaying of the glory of learning in sovereignty that I propound to myself, and not an humour of declaiming in any man's praises. Observe then the speech he used of Diogenes, and see if it tend not to the true state of one of the greatest questions of moral philosophy; whether the enjoying of outward things, or the contemning of them, be the greatest happiness: for when he saw Diogenes so perfectly contented with so little, he said to those that mocked at his condition, 'Were I not Alexander, I would wish to be Diogenes.' But Seneca inverteth it, and saith; *Plus erat, quod hic nollet accipere, quam quod ille posset dare.* 'There were more things which Diogenes would have refused, than those were which Alexander could have given or enjoyed.'

13. Observe again that speech which was usual with him, 'That he felt his mortality chiefly in two things, sleep and lust'; and see if it were not a speech extracted out of the depth of natural philo-

scholastical: following the methods of the 'schools', and therefore pedantic.
Plus erat: Seneca, *On Benefits* V. iv. 4.

sophy, and liker to have comen out of the mouth of Aristotle or Democritus, than from Alexander.

14. See again that speech of humanity and poesy; when upon the bleeding of his wounds, he called unto him one of his flatterers, that was wont to ascribe to him divine honour, and said, 'Look, this is very blood; this is not such a liquor as Homer speaketh of, which ran from Venus' hand, when it was pierced by Diomedes.'

15. See likewise his readiness in reprehension of logic, in the speech he used to Cassander, upon a complaint that was made against his father Antipater: for when Alexander happed to say, 'Do you think these men would have come from so far to complain, except they had just cause of grief?' and Cassander answered, 'Yea, that was the matter, because they thought they should not be disproved'; said Alexander laughing: 'See the subtilties of Aristotle, to take a matter both ways, pro et contra, &c.'

16. But note again how well he could use the same art, which he reprehended, to serve his own humour: when bearing a secret grudge to Callisthenes, because he was against the new ceremony of his adoration, feasting one night where the same Callisthenes was at the table, it was moved by some after supper, for entertainment sake, that Callisthenes, who was an eloquent man, might speak of some theme or purpose at his own choice; which Callisthenes did; choosing the praise of the Macedonian nation for his discourse, and performing the same with so good manner as the hearers were much ravished: whereupon Alexander, nothing pleased, said, 'It was easy to be eloquent upon so good a subject': but saith he 'Turn your style, and let us hear what you can say against us': which Callisthenes presently undertook, and did with that sting and life, that Alexander interrupted him and said, 'The goodness of the cause made him eloquent before, and despite made him eloquent then again.'

17. Consider further, for tropes of rhetoric, that excellent use of a metaphor or translation, wherewith he taxed Antipater, who was an imperious and tyrannous governor: for when one of Antipater's friends commended him to Alexander for his moderation, that he did not degenerate, as his other lieutenants did, into the Persian pride, in use of purple, but kept the ancient habit of Macedon, of black; 'True (saith Alexander), but Antipater is all purple within.'[n] Or that other, when Parmenio came to him in the plain of Arbela, and showed him the innumerable multitude of his enemies, specially

Venus' hand: Homer, *Iliad* v. 340.
reprehension: a term in logic, signifying a refutation or detection of a fallacy.

as they appeared by the infinite number of lights, as it had been a new firmament of stars, and thereupon advised him to assail them by night: whereupon he answered, 'That he would not steal the victory.'

18. For matter of policy, weigh that significant distinction, so much in all ages embraced, that he made between his two friends Hephaestion and Craterus, when he said, 'That the one loved Alexander, and the other loved the king': describing the principal difference of princes' best servants, that some in affection love their person, and other in duty love their crown.

19. Weigh also that excellent taxation of an error, ordinary with counsellors of princes, that they counsel their masters according to the model of their own mind and fortune, and not of their masters'; when upon Darius' great offers Parmenio had said, 'Surely I would accept these offers, were I as Alexander'; saith Alexander, 'So would I were I as Parmenio.'

20. Lastly, weigh that quick and acute reply, which he made when he gave so large gifts to his friends and servants, and was asked what he did reserve for himself, and he answered, *Hope*:[n] weigh, I say, whether he had not cast up his account aright, because *hope* must be the portion of all that resolve upon great enterprises. For this was Caesar's portion[n] when he went first into Gaul, his estate being then utterly overthrown with largesses. And this was likewise the portion of that noble prince, howsoever transported with ambition, Henry Duke of Guise,[n] of whom it was usually said, that he was the greatest usurer in France, because he had turned all his estate into obligations.

21. To conclude therefore: as certain critics are used to say hyperbolically, 'That if all sciences were lost they might be found in Virgil',[n] so certainly this may be said truly, there are the prints and footsteps of learning in those few speeches which are reported of this prince: the admiration of whom, when I consider him not as Alexander the Great, but as Aristotle's scholar, hath carried me too far.

22. As for Julius Caesar, the excellency of his learning needeth not to be argued from his education, or his company, or his speeches; but in a further degree doth declare itself in his writings and works; whereof some are extant and permanent, and some unfortunately perished. For first, we see there is left unto us that excellent history of his own wars, which he intituled only a Commentary, wherein

Caesar: Bacon uses the lives of Caesar by Plutarch and Suetonius.

all succeeding times have admired the solid weight of matter, and the real passages and lively images of actions and persons, expressed in the greatest propriety of words and perspicuity of narration that ever was; which that it was not the effect of a natural gift, but of learning and precept, is well witnessed by that work of his intituled *De Analogia*,[n] being a grammatical philosophy, wherein he did labour to make this same *Vox ad placitum* to become *Vox ad licitum*, and to reduce custom of speech to congruity of speech; and took as it were the pictures of words from the life of reason.[n]

23. So we receive from him, as a monument both of his power and learning, the then reformed computation of the year;[n] well expressing that he took it to be as great a glory to himself to observe and know the law of the heavens, as to give law to men upon the earth.

24. So likewise in that book of his, *Anti-Cato*,[n] it may easily appear that he did aspire as well to victory of wit as victory of war: undertaking therein a conflict against the greatest champion with the pen that then lived, Cicero the orator.

25. So again in his book of Apophthegms[n] which he collected, we see that he esteemed it more honour to make himself but a pair of tables, to take the wise and pithy words of others, than to have every word of his own to be made an apophthegm or an oracle; as vain princes, by custom of flattery, pretend to do. And yet if I should enumerate divers of his speeches, as I did those of Alexander, they are truly such as Salomon noteth, when he saith, *Verba sapientum tanquam aculei et tanquam clavi in altum defixi*: whereof I will only recite three,[n] not so delectable for elegancy, but admirable for vigour and efficacy.

26. As first, it is reason he be thought a master of words, that could with one word appease a mutiny in his army, which was thus. The Romans, when their generals did speak to their army, did use the word *Milites*, but when the magistrates spake to the people, they did use the word *Quirites*. The soldiers were in tumult, and seditiously prayed to be cashiered; not that they so meant, but by expostulation thereof to draw Caesar to other conditions; wherein he being resolute not to give way, after some silence, he began his speech, *Ego Quirites*, which did admit them already cashiered;

real . . . images: i.e. the actual occurrences and the vivid pictures.
Vox: conventional speech to become correct speech.
tables: tablets of wood covered with wax on which the Romans wrote.
Verba: cf. Ecclesiastes 12:11. 'The words of the wise are as goads, and as nails driven deep in.'
Milites: soldiers. *Quirites*, citizens, was used by the Romans in their civil capacity.

wherewith they were so surprised, crossed, and confused, as they would not suffer him to go on in his speech, but relinquished their demands, and made it their suit to be again called by the name of *Milites*.

27. The second speech was thus: Caesar did extremely affect the name of king; and some were set on as he passed by, in popular acclamation to salute him king. Whereupon, finding the cry weak and poor, he put it off thus, in a kind of jest, as if they had mistaken his surname; *Non Rex sum, sed Caesar*; a speech, that if it be searched, the life and fulness of it can scarce be expressed. For, first, it was a refusal of the name, but yet not serious: again, it did signify an infinite confidence and magnanimity, as if he presumed Caesar was the greater title; as by his worthiness it is come to pass till this day. But chiefly it was a speech of great allurement toward his own purpose; as if the state did strive with him but for a name, whereof mean families were vested; for *Rex* was a surname with the Romans, as well as *King* is with us.

28. The last speech which I will mention was used to Metellus: when Caesar, after war declared, did possess himself of the city of Rome; at which time entering into the inner treasury to take the money there accumulate, Metellus being tribune forbade him. Whereto Caesar said, 'That if he did not desist, he would lay him dead in the place.' And presently taking himself up, he added, 'Young man, it is harder for me to speak it than to do it'; Adolescens, durius est mihi hoc dicere quam facere. A speech compounded of the greatest terror and greatest clemency that could proceed out of the mouth of man.

29. But to return and conclude with him, it is evident himself knew well his own perfection in learning, and took it upon him; as appeared when, upon occasion that some spake what a strange resolution it was in Lucius Sulla to resign his dictature; he scoffing at him, to his own advantage, answered, *That Sulla could not skill of letters, and therefore knew not how to dictate*.

30. And here it were fit to leave this point, touching the concurrence of military virtue and learning (for what example should come with any grace after those two of Alexander and Caesar?), were it not in regard of the rareness of circumstance, that I find in one other particular, as that which did so suddenly pass from extreme scorn to extreme wonder: and it is of Xenophon the philosopher,[n] who

affect: aspire to. *Non Rex*: 'I am not King but Caesar.'
dictature: office of dictator. Sulla was dictator in Rome from 82 to 79 B.C.
skill of: understand.

went from Socrates' school into Asia, in the expedition of Cyrus the younger against King Artaxerxes. This Xenophon at that time was very young, and never had seen the wars before; neither had any command in the army, but only followed the war as a voluntary, for the love and conversation of Proxenus his friend. He was present when Falinus came in message from the great king to the Grecians, after that Cyrus was slain in the field, and they a handful of men left to themselves in the midst of the king's territories, cut off from their country by many navigable rivers, and many hundred miles. The message imported that they should deliver up their arms and submit themselves to the king's mercy. To which message before answer was made, divers of the army conferred familiarly with Falinus, and amongst the rest Xenophon happened to say, 'Why, Falinus, we have now but these two things left, our arms and our virtue; and if we yield up our arms, how shall we make use of our virtue?' Whereto Falinus smiling on him said, 'If I be not deceived, young gentleman, you are an Athenian: and I believe you study philosophy, and it is pretty that you say: but you are much abused, if you think your virtue can withstand the king's power.' Here was the scorn; the wonder followed: which was, that this young scholar, or philosopher, after all the captains were murdered in parley by treason, conducted those ten thousand foot, through the heart of all the king's high countries, from Babylon to Grecia in safety, in despite of all the king's forces, to the astonishment of the world, and the encouragement of the Grecians in times succeeding to make invasion upon the kings of Persia; as was after purposed by Jason the Thessalian, attempted by Agesilaus the Spartan, and achieved by Alexander the Macedonian, all upon the ground of the act of that young scholar.

[THE EXCELLENCE OF LEARNING—PRIVATE VIRTUE, POWER OVER MEN'S MINDS, FORTUNE AND ADVANCEMENT, PLEASURE, IMMORTALITY]

VIII. 1. To proceed now from imperial and military virtue to moral and private virtue; first, it is an assured truth, which is contained in the verses,

> Scilicet ingenuas didicisse fideliter artes
> Emollit mores, nec sinit esse feros.

It taketh away the wildness and barbarism and fierceness of men's

Scilicet: Ovid, *Epistles from Pontus* II. ix. 47–8. 'Certainly a faithful study of the liberal arts refines manners and decreases barbarism.'

minds; but indeed the accent had need be upon *fideliter*: for a little superficial learning doth rather work a contrary effect.[n] It taketh away all levity, temerity, and insolency, by copious suggestions of all doubts and difficulties, and acquainting the mind to balance reasons on both sides, and to turn back the first offers and conceits of the mind, and to accept of nothing but examined and tried. It taketh away vain admiration of anything, which is the root of all weakness. For all things are admired either because they are new, or because they are great. For novelty, no man that wadeth in learning or contemplation throughly, but will find that printed in his heart, *Nil novi super terram*. Neither can any man marvel at the play of puppets, that goeth behind the curtain, and adviseth well of the motion. And for magnitude, as Alexander the Great, after that he was used to great armies, and the great conquests of the spacious provinces in Asia, when he received letters out of Greece, of some fights and services there, which were commonly for a passage, or a fort, or some walled town at the most, he said, 'It seemed to him, that he was advertised of the battles of the frogs and the mice, that the old tales went of.' So certainly, if a man meditate much upon the universal frame of nature, the earth with men upon it (the divineness of souls except) will not seem much other than an ant-hill, whereas some ants carry corn, and some carry their young, and some go empty, and all to and fro a little heap of dust. It taketh away or mitigateth fear of death or adverse fortune; which is one of the greatest impediments of virtue, and imperfections of manners. For if a man's mind be deeply seasoned with the consideration of the mortality and corruptible nature of things, he will easily concur with Epictetus, who went forth one day and saw a woman weeping for her pitcher of earth that was broken, and went forth the next day and saw a woman weeping for her son that was dead, and thereupon said, *Heri vidi fragilem frangi, hodie vidi mortalem mori*. And therefore Virgil did excellently and profoundly couple the knowledge of causes and the conquest of all fears together, as *concomitantia*.

> Felix, qui potuit rerum cognoscere causas,
> Quique metus omnes, et inexorabile fatum
> Subjecit pedibus, strepitumque Acherontis avari.[n]

Nil novi: Ecclesiastes 1:9. 'There is no new thing under the sun.'
services: military engagements.
It seemed: told in Plutarch's 'Life' of Agesilaus, 15. The allusion is to a mock-epic, *Batrachomyomachia*, attributed to Homer.
Heri vidi: In Simplicius' commentary on Epictetus' *Enchiridion*, ch. 8 and ch. 33. 'Yesterday I saw a brittle thing broken and today a mortal die.' Epictetus was a Stoic philosopher of the first century A.D.

2. It were too long to go over the particular remedies which learning doth minister to all the diseases of the mind; sometimes purging the ill humours, sometimes opening the obstructions, sometimes helping digestion, sometimes increasing appetite, sometimes healing the wounds and exulcerations thereof, and the like; and therefore I will conclude with that which hath *rationem totius*; which is, that it disposeth the constitution of the mind not to be fixed or settled in the defects thereof, but still to be capable and susceptible of growth and reformation. For the unlearned man knows not what it is to descend into himself, or to call himself to account, nor the pleasure of that *suavissima vita, indies sentire se fieri meliorem.* The good parts he hath he will learn to show to the full, and use them dexterously, but not much to increase them. The faults he hath he will learn how to hide and colour them, but not much to amend them; like an ill mower, that mows on still, and never whets his scythe. Whereas with the learned man it fares otherwise, that he doth ever intermix the correction and amendment of his mind with the use and employment thereof. Nay further, in general and in sum, certain it is that *Veritas* and *Bonitas* differ but as the seal and the print: for Truth prints Goodness, and they be the clouds of error which descend in the storms of passions and perturbations.[n]

3. From moral virtue let us pass on to matter of power and commandment, and consider whether in right reason there be any comparable with that wherewith knowledge investeth and crowneth man's nature. We see the dignity of the commandment is according to the dignity of the commanded: to have commandment over beasts, as herdmen have, is a thing contemptible: to have commandment over children, as schoolmasters have, is a matter of small honour: to have commandment over galley-slaves is a disparagement rather than an honour. Neither is the commandment of tyrants much better, over people which have put off the generosity of their minds: and therefore it was ever holden that honours in free monarchies and commonwealths had a sweetness more than in tyrannies, because the commandment extendeth more over the wills of men, and not only over their deeds and services. And therefore, when Virgil putteth himself forth to attribute to Augustus Caesar the best of human honours, he doth it in these words:

exulcerations: ulcers.
rationem: the essence of the whole.
suavissima: 'that most happy of lives, to feel himself daily get better and better'.
generosity: nobility.
putteth himself forth: endeavours.

> Victorque volentes
> Per populos dat jura, viamque affectat Olympo.

But yet the commandment of knowledge is yet higher than the commandment over the will: for it is a commandment over the reason, belief, and understanding of man, which is the highest part of the mind, and giveth law to the will itself. For there is no power on earth which setteth up a throne or chair of estate in the spirits and souls of men, and in their cogitations, imaginations, opinions, and beliefs, but knowledge and learning. And therefore we see the detestable and extreme pleasure that arch-heretics, and false prophets, and imposters are transported with, when they once find in themselves that they have a superiority in the faith and conscience of men; so great as if they have once tasted of it, it is seldom seen that any torture or persecution can make them relinquish or abandon it. But as this is that which the author of the Revelation calleth the depth or profoundness of Satan, so by argument of contraries, the just and lawful sovereignty over men's understanding, by force of truth rightly interpreted, is that which approacheth nearest to the similitude of the divine rule.

4. As for fortune and advancement, the beneficence of learning is not so confined to give fortune only to states and commonwealths, as it doth not likewise give fortune to particular persons. For it was well noted long ago, that Homer hath given more men their livings, than either Sulla, or Caesar, or Augustus ever did, notwithstanding their great largesses and donatives, and distributions of lands to so many legions. And no doubt it is hard to say whether arms or learning have advanced greater numbers. And in case of sovereignty we see, that if arms or descent have carried away the kingdom, yet learning hath carried the priesthood, which ever hath been in some competition with empire.

5. Again, for the pleasure and delight of knowledge and learning, it far surpasseth all other in nature. For, shall the pleasures of the affections so exceed the pleasure of the sense, as much as the obtaining of desire or victory exceedeth a song or a dinner? and must not of consequence the pleasures of the intellect or understanding exceed the pleasures of the affections? We see in all other pleasures there is satiety, and after they be used, their verdure departeth; which showeth well they be but deceits of pleasure, and not pleasures: and

Victorque: Virgil, *Georgics* iv. 561–2. 'The conqueror gives laws to willing peoples and takes the road to heaven.'
Satan: Revelation 2:24.

that it was the novelty which pleased, and not the quality. And therefore we see that voluptuous men turn friars, and ambitious princes turn melancholy. But of knowledge there is no satiety, but satisfaction and appetite are perpetually interchangeable; and therefore appeareth to be good in itself simply, without fallacy or accident. Neither is that pleasure of small efficacy and contentment to the mind of man, which the poet Lucretius describeth elegantly,

> Suave mari magno, turbantibus aequora ventis, &c.

'It is a view of delight (saith he) to stand or walk upon the shore side, and to see a ship tossed with tempest upon the sea; or to be in a fortified tower, and to see two battles join upon a plain. But it is a pleasure incomparable, for the mind of man to be settled, landed, and fortified in the certainty of truth; and from thence to descry and behold the errors, perturbations, labours, and wanderings up and down of other men.'

6. Lastly, leaving the vulgar arguments, that by learning man excelleth man in that wherein man excelleth beasts; that by learning man ascendeth to the heavens and their motions, where in body he cannot come; and the like; let us conclude with the dignity and excellency of knowledge and learning in that whereunto man's nature doth most aspire, which is immortality or continuance; for to this tendeth generation, and raising of houses and families; to this tend buildings, foundations, and monuments; to this tendeth the desire of memory, fame, and celebration; and in effect the strength of all other human desires. We see then how far the monuments of wit and learning are more durable than the monuments of power or of the hands. For have not the verses of Homer continued twenty-five hundred years, or more, without the loss of a syllable or letter; during which time infinite palaces, temples, castles, cities, have been decayed and demolished? It is not possible to have the true pictures or statuaes of Cyrus, Alexander, Caesar, no nor of the kings or great personages of much later years; for the originals cannot last, and the copies cannot but leese of the life and truth. But the images of men's wits and knowledges remain in books, exempted from the wrong of time and capable of perpetual renovation. Neither are they fitly to be called images, because they generate still, and cast their seeds in the minds of others, provoking and causing infinite actions

fallacy: deception.
Suave mari: Lucretius, *On the Nature of Things* ii. 1–10. Bacon gives a free paraphrase.
infinite: innumerable.

and opinions in succeeding ages. So that if the invention of the ship was thought so noble, which carrieth riches and commodities from place to place, and consociateth the most remote regions in participation of their fruits, how much more are letters to be magnified, which as ships pass through the vast seas of time, and make ages so distant to participate of the wisdom, illuminations, and inventions, the one of the other? Nay further, we see some of the philosophers which were least divine,[n] and most immersed in the senses, and denied generally the immortality of the soul, yet came to this point, that whatsoever motions the spirit of man could act and perform without the organs of the body, they thought might remain after death; which were only those of the understanding, and not of the affection; so immortal and incorruptible a thing did knowledge seem unto them to be. But we, that know by divine revelation that not only the understanding but the affections purified, not only the spirit but the body changed, shall be advanced to immortality, do disclaim in these rudiments of the senses. But it must be remembered, both in this last point, and so it may likewise be needful in other places, that in probation of the dignity of knowledge or learning, I did in the beginning separate divine testimony from human, which method I have pursued, and so handled them both apart.

7. Nevertheless I do not pretend, and I know it will be impossible for me, by any pleading of mine, to reverse the judgement, either of Aesop's cock, that preferred the barley-corn before the gem; or of Midas, that being chosen judge between Apollo, president of the Muses, and Pan, god of the flocks, judged for plenty; or of Paris, that judged for beauty and love against wisdom and power; or of Agrippina, *occidat matrem, modo imperet*,[n] that preferred empire with any condition never so detestable; or of Ulysses, *qui vetulam praetulit immortalitati*,[n] being a figure of those which prefer custom and habit before all excellency; or of a number of the like popular judgements. For these things must continue as they have been: but so will that also continue whereupon learning hath ever relied, and which faileth not: *Justificata est sapientia a filiis suis.*

disclaim in: renounce.
probation: demonstrating and proving.
Aesop: In Phaedrus, *Fables* iii. 12.
Midas: Ovid, *Metamorphoses* xi. 153 f.
Paris: Homer, *Iliad* xxiv. 25–9.
Justificata: Matthew 11:19. 'Wisdom is justified by her children.'

The Second Book of Francis Bacon of the Proficience or Advancement of Learning Divine and Human

TO THE KING

1. It might seem to have more convenience, though it come often otherwise to pass (excellent king), that those which are fruitful in their generations, and have in themselves the foresight of immortality in their descendants, should likewise be more careful of the good estate of future times, unto which they know they must transmit and commend over their dearest pledges.[n] Queen Elizabeth was a sojourner in the world in respect of her unmarried life, and was a blessing to her own times; and yet so as the impression of her good government, besides her happy memory, is not without some effect which doth survive her. But to your Majesty, whom God hath already blessed with so much royal issue, worthy to continue and represent you for ever, and whose youthful and fruitful bed doth yet promise many the like renovations, it is proper and agreeable to be conversant not only in the transitory parts of good government, but in those acts also which are in their nature permanent and perpetual. Amongst the which (if affection do not transport me) there is not any more worthy than the further endowment of the world with sound and fruitful knowledge. For why should a few received authors stand up like Hercules' columns,[n] beyond which there should be no sailing or discovering, since we have so bright and benign a star as your Majesty to conduct and prosper us? To return therefore where we left, it remaineth to consider of what kind those acts are which have been undertaken and performed by kings and others for the increase and advancement of learning: wherein I purpose to speak actively without digressing or dilating.

2. Let this ground therefore be laid, that all works are overcommen by amplitude of reward, by soundness of direction, and by the conjunction of labours. The first multiplieth endeavour, the second preventeth error, and the third supplieth the frailty of man. But the principal of these is direction: for *claudus in via antevertit cursorem*

convenience: to be more in accordance with nature.
affection: inclination or passion.
ground: fundamental principle.
overcommen: accomplished. *supplieth*: assists, relieves.
claudus in via: from St. Augustine's Sermon 169. 'A cripple who keeps to the road outstrips a runner who leaves it.'

extra viam; and Salomon excellently setteth it down, 'If the iron be not sharp, it requireth more strength; but wisdom is that which prevaileth'; signifying that the invention or election of the mean is more effectual than any inforcement or accumulation of endeavours. This I am induced to speak, for that (not derogating from the noble intention of any that have been deservers towards the state of learning) I do observe nevertheless that their works and acts are rather matters of magnificence and memory, than of progression and proficience, and tend rather to augment the mass of learning in the multitude of learned men, than to rectify or raise the sciences themselves.

3. The works or acts of merit towards learning are conversant about three objects; the places of learning, the books of learning, and the persons of the learned. For as water, whether it be the dew of heaven, or the springs of the earth, doth scatter and leese itself in the ground, except it be collected into some receptacle, where it may by union comfort and sustain itself: and for that cause the industry of man hath made and framed springheads, conduits, cisterns, and pools, which men have accustomed likewise to beautify and adorn with accomplishments of magnificence and state, as well as of use and necessity: so this excellent liquor of knowledge, whether it descend from divine inspiration, or spring from human sense, would soon perish and vanish to oblivion, if it were not preserved in books, traditions, conferences, and places appointed, as universities, colleges, and schools, for the receipt and comforting of the same.

4. The works which concern the seats and places of learning are four; foundations and buildings, endowments with revenues, endowments with franchises and privileges, institutions and ordinances for government; all tending to quietness and privateness of life, and discharge of cares and troubles; much like the stations which Virgil prescribeth for the hiving of bees:

> Principio sedes apibus statioque petenda,
> Quo neque sit ventis aditus, &c.

5. The works touching books are two: first, libraries which are as

If the iron: Ecclesiastes 10:10.
mean: means, medium.
derogating: detracting from.
leese: lose.
comfort: strengthen.
accomplishments: adornments.*
franchises: a legal term, signifying legal privileges granted to a body of persons.
Principio: Virgil, *Georgics* iv. 8–9.

> First for thy bees a quiet station find,
> And lodge them under covert of the wind.
> (Dryden)

the shrines where all the relics of the ancient saints, full of true virtue, and that without delusion or imposture, are preserved and reposed; secondly, new editions of authors, with more correct impressions, more faithful translations, more profitable glosses, more diligent annotations, and the like.

6. The work pertaining to the persons of learned men (besides the advancement and countenancing of them in general) are two: the reward and designation of readers in sciences already extant and invented; and the reward and designation of writers and inquirers concerning any parts of learning not sufficiently laboured and prosecuted.

7. These are summarily the works and acts, wherein the merits of many excellent princes and other worthy personages have been conversant. As for any particular commemorations, I call to mind what Cicero said, when he gave general thanks; *Difficile non aliquem, ingratum quenquam praeterire.* Let us rather, according to the scriptures, look unto that part of the race which is before us, than look back to that which is already attained.

8. First therefore, amongst so many great foundations of colleges in Europe, I find strange that they are all dedicated to professions,[n] and none left free to arts and sciences at large. For if men judge that learning should be referred to action, they judge well; but in this they fall into the error described in the ancient fable,[n] in which the other parts of the body did suppose the stomach had been idle, because it neither performed the office of motion, as the limbs do, nor of sense, as the head doth: but yet notwithstanding it is the stomach that digesteth and distributeth to all the rest. So if any man think philosophy and universality to be idle studies, he doth not consider that all professions are from thence served and supplied. And this I take to be a great cause that hath hindered the progression of learning, because these fundamental knowledges have been studied but in passage. For if you will have a tree bear more fruit than it hath used to do, it is not anything you can do to the boughs, but it is the stirring of the earth and putting new mould about the roots that must work it. Neither is it to be forgotten, that this dedicating of foundations and dotations to professory[n] learning

reposed: laid up as in store.
designation: appointment.*
Difficile: Cicero (attrib.), 'Speech to the Senate on his return from exile', §12. 'It is difficult to mention everybody; it would be ungracious to omit anyone.'
scriptures: Philippians 3:13.
universality: i.e. the study of general principles.
in passage: by the way.

hath not only had a malign aspect and influence upon the growth of sciences, but hath also been prejudicial to states and governments. For hence it proceedeth that princes find a solitude in regard of able men to serve them in causes of estate, because there is no education collegiate which is free; where such as were so disposed mought give themselves to histories, modern languages, books of policy and civil discourse, and other the like enablements unto service of estate.

9. And because founders of colleges do plant, and founders of lectures do water, it followeth well in order to speak of the defect which is in public lectures; namely, in the smallness and meanness of the salary[n] or reward which in most places is assigned unto them; whether they be lectures of arts, or of professions. For it is necessary to the progression of sciences that readers be of the most able and sufficient men; as those which are ordained for generating and propagating of sciences, and not for transitory use. This cannot be, except their condition and endowment be such as may content the ablest man to appropriate his whole labour and continue his whole age in that function and attendance; and therefore must have a proportion answerable to that mediocrity or competency of advancement, which may be expected from a profession or the practice of a profession. So as, if you will have sciences flourish, you must observe David's military law, which was, 'That those which staid with the carriage should have equal part with those which were in the action'; else will the carriages be ill attended. So readers in sciences are indeed the guardians of the stores and provisions of sciences, whence men in active courses are furnished, and therefore ought to have equal entertainment with them; otherwise if the fathers in sciences be of the weakest sort or be ill maintained,

Et patrum invalidi referent jejunia nati.

10. Another defect I note, wherein I shall need some alchemist to help me, who call upon men to sell their books, and to build furnaces; quitting and forsaking Minerva and the Muses as barren virgins, and relying upon Vulcan.[n] But certain it is, that unto the

solitude: complete absence.*
free: liberal.
enablements: qualifications (for service to the state).
lectures: i.e. lectureships.†
mediocrity: moderate fortune or condition of life.
That those: In I Samuel 30:24.
Et patrum: Virgil, *Georgics* iii. 128. 'The weakness of the fathers will reappear in the children.'

deep, fruitful, and operative study of many sciences, specially natural philosophy and physic, books be not only the instrumentals; wherein also the beneficence of men hath not been altogether wanting. For we see spheres, globes, astrolabes, maps, and the like, have been provided as appurtenances to astronomy and cosmography, as well as books. We see likewise that some places instituted for physic[n] have annexed the commodity of gardens for simples of all sorts, and do likewise command the use of dead bodies for anatomies. But these do respect but a few things. In general, there will hardly be any main proficience in the disclosing of nature, except there be some allowance for expenses about experiments; whether they be experiments appertaining to Vulcanus or Daedalus,[n] furnace or engine, or any other kind. And therefore as secretaries and spials of princes and states bring in bills for intelligence, so you must allow the spials and intelligencers of nature to bring in their bills; or else you shall be ill advertised.

11. And if Alexander[n] made such a liberal assignation to Aristotle of treasure for the allowance of hunters, fowlers, fishers, and the like, that he mought compile an history of nature, much better do they deserve it that travail in arts of nature.

12. Another defect which I note, is an intermission or neglect, in those which are governors in universities, of consultation, and in princes or superior persons, of visitation: to enter into account and consideration, whether the readings, exercises, and other customs appertaining unto learning, anciently begun and since continued, be well instituted or no; and thereupon to ground an amendment or reformation in that which shall be found inconvenient. For it is one of your Majesty's own most wise and princely maxims, 'That in all usages and precedents, the times be considered wherein they first began; which if they were weak or ignorant, it derogateth from the authority of the usage, and leaveth it for suspect.'[n] And therefore inasmuch as most of the usages and orders of the universities were derived from more obscure times, it is the more requisite they be re-examined. In this kind I will give an instance or two, for example sake, of things that are the most obvious and familiar. The one is a matter, which though it be ancient and general, yet I hold to be an error; which is, that scholars in universities come too soon and too unripe to logic and rhetoric,[n] arts fitter for graduates than children

instrumentals: i.e. books are not the only tools.
commodity: convenience. *anatomies*: dissection.
main: important. *spials*: spies.
advertised: informed. *inconvenient*: unsuitable.

and novices. For these two, rightly taken, are the gravest of sciences, being the arts of arts; the one for judgement, the other for ornament. And they be the rules and directions how to set forth and dispose matter: and therefore for minds empty and unfraught with matter, and which have not gathered that which Cicero calleth *sylva* and *supellex*,[n] stuff and variety, to begin with those arts (as if one should learn to weigh, or to measure, or to paint the wind) doth work but this effect, that the wisdom of those arts, which is great and universal, is almost made contemptible, and is degenerate into childish sophistry and ridiculous affectation. And further, the untimely learning of them hath drawn on by consequence the superficial and unprofitable teaching and writing of them, as fitteth indeed to the capacity of children. Another is a lack I find in the exercises used in the universities, which do make too great a divorce between invention and memory. For their speeches are either premeditate, in *verbis conceptis*, where nothing is left to invention; or merely extemporal, where little is left to memory. Whereas in life and action there is least use of either of these, but rather of intermixtures of premeditation and invention, notes and memory. So as the exercise fitteth not the practice, nor the image the life; and it is ever a true rule in exercises, that they be framed as near as may be to the life of practice; for otherwise they do pervert the motions and faculties of the mind, and not prepare them. The truth whereof is not obscure, when scholars come to the practices of professions, or other actions of civil life; which when they set into, this want is soon found by themselves, and sooner by others. But this part, touching the amendment of the institutions and orders of universities, I will conclude with the clause of Caesar's letter to Oppius and Balbus, *Hoc quemadmodum fieri possit, nonnulla mihi in mentem veniunt, et multa reperiri possunt: de iis rebus rogo vos ut cogitationem suscipiatis.*

13. Another defect which I note, ascendeth a little higher than the precedent. For as the proficience of learning consisteth much in the orders and institutions of universities in the same states and kingdoms, so it would be yet more advanced, if there were more intelligence mutual between the universities of Europe than now there is. We see there be many orders and foundations,[n] which though

verbis conceptis: prepared language.

Hoc quemadmodum: Cicero, *Letters to Atticus* ix. 7. 'I have some ideas as to the way in which this may be brought about, and many others may be thought of; I ask you to reflect on the matter.' Oppius and Balbus were friends and advisers of Julius Caesar.

they be divided under several sovereignties and territories, yet they take themselves to have a kind of contract, fraternity, and correspondence one with the other, insomuch as they have provincials and generals. And surely as nature createth brotherhood in families, and arts mechanical contract brotherhoods in communalties,[n] and the anointment of God superinduceth a brotherhood in kings and bishops, so in like manner there cannot but be a fraternity in learning and illumination, relating to that paternity which is attributed to God, who is called the Father of illuminations or lights.[n]

14. The last defect which I will note is, that there hath not been, or very rarely been, any public designation of writers or inquirers, concerning such parts of knowledge as may apear not to have been already sufficiently laboured or undertaken; unto which point it is an inducement to enter into a view and examination what parts of learning have been prosecuted and what omitted. For the opinion of plenty[n] is amongst the causes of want, and the great quantity of books maketh a show rather of superfluity than lack; which surcharge nevertheless is not to be remedied by making no more books, but by making more good books, which, as the serpent of Moses,[n] mought devour the serpents of the enchanters.

15. The removing of all the defects formerly enumerate, except the last, and of the active part also of the last (which is the designation of writers), are *opera basilica*; towards which the endeavours of a private man may be but as an image in a crossway, that may point at the way, but cannot go it. But the inducing part of the latter (which is the survey of learning) may be set forward by private travail. Wherefore I will now attempt to make a general and faithful perambulation of learning, with an inquiry what parts thereof lie fresh and waste, and not improved and converted by the industry of man; to the end that such a plot made and recorded to memory, may both minister light to any public designation, and also serve to excite voluntary endeavours. Wherein nevertheless my purpose is at this time to note only omissions and deficiences, and not to make any redargution of errors or incomplete prosecutions. For it is one thing to set forth what ground lieth unmanured, and another thing to correct ill husbandry in that which is manured.

designation: appointment. *surcharge*: surfeit.
enumerate: mentioned separately.†
opera basilica: works for a king.
inducing: introductory.*
set forward: promoted. *plot*: plan, outline.
deficiences: *O.E.D.* gives this as the first usage.
unmanured: uncultivated.

In the handling and undertaking of which work I am not ignorant what it is that I do now move and attempt, nor insensible of mine own weakness to sustain my purpose. But my hope is, that if my extreme love to learning carry me too far, I may obtain the excuse of affection; for that 'It is not granted to man to love and to be wise.' But I know well I can use no other liberty of judgement than I must leave to others; and I for my part shall be indifferently glad either to perform myself, or accept from another, that duty of humanity; *Nam qui erranti comiter monstrat viam, &c.* I do foresee likewise that of those things which I shall enter and register as deficiences and omissions, many will conceive and censure that some of them are already done and extant; others to be but curiosities, and things of no great use; and others to be of too great difficulty, and almost impossibility to be compassed and effected. But for the two first, I refer myself to the particulars. For the last, touching impossibility, I take it those things are to be held possible which may be done by some person, though not by every one; and which may be done by many, though not by any one; and which may be done in succession of ages, though not within the hourglass of one man's life; and which may be done by public designation, though not by private endeavour. But notwithstanding, if any man will take to himself rather that of Salomon, *Dicit piger, Leo est in via*, than that of Virgil, *Possunt quia posse videntur*, I shall be content that my labours be esteemed but as the better sort of wishes: for as it asketh some knowledge to demand a question not impertinent, so it requireth some sense to make a wish not absurd.

[TRIPLE DISTRIBUTION OF HUMAN LEARNING. NATURAL HISTORY]

I. 1. The parts of human learning have reference to the three parts of man's understanding, which is the seat of learning: history to his memory, poesy to his imagination, and philosophy to his reason.[n] Divine learning receiveth the same distribution; for the spirit of man

It is not granted: From Publius Syrus, *Sententiae* 15.
Nam qui: Quoted from Ennius in Cicero, *On Duties* i. 16.

> To put a wand'ring traveller in's way
> Is but to light one candle with another:
> I've ne'er the less for what I give.
> (L'Estrange)

censure: judge.
Dicit piger: Proverbs, 22:13. 'The slothful man says there is a lion in the path.'
Possunt: Virgil, *Aeneid* v. 231. 'They can because they think they can.'
impertinent: irrelevant.

is the same, though the revelation of oracle and sense be diverse. So as theology consisteth also of history of the church; of parables, which is divine poesy; and of holy doctrine or precept. For as for that part which seemeth supernumerary, which is prophecy, it is but divine history; which hath that prerogative over human, as the narration may be before the fact as well as after.

2. *Historia Literarum.* History is natural, civil, ecclesiastical,[n] and literary; whereof the three first I allow as extant, the fourth I note as deficient. For no man hath propounded to himself the general state of learning to be described and represented from age to age, as many have done the works of nature, and the state civil and ecclesiastical; without which the history of the world seemeth to me to be as the statua of Polyphemus[n] with his eye out; that part being wanting which doth most show the spirit and life of the person. And yet I am not ignorant that in divers particular sciences, as of the jurisconsults, the mathematicians, the rhetoricians, the philosophers, there are set down some small memorials of the schools, authors, and books; and so likewise some barren relations touching the invention of arts or usages. But a just story of learning,[n] containing the antiquities and originals of knowledges and their sects, their inventions, their traditions, their diverse administrations and managings, their flourishings, their oppositions, decays, depressions, oblivions, removes, with the causes and occasions of them, and all other events concerning learning, throughout the ages of the world, I may truly affirm to be wanting. The use and end of which work I do not so much design for curiosity or satisfaction of those that are the lovers of learning, but chiefly for a more serious and grave purpose, which is this in few words, that it will make learned men wise in the use and administration of learning. For it is not Saint Augustine's nor Saint Ambrose' works[n] that will make so wise a divine, as ecclesiastical history, throughly read and observed; and the same reason is of learning.

3. *Historia Naturae Errantis.* History of nature is of three sorts: of nature in course; of nature erring or varying;[n] and of nature altered or wrought; that is, history of creatures, history of marvels, and history of arts. The first of these no doubt is extant, and that in good perfection: the two latter are handled so weakly and unprofit-

Historia Literarum: History of Letters.
jurisconsults: masters of jurisprudence.*
removes: departures to another place.
Historia Naturae Errantis: History of Erring Nature.
in course: in its due order.

ably, as I am moved to note them as deficient. For I find no sufficient or competent collection of the works of nature which have a digression and deflexion from the ordinary course of generations, productions, and motions; whether they be singularities of place and region, or the strange events of time and chance, or the effects of yet unknown proprieties, or the instances of exception to general kinds. It is true, I find a number of books of fabulous experiments and secrets, and frivolous impostures for pleasure and strangeness; but a substantial and severe collection of the heteroclites or irregulars of nature, well examined and described, I find not: specially not with due rejection of fables and popular errors.[n] For as things now are, if an untruth in nature be once on foot, what by reason of the neglect of examination, and countenance of antiquity, and what by reason of the use of the opinion in similitudes and ornaments of speech,[n] it is never called down.

4. The use of this work, honoured with a precedent in Aristotle,[n] is nothing less than to give contentment to the appetite of curious and vain wits, as the manner of Mirabilaries is to do; but for two reasons, both of great weight; the one to correct the partiality of axioms[n] and opinions, which are commonly framed only upon common and familiar examples; the other because from the wonders of nature is the nearest intelligence[n] and passage towards the wonders of art: for it is no more but by following, and as it were hounding nature in her wanderings, to be able to lead her afterwards to the same place again. Neither am I of opinion, in this history of marvels, that superstitious narrations of sorceries,[n] witchcrafts, dreams, divinations, and the like, where there is an assurance and clear evidence of the fact, be altogether excluded. For it is not yet known in what cases and how far effects attributed to superstition do participate of natural causes: and therefore howsoever the practice of such things is to be condemned, yet from the speculation and consideration of them light may be taken, not only for the discerning of the offences, but for the further disclosing of nature. Neither ought a man to make scruple of entering into these things for inquisition of truth, as your Majesty[n] hath showed in your own example; who with

deflexion: turning aside from the regular path.*
proprieties: natures, essences.
severe: rigidly accurate.*
heteroclites: things that deviate from the ordinary rules.*
called down: decried.
is nothing less than: is by no means intended.
Mirabilaries: i.e. mirabiliaries, collectors of manuals.
speculation: investigation.

the two clear eyes of religion and natural philosophy have looked deeply and wisely into these shadows, and yet proved yourself to be of the nature of the sun, which passeth through pollutions and itself remains as pure as before. But this I hold fit, that these narrations, which have mixture with superstition, be sorted by themselves, and not to be mingled with the narrations which are merely and sincerely natural. But as for the narrations touching the prodigies and miracles of religions, they are either not true, or not natural; and therefore impertinent for the story of nature.

5. *Historia Mechanica.* For history of nature wrought or mechanical, I find some collections made of agriculture, and likewise of manual arts;[n] but commonly with a rejection of experiments familiar and vulgar. For it is esteemed a kind of dishonour unto learning to descend to inquiry or meditation upon matters mechanical, except they be such as may be thought secrets, rarities, and special subtilities; which humour of vain and supercilious arrogancy is justly derided in Plato; where he brings in Hippias, a vaunting sophist, disputing with Socrates, a true and unfeigned inquisitor of truth; where the subject being touching beauty, Socrates, after his wandering manner of inductions, put first an example of a fair virgin, and then of a fair horse, and then of a fair pot well glazed, whereat Hippias was offended, and said, 'More than for courtesy's sake, he did think much to dispute with any that did allege such base and sordid instances.' Whereunto Socrates answereth, 'You have reason, and it becomes you well, being a man so trim in your vestiments, &c.,' and so goeth on in an irony. But the truth is, they be not the highest instances that give the securest information; as may be well expressed in the tale so common of the philosopher,[n] that while he gazed upwards to the stars fell into the water; for if he had looked down he might have seen the stars in the water, but looking aloft he could not see the water in the stars. So it cometh often to pass, that mean and small things discover great, better than great can discover the small: and therefore Aristotle noteth well, 'That the nature of everything is best seen in his smallest portions.' And for that cause he inquireth the nature of a commonwealth, first in a

merely: absolutely, simply.
sincerely: in an absolute manner or degree.
Historia Mechanica: History of Nature wrought.
wrought: worked upon.
Hippias: In Plato, *Hippias Major* 291 a.
think much: hesitate.†
Aristotle: In *Politics* i. 3.
inquireth: investigates.

family, and the simple conjugations of man and wife, parent and child, master and servant, which are in every cottage. Even so likewise the nature of this great city of the world, and the policy thereof, must be first sought in mean concordances and small portions. So we see how that secret of nature, of the turning of iron touched with the loadstone towards the north, was found out in needles of iron, not in bars of iron.

6. But if my judgement be of any weight, the use of history mechanical is of all others the most radical and fundamental towards natural philosophy; such natural philosophy as shall not vanish in the fume of subtile, sublime, or delectable speculation, but such as shall be operative to the endowment and benefit of man's life.[n] For it will not only minister and suggest for the present many ingenious practices in all trades, by a connexion and transferring of the observations of one art to the use of another, when the experiences of several mysteries shall fall under the consideration of one man's mind; but further, it will give a more true and real illumination concerning causes and axioms than is hitherto attained. For like as a man's disposition is never well known till he be crossed, nor Proteus[n] ever changed shapes till he was straitened and held fast; so the passages and variations of nature cannot appear so fully in the liberty of nature as in the trials and vexations of art.

[CIVIL HISTORY]

II. 1. For civil history,[n] it is of three kinds; not unfitly to be compared with the three kinds of pictures or images. For of pictures or images, we see some are unfinished, some are perfect, and some are defaced. So of histories we may find three kinds, memorials, perfect histories, and antiquities; for memorials are history unfinished, or the first or rough draughts of history; and antiquities are history defaced, or some remnants of history which have casually escaped the shipwreck of time.

2. Memorials, or preparatory history, are of two sorts; whereof the one may be termed commentaries, and the other registers. Commentaries are they which set down a continuance of the naked events and actions, without the motives or designs, the counsels, the

conjugations: relationships.*
concordances: instances of agreement.*
mysteries: crafts, trades.
straitened: confined in a narrow space.
vexations: examinations carried out by a process of disturbing.

speeches, the pretexts, the occasions and other passages of action: for this is the true nature of a commentary (though Caesar, in modesty mixed with greatness, did for his pleasure apply the name of a commentary to the best history of the world). Registers are collections of public acts, as decrees of council, judicial proceedings, declarations and letters of estate, orations and the like, without a perfect continuance or contexture of the thread of the narration.

3. Antiquities,[n] or remnants of history, are, as was said, *tanquam tabula naufragii*: when industrious persons, by an exact and scrupulous diligence and observation, out of monuments, names, words, proverbs, traditions, private records and evidences, fragments of stories, passages of books that concern not story, and the like, do save and recover somewhat from the deluge of time.

4. In these kinds of unperfect histories I do assign no deficience, for they are *tanquam imperfecte mista*; and therefore any deficience in them is but their nature. As for the corruptions and moths of history, which are epitomes,[n] the use of them deserveth to be banished, as all men of sound judgement have confessed, as those that have fretted and corroded the sound bodies of many excellent histories, and wrought them into base and unprofitable dregs.

5. History, which may be called just and perfect history, is of three kinds, according to the object which it propoundeth, or pretendeth to represent: for it either representeth a time, or a person, or an action. The first we call chronicles, the second lives, and the third narrations or relations. Of these, although the first be the most complete and absolute kind of history, and hath most estimation and glory, yet the second excelleth it in profit and use, and the third in verity and sincerity. For history of times representeth the magnitude of actions, and the public faces and deportments of persons, and passeth over in silence the smaller passages and motions of men and matters. But such being the workmanship of God, as he doth hang the greatest weight upon the smallest wires, *maxima e minimis suspendens*, it comes therefore to pass, that such histories do rather set forth the pomp of business than the true and inward resorts thereof. But lives, if they be well written, propounding to themselves a person to represent, in whom actions both greater and smaller, public and private, have a commixture, must of necessity contain a more

of the world: i.e. in the world. Caesar's 'Commentaries' are his histories of the Gallic and Civil wars.
tanquam: like the planks of a shipwreck.
tanquam: as it were imperfectly compounded.
resorts: sources, springs.†

true, native, and lively representation. So again narrations and relations of actions, as the war of Peloponnesus,[n] the expedition of Cyrus Minor, the conspiracy of Catiline, cannot but be more purely and exactly true than histories of times, because they may choose an argument comprehensible within the notice and instructions of the writer: whereas he that undertaketh the story of a time, specially of any length, cannot but meet with many blanks and spaces which he must be forced to fill up out of his own wit and conjecture.

6. For the history of times (I mean of civil history), the providence of God hath made the distribution. For it hath pleased God to ordain and illustrate two exemplar states of the world for arms, learning, moral virtue, policy, and laws; the state of Grecia and the state of Rome; the histories whereof, occupying the middle part of time, have more ancient to them histories which may by one common name be termed the antiquities of the world: and after them, histories which may be likewise called by the name of modern history.

7. Now to speak of the deficiences. As to the heathen antiquities of the world, it is in vain to note them for deficient. Deficient they are no doubt, consisting most of fables and fragments; but the deficience cannot be holpen; for antiquity is like fame, *caput inter nubila condit*, her head is muffled from our sight. For the history of the exemplar states it is extant in good perfection. Not but I could wish there were a perfect course of history for Grecia from Theseus to Philopoemen (what time the affairs of Grecia drowned and extinguished in the affairs of Rome), and for Rome from Romulus to Justinianus, who may be truly said to be *ultimus Romanorum*.[n] In which sequences of story the text of Thucydides and Xenophon in the one, and the texts of Livius, Polybius, Sallustius, Caesar, Appianus, Tacitus, Herodianus in the other, to be kept entire without any diminution at all, and only to be supplied and continued. But this is matter of magnificence, rather to be commended than required: and we speak now of parts of learning supplemental and not of supererogation.

8. But for modern histories, whereof there are some few very worthy, but the greater part beneath mediocrity, leaving the care of foreign stories to foreign states, because I will not be *curiosus in aliena republica*, I cannot fail to represent to your Majesty the unworthiness of the history of England in the main continuance there-

exemplar: extraordinary.
holpen: helped.
caput inter: Virgil, *Aeneid* iv. 177. 'She hides her head among the clouds.'
curiosus in: Cicero, *On Duties* i. 34, 'a meddler in the affairs of other states'.

of, and the partiality and obliquity of that of Scotland in the latest and largest author[n] that I have seen: supposing that it would be honour for your Majesty, and a work very memorable, if this island of Great Brittany, as it is now joined in monarchy for the ages to come, so were joined in one history for the times passed; after the manner of the sacred history, which draweth down the story of the ten tribes and of the two tribes as twins together. And if it shall seem that the greatness of this work may make it less exactly performed, there is an excellent period of a much smaller compass of time, as to the story of England; that is to say, from the uniting of the Roses to the uniting of the kingdoms;[n] a portion of time wherein, to my understanding, there hath been the rarest varieties that in like number of successions of any hereditary monarchy hath been known. For it beginneth with the mixed adeption of a crown by arms and title; an entry by battle, an establishment by marriage; and therefore times answerable, like waters after a tempest, full of working and swelling, though without extremity of storm; but well passed through by the wisdom of the pilot, being one of the most sufficient kings of all the number. Then followeth the reign of a king, whose actions, howsoever conducted, had much intermixture with the affairs of Europe, balancing and inclining them variably; in whose time also began that great alteration in the state ecclesiastical, an action which seldom cometh upon the stage. Then the reign of a minor: then an offer of an usurpation (though it was but as *febris ephemera*). Then the reign of a queen matched with a foreigner: then of a queen that lived solitary and unmarried, and yet her government so masculine, as it had greater impression and operation upon the states abroad than it any ways received from thence. And now last, this most happy and glorious event, that this island of Brittany, divided from all the world, should be united in itself: and that oracle of rest given to Aeneas, *antiquam exquirite matrem*, should now be performed and fulfilled upon the nations of England and Scotland, being now reunited in the ancient mother name of Brittany, as a full period of all instability and peregrinations. So that as it cometh to pass in massive bodies, that they have certain trepidations and waverings before they fix and settle, so it seemeth that by the providence of God this monarchy, before it was to settle in your

obliquity: unfairness.
adeption: obtaining.
offer: attempt.
febris: a day-long fever.
operation: influence.*
divided from all the world: cf. Virgil, *Eclogues* i. 67.
antiquam: Virgil, *Aeneid* iii. 96. 'Search out your ancient mother.'
trepidations: vibratory movements.*

majesty and your generations (in which I hope it is now established for ever), it had these prelusive changes and varieties.

9. For lives, I do find strange that these times have so little esteemed the virtues of the times, as that the writings of lives should be no more frequent. For although there be not many sovereign princes or absolute commanders, and that states are most collected into monarchies, yet are there many worthy personages that deserve better than dispersed report or barren elogies. For herein the invention of one of the late poets[n] is proper, and doth well enrich the ancient fiction. For he feigneth that at the end of the thread or web of every man's life there was a little medal containing the person's name, and that Time waited upon the shears, and as soon as the thread was cut, caught the medals, and carried them to the river of Lethe; and about the bank there were many birds flying up and down, that would get the medals and carry them in their beak a little while, and then let them fall into the river. Only there were a few swans, which if they got a name would carry it to a temple where it was consecrate. And although many men, more mortal in their affections than in their bodies, do esteem desire of name and memory but as a vanity and ventosity,

Animi nil magnae laudis egentes;

which opinion cometh from that root, *Non prius laudes contempsimus, quam laudanda facere desivimus*: yet that will not alter Salomon's judgement, *Memoria justi cum laudibus, at impiorum nomen putrescet*: the one flourisheth, the other either consumeth to present oblivion, or turneth to an ill odour. And therefore in that style or addition, which is and hath been long well received and brought in use, *felicis memoriae, piae memoriae, bonae memoriae*, we do acknowledge that which Cicero saith, borrowing it from Demosthenes, that *bona fama propria possessio defunctorum*; which possession I cannot but note that in our times it lieth much waste, and that therein there is a deficience.

prelusive: preliminary.*
find strange: wonder.
elogies: explanatory inscriptions.*
ventosity: an idle conceit.
Animi nil: Virgil, *Aeneid* v. 751. 'Souls that care not for great renown.'
Non prius: quoted (from memory) from Pliny, *Letters* iii. 21. 'Men do not despise praise until they have ceased to do anything that deserves it.'
Memoria justi: Proverbs 10:7. 'The memory of the just is blessed, but the name of the wicked shall rot.'
felicis: 'Of happy, of pious, of good memory.'
bona fama: cf. Cicero, *Philippics* ix. 5. 'Good fame is all that a dead man can possess.'

10. For narrations and relations of particular actions, there were also to be wished a greater diligence therein; for there is no great action but hath some good pen which attends it. And because it is an ability not common to write a good history, as may well appear by the small number of them; yet if particularity of actions memorable were but tolerably reported as they pass, the compiling of a complete history of times mought be the better expected, when a writer should arise that were fit for it: for the collection of such relations mought be as a nursery garden, whereby to plant a fair and stately garden, when time should serve.

11. There is yet another partition of history which Cornelius Tacitus maketh, which is not to be forgotten, specially with that application which he accoupleth it withal, annals and journals: appropriating to the former matters of estate, and to the latter acts and accidents of a meaner nature. For giving but a touch of certain magnificent buildings, he addeth, *Cum ex dignitate populi Romani repertum sit, res illustres annalibus, talia diurnis urbis actis mandare.* So as there is a kind of contemplative heraldry,[n] as well as civil. And as nothing doth derogate from the dignity of a state more than confusion of degrees, so it doth not a little imbase the authority of an history, to intermingle matters of triumph, or matters of ceremony, or matters of novelty, with matters of state. But the use of a journal hath not only been in the history of time[s], but likewise in the history of persons, and chiefly of actions; for princes in ancient time had, upon point of honour and policy both, journals kept, what passed day by day. For we see the chronicle which was read before Ahasuerus, when he could not take rest, contained matter of affairs indeed, but such as had passed in his own time and very lately before. But the journal of Alexander's house[n] expressed every small particularity, even concerning his person and court; and it is yet an use well received in enterprises memorable, as expeditions of war, navigations, and the like, to keep diaries of that which passeth continually.

12. I cannot likewise be ignorant of a form of writing which some grave and wise men have used, containing a scattered history[n] of those actions which they have thought worthy of memory, with politic discourse and observation thereupon: not incorporate into the history, but separately, and as the more principal in their inten-

accoupleth: joins.
giving but a touch: alluding to.
Cum ex: Tacitus, *Annals* xiii. 31. 'It suits the dignity of the Roman people to reserve history for great achievements, and to leave such details to the city's daily register.'
Ahasuerus: Esther 6:1.

tion; which kind of ruminated history I think more fit to place amongst books of policy, whereof we shall hereafter speak, than amongst books of history. For it is the true office of history to represent the events themselves together with the counsels, and to leave the observations and conclusions thereupon to the liberty and faculty of every man's judgement. But mixtures are things irregular, whereof no man can define.

13. So also is there another kind of history manifoldly mixed, and that is history of cosmography:[n] being compounded of natural history, in respect of the regions themselves; of history civil, in respect of the habitations, regiments, and manners of the people; and the mathematics, in respect of the climates and configurations towards the heavens: which part of learning of all others in this latter time hath obtained most proficience. For it may be truly affirmed to the honour of these times, and in a virtuous emulation with antiquity, that this great building of the world had never through-lights made in it, till the age of us and our fathers. For although they had knowledge of the antipodes,

> Nosque ubi primus equis Oriens afflavit anhelis,
> Illic sera rubens accendit lumina Vesper,

yet that mought be by demonstration, and not in fact; and if by travel, it requireth the voyage but of half the globe. But to circle the earth,[n] as the heavenly bodies do, was not done nor enterprised till these later times: and therefore these times may justly bear in their word, not only *plus ultra*, in precedence of the ancient *non ultra*, and *imitabile fulmen*, in precedence of the ancient *non imitabile fulmen*,

> Demens qui nimbos et non imitabile fulmen, &c.

but likewise *imitabile coelum*; in respect of the many memorable

ruminated: meditated.*
regiments: governments. *climates*: regions.
configurations: positions relative to the heavenly bodies.
through-lights: windows on opposite sides of a room.*
Nosque ubi: Virgil, *Georgics* i. 250–1.

> And when on us Morn breathes the living light,
> Red Vesper kindles there the tapers of the night.
> (Dryden)

enterprised: undertaken.
word: Motto. *Plus ultra*: further yet; *non ultra*: no further.
imitabile fulmen: the 'imitable thunder' refers to the invention of gunpowder.
Demens qui: Virgil, *Aeneid* vi. 590. 'Vain fools to mock the clouds, and the bolt we may not imitate.'
imitabile coelum: imitable heaven.

voyages after the manner of heaven about the globe of the earth.

14. And this proficience in navigation and discoveries may plant also an expectation of the further proficience and augmentation of all sciences; because it may seem they are ordained by God to be coevals, that is, to meet in one age. For so the prophet Daniel speaking of the latter times foretelleth, *Plurimi pertransibunt, et multiplex erit scientia*:[n] as if the openness and through-passage of the world and the increase of knowledge were appointed to be in the same ages; as we see it is already performed in great part: the learning of these later times not much giving place to the former two periods[n] or returns of learning, the one of the Grecians, the other of the Romans.

[ECCLESIASTICAL HISTORY]

III. 1. History ecclesiastical receiveth the same divisions with history civil: but further in the propriety thereof may be divided into the history of the church, by a general name; history of prophecy; and history of providence. The first describeth the times of the militant church, whether it be fluctuant, as the ark of Noah, or moveable, as the ark in the wilderness, or at rest, as the ark in the temple: that is, the state of the church in persecution, in remove, and in peace. This part I ought in no sort to note as deficient; only I would the virtue and sincerity of it were according to the mass and quantity. But I am not now in hand with censures, but with omissions.

2. *Historia Prophetica.* The second, which is history of prophecy, consisteth of two relatives, the prophecy, and the accomplishment; and therefore the nature of such a work ought to be, that every prophecy of the scripture be sorted with the event fulfilling the same, throughout the ages of the world: both for the better confirmation of faith, and for the better illumination of the Church touching those parts of prophecies which are yet unfulfilled: allowing nevertheless that latitude which is agreeable and familiar unto divine prophecies; being of the nature of their author, with whom a thousand years are but as one day; and therefore are not fulfilled punctually at once, but have springing and germinant accomplishment

fluctuant: floating on the waves.*
remove: departure to another place.
Historia Prophetica: History of Prophecy.
sorted: classified.
a thousand years: II Peter 3:8.
germinant: slow-growing, like a plant.*

throughout many ages; though the height or fulness of them may refer to some one age. This is a work which I find deficient; but is to be done with wisdom, sobriety, and reverence, or not at all.

3. The third, which is history of providence, containeth that excellent correspondence which is between God's revealed will and his secret will: which though it be so obscure, as for the most part it is not legible to the natural man;[n] no, nor many times to those that behold it from the tabernacle; yet at some times it pleaseth God, for our better establishment and the confuting of those which are as without God in the world, to write it in such text and capital letters, that, as the prophet saith, 'He that runneth by may read it';[n] that is, mere sensual persons, which hasten by God's judgements, and never bend or fix their cogitations upon them, are nevertheless in their passage and race urged to discern it. Such are the notable events and examples of God's judgements, chastisements, deliverances, and blessings: and this is a work which hath passed through the labour of many, and therefore I cannot present as omitted.

4. There are also other parts of learning which are appendices to history. For all the exterior proceedings of man consist of words and deeds; whereof history doth properly receive and retain in memory the deeds, and if words, yet but as inducements and passages to deeds; so are there other books and writings, which are appropriate to the custody and receipt of words only; which likewise are of three sorts; orations, letters, and brief speeches or sayings. Orations are pleadings, speeches of counsel, laudatives, invectives, apologies, reprehensions, orations of formality or ceremony, and the like. Letters are according to all the variety of occasions, advertisements, advices, directions, propositions, petitions, commendatory, expostulatory, satisfactory, of compliment, of pleasure, of discourse, and all other passages of action. And such as are written from wise men are of all the words of man, in my judgement, the best; for they are more natural than orations, and public speeches, and more advised than conferences or present speeches. So again letters of affairs from such as manage them, or are privy to them, are of all others the best instructions for history, and to a diligent reader the best histories in themselves. For apophthegms, it is a great loss of that book of Caesar's; for as his history, and those few letters of his

advertisements: informations.
satisfactory: written to satisfy a debt or obligation.
advised: well-considered.
present: impromptu.
apophthegms: see n. to p. 52 above.

which we have, and those apophthegms which were of his own, excel all men's else, so I suppose would his collection of apophthegms have done. For as for those which are collected by others, either I have no taste in such matters, or else their choice hath not been happy. But upon these three kinds of writings I do not insist, because I have no deficiences to propound concerning them.

5. Thus much therefore concerning history, which is that part of learning which answereth to one of the cells,[n] domiciles, or offices of the mind of man; which is that of the memory.

[POETRY]

IV. 1 Poesy[n] is a part of learning in measure of words for the most part restrained, but in all other points extremely licensed, and doth truly refer to the imagination;[n] which, being not tied to the laws of matter, may at pleasure join that which nature hath severed, and sever that which nature hath joined; and so make unlawful matches and divorces of things; *Pictoribus atque poetis, &c.* It is taken in two senses in respect of words or matter.[n] In the first sense it is but a character of style, and belongeth to arts of speech, and is not pertinent for the present. In the latter it is (as hath been said) one of the principal portions of learning, and is nothing else but feigned history,[n] which may be styled as well in prose as in verse.

2. The use of this feigned history hath been to give some shadow of satisfaction to the mind of man in those points wherein the nature of things doth deny it, the world being in proportion inferior to the soul; by reason whereof there is, agreeable to the spirit of man, a more ample greatness, a more exact goodness, and a more absolute variety, than can be found in the nature of things. Therefore, because the acts or events of true history have not that magnitude which satisfieth the mind of man, poesy feigneth acts and events greater and more heroical.[n] Because true history propoundeth the successes and issues of actions not so agreeable to the merits of virtue and vice, therefore poesy feigns them more just in retribution, and more according to revealed providence. Because true history representeth actions and events more ordinary and less interchanged, therefore poesy endueth them with more rareness, and more unexpected and

restrained: because in metre.
Pictoribus: Horace, *Art of Poetry*, l. 9. 'Painters and poets [have always been allowed to take what liberties they would].'
styled: i.e. may have the title of 'feigned history'.
successes: results of an action, good or bad.
interchanged: varied.

alternative variations. So as it appeareth that poesy serveth and conferreth to magnanimity, morality, and to delectation. And therefore it was ever thought to have some participation of divineness, because it doth raise and erect the mind,[n] by submitting the shows of things to the desires of the mind; whereas reason doth buckle and bow the mind unto the nature of things. And we see that by these insinuations and congruities with man's nature and pleasure, joined also with the agreement and consort it hath with music, it hath had access and estimation in rude times and barbarous regions, where other learning stood excluded.

3. The division of poesy which is aptest in the propriety thereof (besides those divisions which are common unto it with history, as feigned chronicles, feigned lives, and the appendices of history, as feigned epistles, feigned orations, and the rest) is into poesy narrative, representative, and allusive.[n] The narrative is a mere imitation of history, with the excesses before remembered; choosing for subject commonly wars and love, rarely state, and sometimes pleasure or mirth. Representative is as a visible history; and is an image of actions as if they were present, as history is of actions in nature as they are, (that is) past. Allusive or parabolical is a narration applied only to express some special purpose or conceit. Which latter kind of parabolical wisdom was much more in use in the ancient times, as by the fables of Aesop,[n] and the brief sentences of the seven, and the use of hieroglyphics may appear. And the cause was, for that it was then of necessity to express any point of reason which was more sharp or subtile than the vulgar in that manner, because men in those times wanted both variety of examples and subtility of conceit. And as hieroglyphics were before letters, so parables were before arguments: and nevertheless now and at all times they do retain much life and vigour, because reason cannot be so sensible, nor examples so fit.

4. But there remaineth yet another use of poesy parabolical, opposite to that which we last mentioned: for that tendeth to demonstrate and illustrate that which is taught or delivered, and this other to retire and obscure it: that is, when the secrets and mysteries of religion, policy, or philosophy, are involved in fables or parables. Of this in divine poesy[n] we see the use is authorized. In heathen

conferreth: contributes.
magnanimity: high-mindedness, nobility of feeling, nobly ambitious spirit.
insinuations: subtle infusions.
consort: accord.
remembered: mentioned.
vulgar: common, familiar.
conceit: conception.
sensible: effective.
retire: put into obscurity.

poesy we see the exposition of fables doth fall out sometimes with great felicity; as in the fable that the giants being overthrown in their war against the gods, the earth their mother in revenge thereof brought forth Fame:

> Illam terra parens, ira irritata Deorum,
> Extremam, ut perhibent, Coeo Enceladoque sororem
> Progenuit.[n]

Expounded that when princes and monarchs have suppressed actual and open rebels, then the malignity of people (which is the mother of rebellion) doth bring forth libels and slanders, and taxations of the states, which is of the same kind with rebellion, but more feminine. So in the fable that the rest of the gods having conspired to bind Jupiter, Pallas[n] called Briareus with his hundred hands to his aid: expounded that monarchies need not fear any curbing of their absoluteness by mighty subjects, as long as by wisdom they keep the hearts of the people, who will be sure to come in on their side. So in the fable that Achilles[n] was brought up under Chiron the centaur, who was part a man and part a beast, expounded ingeniously but corruptly by Machiavel, that it belongeth to the education and discipline of princes to know as well how to play the part of the lion in violence, and the fox in guile, as of the man in virtue and justice. Nevertheless, in many the like encounters, I do rather think that the fable was first, and the exposition devised, than that the moral was first, and thereupon the fable framed. For I find it was an ancient vanity in Chrysippus,[n] that troubled himself with great contention to fasten the assertions of the Stoics upon the fictions of the ancient poets; but yet that all the fables and fictions of the poets were but pleasure and not figure, I interpose no opinion. Surely of those poets which are now extant, even Homer[n] himself (notwithstanding he was made a kind of scripture by the later schools of the Grecians), yet I should without any difficulty pronounce that his fables had no such inwardness in his own meaning. But what they might have upon a more original tradition, is not easy to affirm; for he was not the inventor of many of them.

5. In this third part[n] of learning, which is poesy, I can report no deficience. For being as a plant that cometh of the lust of the earth, without a formal seed, it hath sprung up and spread abroad more than any other kind. But to ascribe unto it that which is due, for the

taxations: censures. *contention*: effort.
figure: figurative, parabolic.
inwardness: inner meaning.* *lust*: fertility.

expressing of affections, passions, corruptions, and customs, we are beholding to poets[n] more than to the philosophers' works; and for wit and eloquence, not much less than to orators' harangues. But it is not good to stay too long in the theatre. Let us now pass on to the judicial[n] place or palace of the mind, which we are to approach and view with more reverence and attention.

[KNOWLEDGE DIVIDED INTO DIVINITY AND PHILOSOPHY. OF *PHILOSOPHIA PRIMA*]

V. 1. The knowledge of man is as the waters, some descending from above, and some springing from beneath; the one informed by the light of nature, the other inspired by divine revelation. The light of nature consisteth in the notions of the mind and the reports of the senses: for as for knowledge which man receiveth by teaching, it is cumulative and not original; as in a water that besides his own spring-head is fed with other springs and streams. So then, according to these two differing illuminations or originals, knowledge is first of all divided into divinity and philosophy.[n]

2. *Philosophiaprima, sive de fontibus scientiarum.* In philosophy,[n] the contemplations of man do either penetrate unto God, or are circumferred to nature, or are reflected or reverted upon himself. Out of which several inquiries there do arise three knowledges; divine philosophy, natural philosophy, and human philosophy or humanity. For all things are marked and stamped with this triple character, of the power of God, the difference of nature, and the use of man. But because the distributions and partitions of knowledge are not like several lines that meet in one angle, and so touch but in a point; but are like the branches of a tree, that meet in a stem, which hath a dimension and quantity of entireness and continuance, before it come to discontinue and break itself into arms and boughs: therefore it is good, before we enter into the former distribution, to erect and constitute one universal science, by the name of *philosophia prima*,[n] primitive or summary philosophy, as the main and common way, before we come where the ways part and divide themselves; which science whether I should report as deficient or no, I stand doubtful. For I find a certain rhapsody[n] of natural theology, and of divers parts of logic; and of that part of natural philosophy which

beholding: indebted. *a water*: a sea, pond, or lake.
Philosophia: Primary philosophy, or the spring-head of the sciences.
circumferred: carried round.* *reverted*: turned back.
primitive: primary. *summary*: most important.

concerneth the principles, and of that other part of natural philosophy which concerneth the soul or spirit; all these strangely commixed and confused; but being examined, it seemeth to me rather a depredation of other sciences, advanced and exalted unto some height of terms, than anything solid or substantive of itself. Nevertheless I cannot be ignorant of the distinction which is current, that the same things are handled but in several respects. As for example, that logic considereth of many things as they are in notion,[n] and this philosophy as they are in nature; the one in appearance, the other in existence; but I find this difference better made than pursued. For if they had considered quantity, similitude, diversity, and the rest of those extern characters of things, as philosophers, and in nature, their inquiries must of force have been of a far other kind than they are. For doth any of them, in handling quantity, speak of the force of union, how and how far it multiplieth virtue? Doth any give the reason, why some things in nature are so common, and in so great mass, and others so rare, and in so small quantity? Doth any, in handling similitude and diversity, assign the cause why iron should not move to iron, which is more like, but move to the load-stone, which is less like? Why in all diversities of things there should be certain participles in nature,[n] which are almost ambiguous to which kind they should be referred? But there is a mere and deep silence touching the nature and operation of those common adjuncts of things, as in nature: and only a resuming and repeating of the force and use of them in speech or argument. Therefore, because in a writing of this nature I avoid all subtility, my meaning touching this original or universal philosophy is thus, in a plain and gross description by negative: 'That it be a receptacle for all such profitable observations and axioms[n] as fall not within the compass of any of the special parts of philosophy or sciences, but are more common and of a higher stage.'

3. Now that there are many of that kind need not be doubted. For example: is not the rule, *Si inaequalibus aequalia addas, omnia erunt inaequalia*,[n] an axiom as well of justice as of the mathematics? and is there not a true coincidence between commutative and distributive justice, and arithmetical and geometrical proportion? Is not that other rule, *Quae in eodem tertio conveniunt, et inter se conveniunt*,[n] a rule taken from the mathematics, but so potent in logic as all syllogisms are built upon it? Is not the obser-

commixed: mixed. *extern*: external.
virtue: efficacy, superiority in nature or in operation.
mere: absolute. *coincidence*: exact agreement.*

vation, *Omnia mutantur, nil interit,* a contemplation in philosophy thus, that the *quantum* of nature is eternal? in natural theology thus, that it requireth the same omnipotency to make somewhat nothing, which at the first made nothing somewhat? according to the scripture, *Didici quod omnia opera, quae fecit Deus, perseverent in perpetuum; non possumus eis quicquam addere nec auferre.* Is not the ground, which Machiavel wisely and largely discourseth concerning governments, that the way to establish and preserve them, is to reduce them *ad principia,*[n] a rule in religion and nature, as well as in civil administration? Was not the Persian magic a reduction or correspondence of the principles and architectures of nature to the rules and policy of governments? Is not the precept of a musician, to fall from a discord or harsh accord upon a concord or sweet accord, alike true in affection? Is not the trope of music,[n] to avoid or slide from the close or cadence, common with the trope of rhetoric of deceiving expectation? Is not the delight of the quavering upon a stop in music the same with the playing of light upon the water?

Splendet tremulo sub lumine pontus.

Are not the organs of the senses of one kind with the organs of reflection, the eye with a glass, the ear with a cave or strait, determined and bounded? Neither are these only similitudes, as men of narrow observation may conceive them to be, but the same footsteps of nature, treading or printing upon several subjects or matters. This science therefore (as I understand it) I may justly report as deficient: for I see sometimes the profounder sort of wits, in handling some particular argument, will now and then draw a bucket of water out of this well for their present use: but the spring-head thereof seemeth to me not to have been visited; being of so excellent use both for the disclosing of nature and the abridgement of art.

Omnia mutantur: Ovid, *Metamorphoses* xv. 176. 'All things change but nothing is lost.' Cf. *Sylva Sylvarum* i. 100.
Didici quod: Ecclesiastes 3:14. 'I know that whatever God does lasts for ever; to add to it or subtract from it is impossible.'
ground: fundamental principle.
Splendet: Virgil, *Aeneid* vii. 9. 'The sea shines beneath the trembling light.'
glass: mirror.

[DIVINE PHILOSOPHY, OR NATURAL THEOLOGY]

VI. 1. This science being therefore first placed as a common parent like unto Berecynthia, which had so much heavenly issue, *omnes caelicolas, omnes supera alta tenentes*; we may return to the former distribution of the three philosophies, divine, natural, and human. And as concerning divine philosophy or natural theology, it is that knowledge or rudiment of knowledge concerning God, which may be obtained by the contemplation of his creatures; which knowledge may be truly termed divine in respect of the object, and natural in respect of the light. The bounds of this knowledge are, that it sufficeth to convince atheism, but not to inform religion: and therefore there was never miracle wrought by God to convert an atheist, because the light of nature might have led him to confess a God: but miracles have been wrought to convert idolaters and the superstitious, because no light of nature extendeth to declare the will and true worship of God. For as all works do show forth the power and skill of the workman, and not his image, so it is of the works of God, which do show the omnipotency and wisdom of the maker, but not his image. And therefore therein the heathen opinion differeth from the sacred truth; for they supposed the world to be the image of God,[n] and man to be an extract or compendious image of the world; but the scriptures never vouchsafe to attribute to the world that honour, as to be the image of God, but only 'the work of his hands'; neither do they speak of any other image of God, but man. Wherefore by the contemplation of nature to induce and enforce the acknowledgement of God, and to demonstrate his power, providence, and goodness, is an excellent argument, and hath been excellently handled by divers. But on the other side, out of the contemplation of nature, or ground of human knowledges, to induce any verity or persuasion concerning the points of faith, is in my judgement not safe: *Da fidei quae fidei sunt*. For the heathen themselves conclude as much in that excellent and divine fable of the golden chain: 'That men and gods were not able to draw Jupiter down to the earth; but contrariwise Jupiter was able to draw them up to heaven.' So as we ought not to attempt to draw down or to submit the mysteries of God to our reason; but contrariwise to raise and advance our reason to the

omnes: Virgil, *Aeneid* vi. 788. 'All the inhabitants of heaven, all the dwellers in the upper air.'
convince: refute. *inform*: teach. *extract*: summary.
work of his hands: Genesis 1, Psalms 8:3, 6.
Da fidei: Matthew 22:21. 'Give unto faith the things that are faith's.'
golden chain: cf. p. 9 above.

divine truth. So as in this part of knowledge, touching divine philosophy, I am so far from noting any deficience, as I rather note an excess: whereunto I have digressed because of the extreme prejudice which both religion and philosophy hath received and may receive by being commixed together; as that which undoubtedly will make an heretical religion, and an imaginary and fabulous philosophy.

2. Otherwise it is of the nature of angels and spirits,[n] which is an appendix of theology, both divine and natural, and is neither inscrutable nor interdicted. For although the scripture saith, 'Let no man deceive you in sublime discourse touching the worship of angels, pressing into that he knoweth not, &c.,' yet notwithstanding if you observe well that precept, it may appear thereby that there be two things only forbidden, adoration of them, and opinion fantastical of them, either to extol them further than appertaineth to the degree of a creature, or to extol a man's knowledge of them further than he hath ground. But the sober and grounded inquiry, which may arise out of the passages of holy scriptures, or out of the gradations of nature, is not restrained. So of degenerate and revolted spirits, the conversing with them or the employment of them is prohibited, much more any veneration towards them; but the contemplation or science of their nature, their power, their illusions, either by scripture or reason, is a part of spiritual wisdom. For so the apostle saith, 'We are not ignorant of his stratagems.' And it is no more unlawful to inquire into the nature of evil spirits, than to inquire the force of poisons in nature, or the nature of sin and vice in morality. But this part touching angels and spirits I cannot note as deficient, for many have occupied themselves in it; I may rather challenge it, in many of the writers thereof as fabulous and fantastical.

[NATURAL PHILOSOPHY—PHYSICS AND METAPHYSICS]

VII.1. Leaving therefore divine philosophy or natural theology (not divinity or inspired theology, which we reserve for the last of all as the haven and sabbath of all man's contemplations) we will now proceed to natural philosophy. If then it be true that Democritus

touching: concerning. *prejudice*: injury.
Otherwise it is: The case is different with respect to.
Let no man: Colossians 2:4, 18. *grounded*: well founded.
conversing: holding communion.
We are not: II Corinthians 2:11.
Democritus: In Cicero, *Academica* I. xii. 44, and Diogenes Laertius ix. 72.

said, 'that the truth of nature lieth hid in certain deep mines and caves'; and if it be true likewise that the alchemists do so much inculcate, that Vulcan[n] is a second nature, and imitateth that dexterously and compendiously which nature worketh by ambages and length of time; it were good to divide natural philosophy into the mine and the furnace, and to make two professions or occupations of natural philosophers, some to be pioneers and some smiths; some to dig, and some to refine and hammer. And surely I do best allow of a division of that kind, though in more familiar and scholastical terms; namely that these be the two parts of natural philosophy, the inquisition of causes, and the production of effects; speculative, and operative; natural science, and natural prudence. For as in civil matters there is a wisdom of discourse, and a wisdom of direction; so is it in natural. And here I will make a request, that for the latter (or at least for a part thereof) I may revive and reintegrate the misapplied and abused name of natural magic;[n] which in the true sense is but natural wisdom, or natural prudence; taken according to the ancient acception, purged from vanity and superstition. Now although it be true, and I know it well, that there is an intercourse between causes and effects, so as both these knowledges, speculative and operative, have a great connexion between themselves; yet because all true and fruitful natural philosophy hath a double scale or ladder, ascendent and descendent, ascending from experiments to the invention of causes, and descending from causes to the invention of new experiments; therefore I judge it most requisite that these two parts be severally considered and handled.

2. Natural science or theory is divided into physic and metaphysic:[n] wherein I desire it may be conceived that I use the word metaphysic in a differing sense from that that is received. And in like manner, I doubt not but it will easily appear to men of judgement, that in this and other particulars, wheresoever my conception and notion may differ from the ancient, yet I am studious to keep the ancient terms. For hoping well to deliver myself from mistaking, by the order and perspicuous expressing of that I do propound; I am otherwise zealous and affectionate to recede as little from antiquity, either in terms or opinions, as may stand with truth and the proficience of

ambages: circuitous ways.
prudence: practical wisdom (distinguished from theoretical knowledge).
reintegrate: restore.* *acception*: received meaning.
physic: knowledge of the phenomenal universe.
from mistaking: from being misunderstood.
affectionate: eagerly desirous.
stand with: be consistent with.

knowledge. And herein I cannot a little marvel at the philosopher Aristotle,[n] that did proceed in such a spirit of difference and contradiction towards all antiquity: undertaking not only to frame new words of science at pleasure, but to confound and extinguish all ancient wisdom: insomuch as he never nameth or mentioneth an ancient author or opinion, but to confute and reprove; wherein for glory, and drawing followers and disciples, he took the right course. For certainly there cometh to pass, and hath place in human truth, that which was noted and pronounced in the highest truth: *Veni in nomine patris, nec recipitis me; si quis venerit in nomine suo eum recipietis.* But in this divine aphorism (considering to whom it was applied, namely to antichrist, the highest deceiver) we may discern well that the coming in a man's own name, without regard of antiquity or paternity, is no good sign of truth, although it be joined with the fortune and success of an *eum recipietis.* But for this excellent person Aristotle, I will think of him that he learned that humour of his scholar, with whom it seemeth he did emulate; the one to conquer all opinions, as the other to conquer all nations. Wherein nevertheless, it may be, he may at some men's hands, that are of a bitter disposition, get a like title as his scholar did:

> Felix terrarum praedo, non utile mundo
> Editus exemplum, &c.

So,

> Felix doctrinae praedo.

But to me on the other side that do desire as much as lieth in my pen to ground a sociable intercourse between antiquity and proficience, it seemeth best to keep way with antiquity *usque ad aras*; and therefore to retain the ancient terms, though I sometimes alter the uses and definitions, according to the moderate proceeding in civil government; where although there be some alteration, yet that holdeth which Tacitus wisely noteth, *eadem magistratuum vocabula.*

3. To return therefore to the use and acception of the term metaphysic, as I do now understand the word; it appeareth, by that

reprove: refute.
Veni: John 5:43. 'I am come in my Father's name, and you do not receive me; if another comes in his own name, you will receive him.'
his scholar: Alexander the Great.
Felix: Adapted from Lucan, *Pharsalia* x. 20 f. 'A fortunate robber of territory, who was a bad example to the world.' So, 'A fortunate robber of learning'.
keep way with: stay with.
usque ad aras: so far as is consistent with religious obligations.
eadem: Tacitus, *Annals* i. 3. 'The names of the magistracies are not changed.'

which hath been already said, that I intend *philosophia prima*, summary philosophy and metaphysic, which heretofore have been confounded as one, to be two distinct things. For the one I have made as a parent or common ancestor to all knowledge; and the other I have now brought in as a branch or descendant of natural science. It appeareth likewise that I have assigned to summary philosophy the common principles and axioms which are promiscuous and indifferent to several sciences: I have assigned unto it likewise the inquiry touching the operation of the relative and adventive characters of essences, as quantity, similitude, diversity, possibility, and the rest: with this distinction and provision; that they be handled as they have efficacy in nature, and not logically. It appeareth likewise that natural theology, which heretofore hath been handled confusedly with metaphysic, I have inclosed and bounded by itself. It is therefore now a question what is left remaining for metaphysic;[n] wherein I may without prejudice preserve thus much of the conceit of antiquity, that physic should contemplate that which is inherent in matter, and therefore transitory; and metaphysic that which is abstracted and fixed. And again, that physic should handle that which supposeth in nature only a being and moving; and metaphysic should handle that which supposeth further in nature a reason, understanding, and platform. But the difference, perspicuously expressed, is most familiar and sensible. For as we divided natural philosophy in general into the inquiry of causes, and productions of effects: so that part which concerneth the inquiry of causes we do subdivide according to the received and sound division of causes. The one part, which is physic, inquireth and handleth the material and efficient causes; and the other, which is metaphysic, handleth the formal and final causes.

4. Physic (taking it according to the derivation, and not according to our idiom for medicine) is situate in a middle term or distance between natural history and metaphysic. For natural history describeth the variety of things; physic the causes, but variable or respective causes; and metaphysic the fixed and constant causes.

> Limus ut hic durescit, et haec ut cera liquescit,
> Uno eodemque igni.

promiscuous: mixed indiscriminately.*
adventive: Adventitious, not essentially inherent.*
essences: things in themselves.†
platform: design.
respective: particular, relative.
Limus ut: Virgil, *Eclogues* viii. 80. 'The same fire which makes the soft clay hard makes hard wax soft.'

Fire is the cause of induration, but respective to clay; fire is the cause of colliquation, but respective to wax. But fire is no constant cause either of induration or colliquation: so then the physical causes are but the efficient and the matter. Physic hath three parts, whereof two respect nature united or collected, the third contemplateth nature diffused or distributed. Nature is collected either into one entire total, or else into the same principles or seeds. So as the first doctrine is touching the contexture or configuration of things, as *de mundo, de universitate rerum.* The second is the doctrine concerning the principles or originals of things. The third is the doctrine concerning all variety and particularity of things; whether it be of the differing substances, or their differing qualities and natures; whereof there needeth no enumeration, this part being but as a gloss or paraphrase that attendeth upon the text of natural history. Of these three I cannot report any as deficient. In what truth or perfection they are handled, I make not now any judgement; but they are parts of knowledge not deserted by the labour of man.

5. *Metaphysica sive de formis et finibus rerum.* For Metaphysic, we have assigned unto it the inquiry of formal and final causes; which assignation, as to the former of them, may seem to be nugatory and void, because of the received and inveterate opinion, that the inquisition of man is not competent to find out essential forms or true differences: of which opinion we will take this hold, that the invention of forms[n] is of all other parts of knowledge the worthiest to be sought, if it be possible to be found. As for the possibility, they are ill discoverers that think there is no land, when they can see nothing but sea. But it is manifest that Plato, in his opinion of ideas, as one that had a wit of elevation situate as upon a cliff, did descry 'that forms were the true object of knowledge'; but lost the real fruit of his opinion, by considering of forms as absolutely abstracted from matter, and not confined and determined by matter; and so turning his opinion upon theology, wherewith all his natural philosophy is infected. But if any man shall keep a continual watchful and severe eye upon action, operation, and the use of knowledge, he may advise and take notice what are the forms, the disclosures whereof are fruitful and important to the state of man. For as to the

colliquation: melting.*
de mundo: of the world, of the universe.
Metaphysica: Metaphysics or of forms and final causes.
nugatory: invalid, futile.*
inveterate: firmly established by long continuance.
advise: consider.

forms of substances (man only except, of whom it is said, *Formavit hominem di limo terrae, et spiravit in faciem ejus spiraculum vitae*, and not as of all other creatures, *Producant aquae, producat terra*), the forms of substances I say (as they are now by compounding and transplanting multiplied) are so perplexed, as they are not to be inquired; no more than it were either possible or to purpose to seek in gross the forms of those sounds which make words, which by composition and transposition of letters are infinite. But on the other side to inquire the form of those sounds or voices which make simple letters is easily comprehensible; and being known induceth and manifesteth the forms of all words, which consist and are compounded of them. In the same manner to inquire the form of a lion, of an oak, of gold; nay, of water, of air, is a vain pursuit: but to inquire the forms of sense, of voluntary motion, of vegetation, of colours, of gravity and levity, of density, of tenuity, of heat, of cold, and all other natures and qualities, which, like an alphabet, are not many, and of which the essences (upheld by matter) of all creatures do consist; to inquire, I say, the true forms of these, is that part of metaphysic which we now define of. Not but that physic doth make inquiry and take consideration of the same natures: but how? Only as to the material and efficient causes of them, and not as to the forms. For example, if the cause of whiteness in snow or froth be inquired, and it be rendered thus, that the subtile intermixture of air and water is the cause, it is well rendered, but nevertheless is this the form of whiteness? No; but it is the efficient, which is ever but *vehiculum formae*. This part of metaphysic I do not find laboured and performed: whereat I marvel not: because I hold it not possible to be invented by that course of invention which hath been used; in regard that men (which is the root of all error) have made too untimely a departure and too remote a recess from particulars.

6. But the use of this part of metaphysic, which I report as deficient, is of the rest the most excellent in two respects: the one, because it is the duty and virtue of all knowledge to abridge the infinity of individual experience, as much as the conception of truth will permit, and to remedy the complaint of *vita brevis, ars longa*; which is performed by uniting the notions and conceptions of

Formavit: Genesis 2:7. 'God formed man from the dust of the ground and breathed into his nostrils the breath of life.'
Producant: Genesis 1:20, 24. 'Let the waters bring forth; let the earth bring forth.'
vehiculum: the carrier of the form. Cf. *Novum Organum* ii. 23.
recess: withdrawal.
vita brevis: life is short and art is long.

sciences. For knowledges are as pyramides, whereof history is the basis. So of natural philosophy, the basis is natural history; the stage next the basis is physic; the stage next the vertical point is metaphysic. As for the vertical point, *opus quod operatur Deus a principio usque ad finem*, the summary law of nature, we know not whether man's inquiry can attain unto it. But these three be the true stages of knowledge, and are to them that are depraved[n] no better than the giants' hills:

> Ter sunt conati imponere Pelio Ossam,
> Scilicet, atque Ossae frondosum involvere Olympum.

But to those which refer all things to the glory of God, they are as the three acclamations, *Sancte, sancte, sancte!* holy in the description or dilatation of his works; holy in the connexion or concatenation of them; and holy in the union of them in a perpetual and uniform law. And therefore the speculation was excellent in Parmenides[n] and Plato, although but a speculation in them, that all things by scale did ascend to unity. So then always that knowledge is worthiest which is charged with least multiplicity, which appeareth to be metaphysic; as that which considereth the simple forms or differences of things, which are few in number, and the degrees and co-ordinations whereof make all this variety. The second respect, which valueth and commendeth this part of metaphysic, is that it doth enfranchise the power of man unto the greatest liberty and possibility of works and effects. For physic carrieth men in narrow and restrained ways, subject to many accidents of impediments, imitating the ordinary flexous courses of nature. But *latae undique sunt sapientibus viae*: to sapience (which was anciently defined to be *rerum divinarum et humanarum scientia*) there is ever choice of means. For physical causes give light to new invention in *simili materia*. But whosoever knoweth any form, knoweth the utmost possibility of superinducing that nature upon any variety of matter;

opus quod: Ecclesiastes 3:11. 'The work which God worketh from beginning to end.'
Ter sunt: Virgil, *Georgics* i. 281–2. 'Mountain on mountain thrice they strove to heap, / Olympus, Ossa, piled on Pelion's steep.'
Sancte: Revelation 4:8. 'Holy, holy, holy!'
dilatation: amplification.
by scale: by degrees.
valueth: gives greater value to.
flexuous: winding.*
latae undique: To the wise there are broad paths every way.
rerum: Cicero, *On Duties* i. 43. 'Knowledge of things divine and human.'
in simili: in similar matter.

and so is less restrained in operation, either to the basis of the matter, or the condition of the efficient; which kind of knowledge Salomon likewise, though in a more divine sense, elegantly describeth; *non arctabuntur gressus tui, et currens non habebis offendiculum.* The ways of sapience are not much liable either to particularity or chance.

7. The second part of metaphysic is the inquiry of final causes,[n] which I am moved to report not as omitted but as misplaced. And yet if it were but a fault in order, I would not speak of it: for order is matter of illustration, but pertaineth not to the substance of sciences. But this misplacing hath caused a deficience, or at least a great improficience in the sciences themselves. For the handling of final causes, mixed with the rest in physical inquiries, hath intercepted the severe and diligent inquiry of all real and physical causes, and given men the occasion to stay upon these satisfactory and specious causes, to the great arrest and prejudice of further discovery. For this I find done not only by Plato, who ever anchoreth upon that shore, but by Aristotle, Galen, and others which do usually likewise fall upon these flats of discoursing causes. For to say that 'the hairs of the eye-lids are for a quickset and fence about the sight'; or that 'the firmness of the skins and hides of living creatures is to defend them from the extremities of heat or cold'; or that 'the bones are for the columns or beams, whereupon the frames of the bodies of living creatures are built': or that 'the leaves of trees are for protecting of the fruit'; or that 'the clouds are for watering of the earth'; or that 'the solidness of the earth is for the station and mansion of living creatures', and the like, is well inquired and collected in metaphysic, but in physic they are impertinent. Nay, they are indeed but *remoraes*[n] and hindrances to stay and slug the ship from further sailing; and have brought this to pass, that the search of the physical causes hath been neglected and passed in silence. And therefore the natural philosophy of Democritus[n] and some others, who did not suppose a mind or reason in the frame of things, but attributed the form thereof able to maintain itself to infinite essays or proofs of nature, which they term fortune, seemeth to me (as far as I can judge by the recital and fragments which remain unto us) in particularities of physical causes more real and

non arctabuntur: Proverbs 4:12. 'As you walk you will not slip, and, if you run, nothing will bring you down.'
improficience: lack of progress.* *intercepted*: put a stop to.
satisfactory: merely plausible. *quickset*: a hedge grown from cuttings.
slug: retard.* *proofs*: experiments, tests.

better inquired than that of Aristotle and Plato; whereof both intermingled final causes, the one as a part of theology, and the other as a part of logic, which were the favourite studies respectively of both those persons. Not because those final causes are not true, and worthy to be inquired, being kept within their own province; but because their excursions into the limits of physical causes hath bred a vastness and solitude in that tract. For otherwise, keeping their precincts and borders, men are extremely deceived if they think there is an enmity or repugnancy at all between them. For the cause rendered, that 'the hairs about the eye-lids are for the safeguard of the sight', doth not impugn the cause rendered, that 'pilosity is incident to orifices of moisture; muscosi fontes', &c. Nor the cause rendered, that 'the firmness of hides is for the armour of the body against extremities of heat or cold', does not impugn the cause rendered, that 'contraction of pores is incident to the outwardest parts, in regard of their adjacence to foreign or unlike bodies': and so of the rest: both causes being true and compatible, the one declaring an intention, the other a consequence only. Neither doth this call in question, or derogate from divine providence, but highly confirm and exalt it. For as in civil actions he is the greater and deeper politique, that can make other men the instruments of his will and ends, and yet never acquaint them with his purpose, so as they shall do it and yet not know what they do, than he that imparteth his meaning to those he employeth; so is the wisdom of God more admirable, when nature intendeth one thing, and providence draweth forth another, than if he had communicated to particular creatures and motions the characters and impressions of his providence. And thus much for metaphysic: the latter part whereof I allow as extant, but wish it confined to his proper place.

[NATURAL PHILOSOPHY—MATHEMATICS AND NATURAL PRUDENCE (EXPERIMENTAL, PHILOSOPHICAL, AND MAGICAL)]

VIII. 1. Nevertheless there remaineth yet another part of natural philosophy, which is commonly made a principal part, and holdeth rank with physic special and metaphysic, which is mathematic;[n] but I think it more agreeable to the nature of things, and to the light of order, to place it as a branch of metaphysic. For the subject of it

vastness: desolation.*
pilosity: hairiness.*
muscosi fontes: Virgil, *Eclogues* vii. 45. 'Ye mossy fountains.'
adjacence: contiguity.*

being quantity, not quantity indefinite, which is but a relative, and belongeth to *philosophia prima* (as hath been said), but quantity determined or proportionable, it appeareth to be one of the essential forms of things, as that that is causative in nature of a number of effects; insomuch as we see in the schools both of Democritus and of Pythagoras, that the one did ascribe figure to the first seeds of things, and the other did suppose numbers to be the principles and originals of things. And it is true also that of all other forms (as we understand forms) it is the most abstracted and separable from matter, and therefore most proper to metaphysic; which hath likewise been the cause why it hath been better laboured and inquired than any of the other forms, which are more immersed into matter. For it being the nature of the mind of man (to the extreme prejudice of knowledge) to delight in the spacious liberty of generalities, as in a champain region, and not in the inclosures of particularity, the mathematics of all other knowledge were the goodliest fields to satisfy that appetite. But for the placing of this science, it is not much material: only we have endeavoured in these our partitions to observe a kind of perspective, that one part may cast light upon another.

2. The mathematics are either pure or mixed. To the pure mathematics are those sciences belonging which handle quantity determinate, merely severed from any axioms of natural philosophy; and these are two, geometry and arithmetic; the one handling quantity continued, and the other dissevered. Mixed hath for subject some axioms or parts of natural philosophy, and considereth quantity determined, as it is auxiliary and incident unto them. For many parts of nature can neither be invented with sufficient subtilty, nor demonstrated with sufficient perspicuity, nor accommodated unto use with sufficient dexterity, without the aid and intervening of the mathematics; of which sort are perspective, music, astronomy, cosmography, architecture, enginery, and divers others. In the mathematics I can report no deficience, except it be that men do not sufficiently understand the excellent use of the pure mathematics, in that they do remedy and cure many defects in the wit and faculties intellectual. For if the wit be too dull, they sharpen it; if too wandering, they fix it; if too inherent in the sense, they abstract it. So that as tennis is a game of no use in itself, but of great use in respect it

champain: open and unenclosed, as common land.
merely severed: wholly abstracted from.
invented: discovered.
enginery: the art of constructing engines.*

maketh a quick eye and a body ready to put itself into all postures; so in the mathematics, that use which is collateral and intervenient is no less worthy than that which is principal and intended. And as for the mixed mathematics, I may only make this prediction, that there cannot fail to be more kinds of them, as nature grows further disclosed. Thus much of natural science, or the part of nature speculative.

3. For natural prudence, or the part operative of natural philosophy, we will divide it into three parts, experimental, philosophical, and magical: which three parts active have a correspondence and analogy with the three parts speculative, natural history, physic, and metaphysic. For many operations have been invented, sometime by a casual incidence and occurrence, sometimes by a purposed experiment: and of those which have been found by an intentional experiment, some have been found out by varying or extending the same experiment, some by transferring and compounding divers experiments the one into the other, which kind of invention an empiric may manage. Again by the knowledge of physical causes there cannot fail to follow many indications and designations of new particulars, if men in their speculation will keep one eye upon use and practice. But these are but coastings along the shore *premendo littus iniquum*: for it seemeth to me there can hardly be discovered any radical or fundamental alterations and innovations in nature, either by the fortune and essays of experiments, or by the light and direction of physical causes.

Naturalis Magia sive Physica Operativa major. If therefore we have reported metaphysic deficient, it must follow that we do the like of natural magic, which hath relation thereunto. For as for the natural magic[n] whereof now there is mention in books, containing certain credulous and superstitious conceits and observations of sympathies and antipathies, and hidden proprieties, and some frivolous experiments, strange rather by disguisement than in themselves, it is as far differing in truth of nature from such a knowledge as we require, as the story of King Arthur of Britain, or Hugh of Bourdeaux, differs from Caesar's Commentaries in truth of story. For it is manifest that Caesar did greater things *de vero* than those

intervenient: incidental.*
incidence: coincidence.
divers: several, different.
empiric: a person versed only in practical skills.
premendo: Horace, *Odes* II. x. 3. 'Hugging the hostile shore.'
Naturalis: Natural Magic or the major Operative Physics.
de vero: in reality.

imaginary heroes were feigned to do. But he did them not in that fabulous manner. Of this kind of learning the fable of Ixion was a figure, who designed to enjoy Juno, the goddess of power; and instead of her had copulation with a cloud, of which mixture were begotten centaurs and chimeras. So whosoever shall entertain high and vaporous imaginations, instead of a laborious and sober inquiry of truth, shall beget hopes and beliefs of strange and impossible shapes. And therefore we may note in these sciences which hold so much of imagination and belief, as this degenerate natural magic, alchemy, astrology, and the like, that in their propositions the description of the means is ever more monstrous than the pretence or end. For it is a thing more probable, that he that knoweth well the natures of weight, of colour, of pliant and fragile in respect of the hammer, of volatile and fixed in respect of the fire, and the rest, may superinduce upon some metal the nature and form of gold by such mechanique as longeth to the production of the natures afore rehearsed, than that some grains of the medicine projected should in a few moments of time turn a sea of quicksilver or other material into gold. So it is more probable that he that knoweth the nature of arefaction, the nature of assimilation of nourishment to the thing nourished, the manner of increase and clearing of spirits, the manner of the depredations which spirits[n] make upon the humours and solid parts, shall by ambages of diets, bathings, anointings, medicines, motions, and the like, prolong life, or restore some degree of youth or vivacity, than that it can be done with the use of a few drops or scruples of a liquor or receipt.

Inventarium opum humanarum. To conclude therefore, the true natural magic, which is that great liberty and latitude of operation which dependeth upon the knowledge of forms, I may report deficient, as the relative thereof is. To which part, if we be serious and incline not to vanities and plausible discourse, besides the deriving and deducing the operations themselves from metaphysic, there are pertinent two points of much purpose, the one by way of preparation, the other by way of caution. The first is, that there be made a kalendar,[n] resembling an inventory of the estate of man, containing all the inventions (being the works of fruits of nature or art) which are now extant, and whereof man is already possessed; out of which

Ixion: Pindar, *Pythian Odes* ii. 21 f. *pretence*: expressed aim.
mechanique: mechanism. Only used by Bacon. *longeth*: belongs.
arefaction: the process of growing dry. *ambages*: circuitous ways.
motions: bodily exercises. *scruples*: very small amounts.
Inventarium: Inventory of Man's Inventions.
relative: i.e. metaphysics.

doth naturally result a note, what things are yet held impossible, or not invented: which kalendar will be the more artificial and serviceable, if to every reputed impossibility you add what thing is extant which cometh the nearest in degree to that impossibility; to the end that by these optatives and potentials man's inquiry may be the more awake in deducing direction of works from the speculation of causes. And secondly, that those experiments be not only esteemed which have an immediate and present use, but those principally which are of most universal consequence for invention of other experiments, and those which give most light to the invention of causes. For the invention of the mariner's needle, which giveth the direction, is of no less benefit for navigation than the invention of the sails which give the motion.

4. Thus have I passed through natural philosophy and the deficiences thereof; wherein if I have differed from the ancient and received doctrines, and thereby shall move contradiction, for my part, as I affect not to dissent, so I purpose not to contend. If it be truth,

Non canimus surdis, respondent omnia sylvae;

the voice of nature will consent, whether the voice of man do or no. And as Alexander Borgia[n] was wont to say of the expedition of the French for Naples, that they came with chalk in their hands to mark up their lodgings, and not with weapons to fight; so I like better that entry of truth which cometh peaceably with chalk to mark up those minds which are capable to lodge and harbour it, than that which cometh with pugnacity and contention.

5. *Continuatic Problematum in natura. Catalogus falsitatum grassantium in historia naturae.* But there remaineth a division of natural philosophy according to the report of the inquiry, and nothing concerning the matter or subject: and that is positive and considerative; when the inquiry reporteth either an assertion or a doubt. These doubts or *non liquets*[n] are of two sorts, particular and total. For the first, we see a good example thereof in Aristotle's Problems,[n] which deserved to have had a better continuance; but so nevertheless as there is one point whereof warning is to be given and taken. The

artificial: skilfully made.
optatives: things to be desired.*
move: excite.
Non canimus: Virgil, *Eclogues* x. 8. 'We sing not to the deaf; the woods reply to all.'
Continuatio: Continuation of the Problems in nature. Catalogue of the falsehoods lurking in natural history.
considerative: requiring consideration.

registering of doubts hath two excellent uses: the one, that it saveth philosophy from errors and falsehoods; when that which is not fully appearing is not collected into assertion, whereby error might draw error, but reserved in doubt: the other, that the entry of doubts are as so many suckers or sponges to draw use of knowledge; insomuch as that which, if doubts had not preceded, a man should never have advised, but passed it over without note, by the suggestion and solicitation of doubts is made to be attended and applied. But both these commodities do scarcely countervail an inconvenience, which will intrude itself if it be not debarred; which is, that when a doubt is once received, men labour rather how to keep it a doubt still, than how to solve it; and accordingly bend their wits. Of this we see the familiar example in lawyers and scholars, both which, if they have once admitted a doubt, it goeth ever after authorized for a doubt. But that use of wit and knowledge is to be allowed, which laboureth to make doubtful things certain, and not those which labour to make certain things doubtful. Therefore these kalendars of doubts I commend as excellent things; so that there be this caution used, that when they be throughly sifted and brought to resolution, they be from thenceforth omitted, decarded, and not continued to cherish and encourage men in doubting. To which kalendar of doubts or problems, I advise be annexed another kalendar, as much or more material, which is a kalendar of popular errors:[n] I mean chiefly in natural history, such as pass in speech and conceit, and are nevertheless apparently detected and convicted of untruth; that man's knowledge be not weakened nor imbased by such dross and vanity.

De antiquis philosophiis. As for the doubts or *non liquets* general or in total, I understand those differences of opinions touching the principles of nature, and the fundamental points of the same, which have caused the diversity of sects, schools, and philosophies, as that of Empedocles, Pythagoras, Democritus, Parmenides, and the rest.[n] For although Aristotle, as though he had been of the race of the Ottomans, thought he could not reign except the first thing he did he killed all his brethren; yet to those that seek truth and not magistrality, it cannot but seem a matter of great profit, to see before them the several opinions touching the foundations of nature. Not

use: increase. *countervail*: equal.
accordingly: in accordance therewith.
decarded: set aside.*
De antiquis: Concerning the ancient philosophies.
magistrality: dogmatism.

for any exact truth that can be expected in those theories; for as the same phenomena in astronomy are satisfied by the received astronomy of the diurnal motion, and the proper motions of the planets, with their eccentrics and epicycles, and likewise by the theory of Copernicus,[n] who supposed the earth to move, and the calculations are indifferently agreeable to both, so the ordinary face and view of experience is many times satisfied by several theories and philosophies; whereas to find the real truth requireth another manner of severity and attention. For as Aristotle saith, that children at the first will call every woman mother, but afterward they come to distinguish according to truth; so experience, if it be in childhood, will call every philosophy mother, but when it cometh to ripeness it will discern the true mother. So as in the mean time it is good to see the several glosses and opinions upon nature, whereof it may be every one in some one point hath seen clearer than his fellows, therefore I wish some collection to be made painfully and understandingly *de antiquis philosophiis*, out of all the possible light which remaineth to us of them: which kind of work I find deficient. But here I must give warning, that it be done distinctly and severedly; the philosophies of every one throughout by themselves, and not by titles packed and faggoted up together, as hath been done by Plutarch.[n] For it is the harmony of a philosophy in itself which giveth it light and credence; whereas if it be singled and broken, it will seem more foreign and dissonant. For as when I read in Tacitus[n] the actions of Nero or Claudius, with circumstances of times, inducements, and occasions, I find them not so strange; but when I read them in Suetonius Tranquillus, gathered into titles and bundles and not in order of time, they seem more monstrous and incredible: so is it of any philosophy reported entire, and dismembered by articles. Neither do I exclude opinions of latter times to be likewise represented in this kalendar of sects of philosophy, as that of Theophrastus Paracelsus,[n] eloquently reduced into an harmony by the pen of Severinus the Dane; and that of Tilesius,[n] and his scholar Donius,[n] being as a pastoral philosophy, full of sense, but of no great depth; and that of Fracastorius,[n] who, though he pretended not to make any new philosophy, yet did use the absoluteness of his own sense upon the old; and that of Gilbertus[n] our countryman, who revived, with some alterations and demonstrations, the opinions of Xenophanes; and any other worthy to be admitted.

that children: Aristotle, *Physics* i. 1.
understandingly: intelligently.
inducements: introductions.*
painfully: diligently.
severedly: separately.*
absoluteness: independence.

6. Thus have we now dealt with two of the three beams of man's knowledge; that is *radius directus*, which is referred to nature, *radius refractus*, which is referred to God, and cannot report truly because of the inequality of the medium. There resteth *radius reflexus*, whereby man beholdeth and contemplateth himself.

[HUMAN PHILOSOPHY, OR THE KNOWLEDGE OF OURSELVES]

IX. 1. We come therefore now to that knowledge whereunto the ancient oracle[n] directeth us, which is the knowledge of ourselves; which deserveth the more accurate handling, by how much it toucheth us more nearly. This knowledge, as it is the end and term of natural philosophy in the intention of man, so notwithstanding it is but a portion of natural philosophy in the continent of nature. And generally let this be a rule, that all partitions of knowledges[n] be accepted rather for lines and veins than for sections and separations; and that the continuance and entireness of knowledge be preserved. For the contrary hereof hath made particular sciences to become barren, shallow, and erroneous, while they have not been nourished and maintained from the common fountain. So we see Cicero[n] the orator complained of Socrates and his school, that he was the first that separated philosophy and rhetoric; whereupon rhetoric became an empty and verbal art. So we may see that the opinion of Copernicus[n] touching the rotation of the earth, which astronomy itself cannot correct, because it is not repugnant to any of the *phainomena*, yet natural philosophy may correct. So we see also that the science of medicine if it be destituted and forsaken by natural philosophy, it is not much better than an empirical practice. With this reservation therefore we proceed to human philosophy or humanity, which hath two parts: the one considereth man segregate or distributively; the other congregate, or in society. So as human philosophy is either simple and particular, or conjugate and civil. Humanity particular consisteth of the same parts whereof man consisteth; that is, of knowledges which respect the body, and of knowledges that respect the mind. But before we distribute so far, it is good to constitute. For I do take the consideration in general, and at large, of human nature to be fit to be emancipate and made a knowledge by itself: not so much in regard of those delightful and elegant discourses which have been made of the dignity of man,[n] of his miseries, of

radius: the direct beam, the refracted beam, the reflected beam.
destituted: abandoned. *constitute*: establish. *emancipate*: set free.*

his state and life, and the like adjuncts of his common and undivided nature; but chiefly in regard of the knowledge concerning the sympathies and concordances between the mind and body, which being mixed cannot be properly assigned to the sciences of either.

2. *Pars Physiognomiae, de gestu sive motu corporis.* This knowledge hath two branches: for as all leagues and amities consist of mutual intelligence and mutual offices, so this league of mind and body hath these two parts; how the one discloseth the other, and how the one worketh upon the other; discovery and impression. The former of these hath begotten two arts, both of prediction or prenotion; whereof the one is honoured with the inquiry of Aristotle, and the other of Hippocrates.[n] And although they have of later time been used to be coupled with superstitious and fantastical arts, yet being purged and restored to their true state, they have both of them a solid ground in nature, and a profitable use in life. The first is physiognomy, which discovereth the disposition of the mind by the lineaments of the body. The second is the exposition of natural dreams, which discovereth the state of the body by the imaginations of the mind. In the former of these I note a deficience. For Aristotle hath very ingeniously and diligently handled the factures of the body, but not the gestures of the body, which are no less comprehensible by art, and of greater use and advantage. For the lineaments of the body do disclose the disposition and inclination of the mind in general; but the motions of the countenance and parts do not only so, but do further disclose the present humour and state of the mind and will. For as your majesty saith most aptly and elegantly, 'As the tongue speaketh to the ear so the gesture speaketh to the eye.'[n] And therefore a number of subtile persons, whose eyes do dwell upon the faces and fashions of men, do well know the advantage of this observation, as being most part of their ability; neither can it be denied, but that it is a great discovery of dissimulations, and a great direction in business.

3. The latter branch, touching impression, hath not been collected into art, but hath been handled dispersedly; and it hath the same relation or *antistrophe* that the former hath. For the consideration is double: either, how and how far the humours and affects of the

Pars: The Parts of Physiognomy, concerning the gesture or motion of the body.
intelligence: interchange of information.
offices: obligations.
prenotion: foreknowledge.
factures: shapes, structures.
antistrophe: correspondence.*
affects: dispositions.

body do alter or work upon the mind; or again, how and how far the passions or apprehensions of the mind do alter or work upon the body. The former of these hath been inquired and considered as a part and appendix of medicine,[n] but much more as a part of religion or superstition. For the physician prescribeth cures of the mind in phrensies and melancholy passions; and pretendeth also to exhibit medicines to exhilarate the mind, to confirm the courage, to clarify the wits, to corroborate the memory, and the like: but the scruples and superstitions of diet[n] and other regiment of the body in the sect of the Pythagoreans, in the heresy of the Manichees, and in the law of Mahomet, do exceed. So likewise the ordinances in the ceremonial law, interdicting the eating of the blood and the fat, distinguishing between beasts clean and unclean for meat, are many and strict. Nay the faith itself being clear and serene from all clouds of ceremony, yet retaineth the use of fastings, abstinences, and other macerations and humiliations of the body, as things real, and not figurative. The root and life of all which prescripts is (besides the ceremony) the consideration of that dependency which the affections of the mind are submitted unto upon the state and disposition of the body. And if any man of weak judgement do conceive that this suffering of the mind from the body doth either question the immortality, or derogate from the sovereignty of the soul, he may be taught in easy instances, that the infant in the mother's womb is compatible with the mother and yet separable; and the most absolute monarch is sometimes led by his servants and yet without subjection. As for the reciprocal knowledge, which is the operation of the conceits and passions of the mind upon the body, we see all wise physicians, in the prescriptions of their regiments to their patients, do ever consider *accidentia animi* as of great force to further or hinder remedies or recoveries: and more specially it is an inquiry of great depth and worth concerning imagination,[n] how and how far it altereth the body proper of the imaginant. For although it hath a manifest power to hurt, it followeth not it hath the same degree of power to help. No more than a man can conclude, that because there be pestilent airs, able suddenly to kill a man in health, therefore there should be sovereign airs, able suddenly to cure a man in sickness. But the

exhibit: administer. *corroborate*: strengthen.
exceed: go too far. *macerations*: mortifications.
prescripts: rules. *question*: make doubtful.†
compatible: participating in.
regiments: rule of diet, or mode of living.
accidentia: the symptoms of the mind.
imaginant: one who imagines.*

inquisition of this part is of great use, though it needeth, as Socrates said, *a Delian diver*,[n] being difficult and profound. But unto all this knowledge *de communi vinculo*, of the concordances between the mind and the body, that part of inquiry is most necessary, which considereth of the seats and domiciles which the several faculties of the mind do take and occupate in the organs of the body; which knowledge hath been attempted, and is controverted, and deserveth to be much better inquired. For the opinion of Plato,[n] who placed the understanding in the brain, animosity (which he did unfitly call anger, having a greater mixture with pride) in the heart, and concupiscence or sensuality in the liver, deserveth not to be despised; but much less to be allowed. So then we have constituted (as in our own wish and advice) the inquiry touching human nature entire, as a just portion of knowledge to be handled apart.

[ARTS CONCERNING THE BODY—MEDICINE, COSMETICS, ATHLETICS, VOLUPTUARY, AND JOCULARY]

X. 1. The knowledge that concerneth man's body is divided as the good of man's body is divided, unto which it referreth. The good of man's body is of four kinds, health, beauty, strength, and pleasure: so the knowledges are medicine, or art of cure: art of decoration, which is called cosmetic; art of activity, which is called athletic; and art voluptuary, which Tacitus truly calleth *eruditus luxus*. This subject of man's body is of all other things in nature most susceptible of remedy; but then that remedy is most susceptible of error. For the same subtility of the subject doth cause large possibility and easy failing; and therefore the inquiry ought to be the more exact.

2. To speak therefore of medicine, and to resume that we have said, ascending a little higher: the ancient opinion that man was *microcosmus*,[n] an abstract or model of the world, hath been fantastically strained by Paracelsus and the alchemists, as if there were to be found in man's body certain correspondences and parallels, which should have respect to all varieties of things, as stars, planets, minerals, which are extant in the great world. But thus much is evidently true, that of all substances which nature hath produced, man's body is the most extremely compounded. For we see herbs

de communi vinculo: of the common tie.
occupate: occupy.
cosmetic: the art of beautifying the body.*
eruditus luxus: Tacitus, *Annals* xvi. 18. 'Refined luxury.'

and plants are nourished by earth and water; beasts for the most part by herbs and fruits; man by the flesh of beasts, birds, fishes, herbs, grains, fruits, water, and the manifold alterations, dressings and preparations of these several bodies, before they come to be his food and aliment. Add hereunto that beasts have a more simple order of life, and less change of affections to work upon their bodies; whereas man in his mansion, sleep, exercise, passions, hath infinite variations: and it cannot be denied but that the body of man of all other things is of the most compounded mass. The soul on the other side is the simplest of substances, as is well expressed:

> Purumque reliquit
> Æthereum sensum atque auraï simplicis ignem.

So that it is no marvel though the soul so placed enjoy no rest, if that principle be true, that *Motus rerum est rapidus extra locum, placidus in loco*. But to the purpose: this variable composition of man's body hath made it as an instrument easy to distemper; and therefore the poets did well to conjoin music and medicine in Apollo,[n] because the office of medicine is but to tune this curious harp of man's body and to reduce it to harmony. So then the subject being so variable, hath made the art by consequent more conjectural; and the art being conjectural hath made so much the more place to be left for imposture. For almost all other arts and sciences are judged by acts or masterpieces, as I may term them, and not by the successes and events. The lawyer is judged by the virtue of his pleading, and not by the issue of the cause. The master in the ship is judged by the directing his course aright, and not by the fortune of the voyage. But the physician, and perhaps the politique, hath no particular acts demonstrative of his ability, but is judged most by the event; which is ever but as it is taken: for who can tell, if a patient die or recover, or if a state be preserved or ruined, whether it be art or accident? And therefore many times the impostor is prized, and the man of virtue taxed. Nay, we see [the] weakness and credulity of men is such, as they will often prefer a mountebank or witch before a learned physician. And therefore the poets were

mansion: the act of staying in one place.
Purumque: Virgil, *Aeneid* vi. 747. 'Pure and unmixed the aetherial sense is left, mere air and fire.'
Motus rerum: The motion of things is rapid out of their place, quiet in their place.
distemper: disorder.
curious: made with care.
by consequent: in consequence.
mountebank: an itinerant quack doctor.

clear-sighted in discerning this extreme folly, when they made Aesculapius and Circe brother and sister, both children of the sun, as in the verses,

> Ipse repertorem medicinae talis et artis
> Fulmine Phoebigenam Stygias detrusit ad undas:[n]

And again,

> Dives inaccessos ubi Solis filia lucos, &c.

For in all times, in the opinion of the multitude, witches and old women and impostors have had a competition with physicians. And what followeth? Even this, that physicians say to themselves, as Salomon expresseth it upon an higher occasion, 'If it befal to me as befalleth to the fools, why should I labour to be more wise?' And therefore I cannot much blame physicians, that they use commonly to intend some other art or practice, which they fancy, more than their profession. For you shall have of them antiquaries, poets, humanists, statesmen, merchants, divines, and in every of these better seen than in their profession; and no doubt upon this ground, that they find that mediocrity and excellency in their art maketh no difference in profit or reputation towards their fortune; for the weaknes of patients, and sweetness of life, and nature of hope, maketh men depend upon physicians with all their defects. But nevertheless these things which we have spoken of are courses begotten between a little occasion, and a great deal of sloth and default; for if we will excite and awake our observation, we shall see in familiar instances what a predominant faculty the subtilty of spirit hath over the variety of matter or form. Nothing more variable than faces and countenances: yet men can bear in memory the infinite distinctions of them; nay, a painter with a few shells of colours, and the benefit of his eye, and habit of his imagination, can imitate them all that ever have been, are, or may be, if they were brought before him. Nothing more variable than voices; yet men can likewise discern them personally: nay, you shall have a *buffon* or *pantomimus*, will express as many as he pleaseth. Nothing more variable than the differing sounds of words: yet men have found the way to reduce them to a few simple letters. So that it is not the insufficiency or incapacity of man's mind, but it is the remote standing or placing thereof, that breedeth these mazes and incomprehensions. For as the sense afar off

If it befal: Ecclesiastes 2:15.
intend: occupy themselves with.
every: each.
seen: skilled.
faculty: influence.
pantomimus: a mimic.
incomprehensions: failures to comprehend.*

is full of mistaking, but is exact at hand, so is it of the understanding: the remedy whereof is, not to quicken or strengthen the organ, but to go nearer to the object; and therefore there is no doubt but if the physicians will learn and use the true approaches and avenues of nature, they may assume as much as the poet saith:

> Et quoniam variant morbi, variabimus artes;
> Mille mali species, mille salutis erunt.

Which that they should do, the nobleness of their art doth deserve; well shadowed by the poets, in that they made Aesculapius to be the son of [the] sun, the one being the fountain of life, the other as the second stream: but infinitely more honoured by the example of our Saviour, who made the body of man the object of his miracles, as the soul was the object of his doctrine. For we read not that ever he vouchsafed to do any miracle about honour or money (except that one for giving tribute to Caesar), but only about the preserving, sustaining, and healing the body of man.

3. Medicine is a science which hath been (as we have said) more professed than laboured, and yet more laboured than advanced; the labour having been, in my judgement, rather in circle than in progression. For I find much iteration, but small addition. It considereth causes of diseases, with the occasions or impulsions; the diseases themselves, with the accidents; and the cures, with the preservations. The deficiencies which I think good to note, being a few of many, and those such as are of a more open and manifest nature, I will enumerate and not place.

4. *Narrationes medicinales.* The first is the discontinuance of the ancient and serious diligence of Hippocrates,[n] which used to set down a narrative of the special cases of his patients, and how they proceeded, and how they were judged by recovery or death. Therefore having an example proper in the father of the art, I shall not need to allege an example foreign, of the wisdom of the lawyers, who are careful to report new cases and decisions, for the direction of future judgements. This continuance of medicinal history I find deficient; which I understand neither to be so infinite as to extend

Et quoniam: Adapted from Ovid, *Remedies of Love*, ll. 525–6. 'And since diseases vary, we will vary our arts. There are a thousand forms of disease; there shall be a thousand methods of healing.'
money: Matthew 17:24–27.
impulsions: impelling causes.*
accidents: symptoms.
Narrationes: Medical histories.
judged: decided.
example foreign: i.e. from another area, the law.

to every common case, nor so reserved as to admit none but wonders: for many things are new in the manner, which are not new in the kind; and if men will intend to observe, they shall find much worthy to observe.

5. *Anatomia comparata.* In the inquiry which is made by anatomy,[n] I find much deficience: for they inquire of the parts, and their substances, figures, and collocations; but they inquire not of the diversities of the parts, the secrecies of the passages, and the seats or nestling of the humours, nor much of the footsteps and impressions of diseases. The reason of which omission I suppose to be, because the first inquiry may be satisfied in the view of one or a few anatomies: but the latter, being comparative and casual, must arise from the view of many. And as to the diversity of parts, there is no doubt but the facture or framing of the inward parts is as full of difference as the outward, and in that is the cause continent[n] of many diseases; which not being observed, they quarrel many times with the humours, which are not in fault; the fault being in the very frame and mechanique of the part, which cannot be removed by medicine alterative, but must be accommodate and palliate by diets and medicines familiar. And for the passages and pores, it is true which was anciently noted, that the more subtile of them appear not in anatomies, because they are shut and latent in dead bodies, though they be open and manifest in live: which being supposed, though the inhumanity of *anatomia vivorum* was by Celsus[n] justly reproved, yet in regard of the great use of this observation, the inquiry needed not by him so slightly to have been relinquished altogether, or referred to the casual practices of surgery; but mought have been well diverted upon the dissection of beasts alive, which notwithstanding the dissimilitude of their parts may sufficiently satisfy this inquiry. And for the humours, they are commonly passed over in anatomies as purgaments; whereas it is most necessary to observe, what cavities, nests, and receptacles the humours do find

intend: occupy themselves.
Anatomia: Comparative anatomy.
collocations: arrangements.*
nestling: i.e. where the humours breed.*
anatomies: dissections.
casual: uncertain.
facture: structure.
mechanique: mechanism.*
alterative: tending to produce alteration.*
accommodate: adjusted.
anatomia: the dissection of living creatures.
purgaments: excrements.

in the parts, with the differing kind of the humour so lodged and received. And as for the footsteps of diseases, and their devastations of the inward parts, impostumations, exulcerations, discontinuations, putrefactions, consumptions, contractions, extensions, convulsions, dislocations, obstructions, repletions, together with all preternatural substances, as stones, carnosities, excrescences, worms, and the like; they ought to have been exactly observed by multitude of anatomies, and the contribution of men's several experiences, and carefully set down both historically according to the appearances, and artificially with a reference to the diseases and symptoms which resulted from them, in case where the anatomy is of a defunct patient; whereas now upon opening of bodies they are passed over slightly and in silence.

6. *Inquisitio ulterior de morbis insanabilibus.* In the inquiry of diseases, they do abandon the cures of many, some as in their nature incurable, and others as passed the period of cure; so that Sulla[n] and the Triumvirs never proscribed so many men to die, as they do by their ignorant edicts: whereof numbers do escape with less difficulty than they did in the Roman proscriptions. Therefore I will not doubt to note as a deficience, that they inquire not the perfect cures of many diseases, or extremities of diseases; but pronouncing them incurable do enact a law of neglect, and exempt ignorance from discredit.

7. *De Euthanasia exteriore.* Nay further, I esteem it the office of a physician not only to restore health, but to mitigate pain and dolors; and not only when such mitigation may conduce to recovery, but when it may serve to make a fair and easy passage. For it is no small felicity which Augustus Caesar was wont to wish to himself, that same *Euthanasia*;[n] and which was specially noted in the death of Antoninus Pius whose death was after the fashion and semblance of a kindly and pleasant sleep. So it is written of Epicurus, that after his disease was judged desperate, he drowned his stomach and senses with a large draught and ingurgitation of wine; whereupon the epigram was made, *Hinc Stygias ebrius hausit aquas*;[n] he was not sober enough to taste any bitterness of the Stygian water. But the physicians contrariwise do make a kind of scruple and religion to

impostumations: abscesses. *exulcerations*: ulcers.
discontinuations: want of cohesion of parts.†
carnosities: morbid fleshy growths.
Inquisitio: Further investigation into incurable diseases.
proscriptions: decrees condemning to death or banishment.
De Euthanasia: Of Euthanasia at the end.
dolors: physical sufferings. *kindly*: natural.

stay with the patient after the disease is deplored; whereas in my judgement they ought both to inquire the skill, and to give the attendances, for the facilitating and assuaging of the pains and agonies of death.

8. *Medicinae experimentales.* In the consideration of the cures of diseases, I find a deficience in the receipts of propriety, respecting the particular cures of diseases: for the physicians have frustrated the fruit of tradition and experience by their magistralities, in adding and taking out and changing *quid pro quo* in their receipts, at their pleasures; commanding so over the medicine, as the medicine cannot command over the disease. For except it be treacle[n] and *mithridatum*, and of late *diascordium*, and a few more, they tie themselves to no receipts severely and religiously. For as to the confections of sale which are in the shops, they are for readiness and not for propriety.[n] For they are upon general intentions of purging, opening, comforting, altering, and not much appropriate to particular diseases. And this is the cause why empirics and old women are more happy many times in their cures than learned physicians, because they are more religious in holding their medicines. Therefore here is the deficience which I find, that physicians have not, partly out of their own practice, partly out of the constant probations reported in books, and partly out of the traditions of empirics, set down and delivered over certain experimental medicines for the cure of particular diseases, besides their own conjectural and magistral descriptions. For as they were the men of the best composition in the state of Rome, which either being consuls inclined to the people, or being tribunes inclined to the senate; so in the matter we now handle, they be the best physicians, which being learned incline to the traditions of experience, or being empirics incline to the methods of learning.

9. *Imitatio naturae in balneis, et aquis medicinalibus.* In preparation of medicines I do find strange, specially considering how mineral medicines[n] have been extolled, and that they are safer for the outward than inward parts, that no man hath sought to make an imi-

deplored: given up as hopeless.
Medicinae: Experimental medicines.
receipts of propriety: receipts proper to a particular disease.
magistralities: special prescriptions for individual cases.
religious: scrupulous, exact.
probations: tests.
magistral: used of a remedy devised for a particular case.
Imitatio: Imitation of nature in baths and medicinal waters.
find strange: wonder.

tation by art of natural baths and medicinal fountains: which nevertheless are confessed to receive their virtues from minerals: and not so only, but discerned and distinguished from what particular mineral they receive tincture, as sulphur, vitriol, steel, or the like: which nature, if it may be reduced to compositions of art, both the variety of them will be increased, and the temper of them will be more commanded.

10. *Filum medicinale, sive de vicibus medicinarum.* But lest I grow to be more particular than is agreeable either to my intention or to proportion, I will conclude this part with the note of one deficience more, which seemeth to me of greatest consequence; which is, that the prescripts in use are too compendious to attain their end: for, to my understanding, it is a vain and flattering opinion to think any medicine can be so sovereign or so happy, as that the receipt or use of it can work any great effect upon the body of man. It were a strange speech which spoken, or spoken oft, should reclaim a man from a vice to which he were by nature subject. It is order, pursuit, sequence, and interchange of application, which is mighty in nature; which although it require more exact knowledge in prescribing, and more precise obedience in observing, yet is recompensed with the magnitude of effects. And although a man would think, by the daily visitations of the physicians, that there were a pursuance in the cure: yet let a man look into their prescripts and ministrations, and he shall find them but inconstancies and every day's devices, without any settled providence or project. Not that every scrupulous or superstitious prescript is effectual, no more than every straight way is the way to heaven; but the truth of the direction must precede severity of observance.

11. For cosmetic, it hath parts civil, and parts effeminate: for cleanness of body was ever esteemed to proceed from a due reverence to God, to society, and to ourselves. As for artificial decoration, it is well worthy of the deficiences which it hath; being neither fine enough to deceive, nor handsome to use, nor wholesome to please.

12. For athletic, I take the subject of it largely, that is to say, for any point of ability whereunto the body of man may be brought, whether it be of activity, or of patience; whereof activity hath two parts, strength and swiftness; and patience likewise hath two parts, hardness against wants and extremities, and endurance of pain or tor-

Filum: The Physician's Clue, or the sequence of remedies.
prescripts: prescriptions. *pursuit*: sequence.*
pursuance: sequence.* *largely*: broadly.

ment; whereof we see the practices in tumblers, in savages, and in those that suffer punishment. Nay, if there be any other faculty which falls not within any of the former divisions, as in those that dive, that obtain a strange power of containing respiration, and the like, I refer it to this part. Of these things the practices are known, but the philosophy which concerneth them is not much inquired; the rather, I think, because they are supposed to be obtained, either by an aptness of nature, which cannot be taught, or only by continual custom, which is soon prescribed: which though it be not true, yet I forbear to note any deficiences: for the Olympian games are down long since, and the mediocrity of these things is for use; as for the excellency of them it serveth for the most part but for mercenary ostentation.

13. For arts of pleasure sensual, the chief deficience in them is of laws to repress them. For as it hath been well observed, that the arts which flourish in times while virtue is in growth, are military; and while virtue is in state, are liberal; and while virtue is in declination, are voluptuary: so I doubt that this age of the world is somewhat upon the descent of the wheel. With arts voluptuary I couple practices joculary;[n] for the deceiving of the senses is one of the pleasures of the senses. As for games of recreation, I hold them to belong to civil life and education. And thus much of that particular human philosophy which concerns the body, which is but the tabernacle of the mind.

[HUMAN PHILOSOPHY AS IT CONCERNS THE MIND—THE NATURE OF THE SOUL OR MIND, DIVINATION, FASCINATION]

XI. 1. For human knowledge which concerns the mind, it hath two parts; the one that inquireth of the substance or nature of the soul or mind, the other that inquireth of the faculties or functions thereof. Unto the first of these, the considerations of the original of the soul, whether it be native or adventive,[n] and how far it is exempted from laws of matter, and of the immortality thereof, and many other points, do appertain: which have been not more laboriously inquired than variously reported; so as the travail therein taken seemeth to have been rather in a maze than in a way. But although I am of opinion that this knowledge may be more really and soundly inquired, even in nature, than it hath been: yet I hold that in the end

down: have long since ceased, and a mediocrity in such things is adequate.
declination: decline.

it must be bounded by religion, or else it will be subject to deceit and delusion. For as the substance of the soul in the creation was not extracted out of the mass of heaven and earth by the benediction of a *producat*, but was immediately inspired from God, so it is not possible that it should be (otherwise than by accident) subject to the laws of heaven and earth, which are the subject of philosophy; and therefore the true knowledge of the nature and state of the soul must come by the same inspiration that gave the substance. Unto this part of knowledge touching the soul there be two appendices; which, as they have been handled, have rather vapoured forth fables than kindled truth; divination and fascination.

2. Divination[n] hath been anciently and fitly divided into artificial and natural; whereof artificial is, when the mind maketh a prediction by argument, concluding upon signs and tokens; natural is, when the mind hath a presention by an internal power, without the inducement of a sign. Artificial is of two sorts; either when the argument is coupled with a derivation of causes, which is rational; or when it is only grounded upon a coincidence of the effect, which is experimental: whereof the latter for the most part is superstitious; such as were the heathen observations upon the inspection of sacrifices, the flights of birds, the swarming of bees; and such as was the Chaldean astrology, and the like. For artificial divination, the several kinds thereof are distributed amongst particular knowledges. The astronomer hath his predictions, as of conjunctions, aspects, eclipses, and the like. The physician hath his predictions, of death, of recovery, of the accidents and issues of diseases. The politique hath his predictions; *O urbem venalem, et cito perituram, si emptorem invenerit!* which stayed not long to be performed, in Sulla first, and after in Caesar. So as these predictions are now impertinent, and to be referred over. But the divination which springeth from the internal nature of the soul, is that which we now speak of; which hath been made to be of two sorts, primitive and by influxion. Primitive is grounded upon the supposition, that the mind, when it is withdrawn and collected into itself, and not diffused into the organs of the body, hath some extent and latitude of prenotion; which therefore appeareth most in sleep, in ecstasies, and near death, and more rarely in waking apprehensions; and is induced and furthered

fascination: sorcery.* *presentation*: presentiment.
O urbem: Sallust, *War with Jugurtha* xxxv. 'A city for sale, and doomed to speedy destruction, if it finds a purchaser.'
impertinent: irrelevant. *referred over*: deferred.
influxion: influence.* *prenotion*: foreknowledge.

by those abstinences and observances which make the mind most to consist in itself. By influxion, is grounded upon the conceit that the mind, as a mirror or glass, should take illumination from the foreknowledge of God and spirits: unto which the same regiment doth likewise conduce. For the retiring of the mind within itself is the state which is most susceptible of divine influxions; save that it is accompanied in this case with a fervency and elevation (which the ancients noted by fury), and not with a repose and quiet, as it is in the other.

3. Fascination[n] is the power and act of imagination intensive upon other bodies than the body of the imaginant, for of that we spake in the proper place. Wherein the school of Paracelsus, and the disciples of pretended magic have been so intemperate, as they have exalted the power of the imagination to be much one with the power of miracle-working faith. Others, that draw nearer to probability, calling to their view the secret passages of things, and specially of the contagion that passeth from body to body, do conceive it should likewise be agreeable to nature, that there should be some transmissions and operations from spirit to spirit without the mediation of the senses; whence the conceits have grown (now almost made civil) of the mastering spirit, and the force of confidence and the like. Incident unto this is the inquiry how to raise and fortify the imagination: for if the imagination fortified have power, then it is material to know how to fortify and exalt it. And herein comes in crookedly and dangerously a palliation of a great part of ceremonial magic. For it may be pretended that ceremonies, characters, and charms do work, not by any tacit or sacramental contract with evil spirits, but serve only to strengthen the imagination of him that useth it; as images are said by the Roman church to fix the cogitations and raise the devotions of them that pray before them. But for mine own judgement, if it be admitted that imagination hath power, and that ceremonies fortify imagination, and that they be used sincerely and intentionally for that purpose; yet I should hold them unlawful, as opposing to that first edict which God gave unto man, *In sudore vultus comedes panem tuum*. For they propound those noble effects, which God hath set forth unto man to be bought at the price of

consist in: inhere in. *noted*: denoted.
intensive: strenuously directed upon.*
fortify: render more effective.
palliation: excuse. *characters*: magical signs.
sacramental: bound by an oath.
opposing to: conflicting with.†
In sudore: Genesis 3:19. 'In the sweat of thy brow shalt thou eat thy bread.'

labour, to be attained by a few easy and slothful observances. Deficiences in these knowledges I will report none, other than the general deficience, that it is not known how much of them is verity, and how much vanity.

[DIVISION OF KNOWLEDGE INTO INTELLECTUAL AND MORAL—THE FACULTIES]

XII. 1. The knowledge which respecteth the faculties[n] of the mind of man is of two kinds; the one respecting his understanding and reason, and the other his will, appetite, and affection; whereof the former produceth position or decree, the latter action or execution. It is true that the imagination is an agent or *nuncius*, in both provinces, both the judicial and the ministerial. For sense sendeth over to imagination before reason have judged: and reason sendeth over to imagination before the decree can be acted. For imagination ever precedeth voluntary motion. Saving that this Janus of imagination hath differing faces: for the face towards reason hath the print of truth, but the face towards action hath the print of good; which nevertheless are faces,

Quales decet esse sororum.

Neither is the imagination simply and only a messenger; but is invested with, or at least wise usurpeth no small authority in itself, besides the duty of the message. For it was well said by Aristotle, 'That the mind hath over the body that commandment, which the lord hath over a bondman; but that reason hath over the imagination that commandment which a magistrate hath over a free citizen';[n] who may come also to rule in his turn. For we see that, in matters of faith and religion, we raise our imagination above our reason; which is the cause why religion sought ever access to the mind by similitudes, types, parables, visions, dreams. And again, in all persuasions that are wrought by eloquence,[n] and other impressions of like nature, which do paint and disguise the true appearance of things, the chief recommendation unto reason is from the imagination. Nevertheless, because I find not any science that doth properly or fitly pertain to the imagination, I see no cause to alter the former division.[n] For as for poesy, it is rather a pleasure or play of imagin-

position: the laying down of a law.
nuncius: ambassador.
Quales: Ovid, *Metamorphoses* ii. 14. 'Such as sisters' faces should be.'
at least wise: at least.
former division: i.e. at II. i. 1 into Memory, Imagination, and Reason.

ation, than a work or duty thereof. And if it be a work, we speak not now of such parts of learning as the imagination produceth, but of such sciences as handle and consider of the imagination. No more than we shall speak now of such knowledges as reason produceth (for that extendeth to all philosophy), but of such knowledges as do handle and inquire of the faculty of reason: so as poesy had his true place. As for the power of the imagination in nature, and the manner of fortifying the same, we have mentioned it in the doctrine *De Anima*, whereunto most fitly it belongeth. And lastly, for imaginative or insinuative reason, which is the subject of rhetoric, we think it best to refer it to the arts of reason. So therefore we content ourselves with the former division, that human philosophy, which respecteth the faculties of the mind of man, hath two parts, rational and moral.

2. The part of human philosophy which is rational, is of all knowledges, to the most wits, the least delightful, and seemeth but a net of subtility and spinosity. For as it was truly said, that knowledge is *pabulum animi*; so in the nature of men's appetite to this food, most men are of the taste and stomach of the Israelites in the desert, that would fain have returned *ad ollas carnium*, and were weary of manna; which, though it were celestial, yet seemed less nutritive and comfortable. So generally men taste well knowledges that are drenched in flesh and blood, civil history, morality, policy, about the which men's affections, praises, fortunes do turn and are conversant. But this same *lumen siccum* doth parch and offend most men's watery and soft natures. But to speak truly of things as they are in worth, rational knowledges are the keys of all other arts: for as Aristotle saith aptly and elegantly, 'That the hand is the instrument of instruments, and the mind is the form of forms'; so these be truly said to be the art of arts. Neither do they only direct, but likewise confirm and strengthen: even as the habit of shooting doth not only enable to shoot a nearer shoot, but also to draw a stronger bow.

3. The arts intellectual are four in number; divided according to the ends whereunto they are referred: for man's labour is to

De Anima: when discussing the Soul, at II. xi. 3.
former division: i.e. at the beginning of the paragraph, into Logic and Ethics.
spinosity: the quality of being thorny.*
pabulum: food of the mind.
ad ollas: Numbers 11:4–6. 'To the flesh-pots [of Egypt].'
comfortable: strengthening.
lumen siccum: dry light, cf. p. 8 above.
That the hand: Aristotle, *De Anima* iii. 8.
shoot: i.e. nearer the mark.

invent that which is sought or propounded; or to judge that which is invented; or to retain that which is judged; or to deliver over that which is retained. So as the arts must be four: art of inquiry or invention: art of examination or judgement: art of custody or memory: and art of elocution or tradition.

[OF INVENTION: OF SCIENCES AND OF SPEECH]

XIII. 1. Invention[n] is of two kinds much differing: the one of arts and sciences, and the other of speech and arguments. The former of these I do report deficient; which seemeth to me to be such a deficience as if, in the making of an inventory touching the state of a defunct, it should be set down that there is no ready money. For as money will fetch all other commodities, so this knowledge is that which should purchase all the rest. And like as the West Indies had never been discovered if the use of the mariner's needle had not been first discovered, though the one be vast regions, and the other a small motion; so it cannot be found strange if sciences be no further discovered, if the art itself of invention and discovery hath been passed over.

2. That this part of knowledge is wanting, to my judgement standeth plainly confessed; for first, logic doth not pretend to invent sciences, or the axioms of sciences, but passeth it over with a *cuique in sua arte credendum.*[n] And Celsus acknowledgeth it gravely, speaking of the empirical and dogmatical sects of physicians, 'That medicines and cures were first found out, and then after the reasons and causes were discoursed; and not the causes first found out, and by light from them the medicines and cures discovered.' And Plato[n] in his Theaetetus noteth well, 'That particulars are infinite, and the higher generalities give no sufficient direction: and that the pith of all sciences, which maketh the artsman differ from the inexpert, is in the middle propositions, which in every particular knowledge are taken from tradition and experience.' And therefore we see, that they which discourse of the inventions and originals of things refer them rather to chance than to art, and rather to beasts, birds, fishes, serpents, than to men.

> Dictamnum genetrix Cretaea carpit ab Ida,
> Puberibus caulem foliis et flore comantem

invent: find out.
tradition: the delivery of information.
state: the estate of one deceased.
fetch: purchase.* *found strange*: wondered at.

Purpureo; non illa feris incognita capris
Gramina, cum tergo volucres haesere sagittae.[n]

So that it was no marvel (the manner of antiquity being to consecrate inventors) that the Egyptians had so few human idols in their temples, but almost all brute:

Omnigenumque Deum monstra, et latrator Anubis,
Contra Neptunum, et Venerem, contraque Minervam, &c.[n]

And if you like better the tradition of the Grecians, and ascribe the first inventions[n] to men, yet you will rather believe that Prometheus first stroke the flints, and marvelled at the spark, than that when he first stroke the flints he expected the spark: and therefore we see the West Indian Prometheus had no intelligence with the European, because of the rareness with them of flint, that gave the first occasion. So as it should seem, that hitherto men are rather beholden to a wild goat for surgery, or to a nightingale for music, or to the ibis for some part of physic, or to the pot-lid that flew open for artillery, or generally to chance or anything else than to logic for the invention of arts and sciences. Neither is the form of invention which Virgil describeth much other:

Ut varias usus meditando extunderet artes
Paulatim.

For if you observe the words well, it is no other method than that which brute beasts are capable of, and do put in ure; which is a perpetual intending or practising some one thing, urged and imposed by an absolute necessity of conservation of being. For so Cicero saith very truly, *Usus uni rei deditus et naturam et artem saepe vincit*. And therefore if it be said of men.

Labor omnia vincit
Improbus, et duris urgens in rebus egestas,

it is likewise said of beasts, *Quis psittaco docuit suum χαῖρε?*[n] Who taught the raven in a drowth to throw pebbles into an hollow tree, where she spied water, that the water might rise so as she might

stroke: struck.
Ut varias: Virgil, *Georgics* i. 133. 'That practice, with meditation, might by degrees hammer out the arts.'
ure: use.
Usus uni: Cicero, *Pro Balbo* 20. 'Practice applied to one thing often accomplishes more than nature and art.'
Labor: Virgil, *Georgics* i. 145–6. 'Incessant labour and want, in hardships urgent, surmounted every obstacle.'

come to it? Who taught the bee to sail through such a vast sea of air, and to find the way from a field in flower a great way off to her hive? Who taught the ant to bite every grain of corn that she burieth in her hill, lest it should take root and grow? Add then the word *extundere*, which importeth the extreme difficulty, and the word *paulatim*, which importeth the extreme slowness, and we are where we were, even amongst the Egyptians' gods; there being little left to the faculty of reason, and nothing to the duty of art, for matter of invention.

3. Secondly, the induction[n] which the logicians speak of, and which seemeth familiar with Plato, whereby the principles of sciences may be pretended to be invented, and so the middle propositions by derivation from the principles; their form of induction, I say, is utterly vicious and incompetent: wherein their error is the fouler, because it is the duty of art to perfect and exalt nature; but they contrariwise have wronged, abused, and traduced nature. For he that shall attentively observe how the mind doth gather this excellent dew of knowledge, like unto that which the poet speaketh of, *Aërei mellis caelestia dona,* distilling and contriving it out of particulars natural and artificial, as the flowers of the field and garden, shall find that the mind of herself by nature doth manage and act an induction much better than they describe it. For to conclude upon an enumeration of particulars, without instance contradictory, is no conclusion, but a conjecture; for who can assure (in many subjects) upon those particulars which appear of a side, that there are not other on the contrary side which appear not? As if Samuel should have rested upon those sons of Issay which were brought before him, and failed of David which was in the field. And this form (to say truth) is so gross, as it had not been possible for wits so subtile as have managed these things to have offered it to the world, but that they hasted to their theories and dogmaticals, and were imperious and scornful towards particulars; which their manner was to use but as *lictores*[n] and *viatores*, for sergeants and whifflers, *ad summovendam turbam*, to make way and make room for their opinions, rather than in their true use and service. Certainly it is a thing may touch a man with a religious wonder, to see how the footsteps of seducement are the very same in divine and human truth: for as in divine

extundere: hammer out. *paulatim*: by degrees.
traduced: dishonoured.
Aërei mellis: Virgil, *Georgics* iv. 1. 'The heavenly gift, celestial honey.'
assure: guarantee.
Issay: Jesse. See I Samuel 16:1–13.
dogmaticals: systems of dogma.*

truth man cannot endure to become as a child;[n] so in human, they reputed the attending the inductions (whereof we speak) as if it were a second infancy or childhood.

4. Thirdly, allow some principles or axioms were rightly induced, yet nevertheless certain it is that middle propositions[n] cannot be deduced from them in subject of nature by syllogism, that is, by touch and reduction of them to principles in a middle term. It is true that in sciences popular, as moralities, laws, and the like, yea, and divinity (because it pleaseth God to apply himself to the capacity of the simplest), that form may have use; and in natural philosophy likewise, by way of argument or satisfactory reason, *Quae assensum parit, operis effoeta est*: but the subtilty of nature and operations will not be enchained in those bonds. For arguments consist of propositions, and propositions of words, and words are but the current tokens or marks of popular notions of things;[n] which notions, if they be grossly and variably collected out of particulars, it is not the laborious examination either of consequences or arguments, or of the truth of propositions, that can ever correct that error, being (as the physicians speak) in the first digestion. And therefore it was not without cause, that so many excellent philosophers became Sceptics[n] and Academics, and denied any certainty of knowledge or comprehension; and held opinion that the knowledge of man extended only to appearances and probabilities. It is true that in Socrates it was supposed to be but a form of irony, *Scientiam dissimulando simulavit*:[n] for he used to disable his knowledge, to the end to enhance his knowledge: like the humour of Tiberius in his beginnings, that would reign, but would not acknowledge so much. And in the later Academy, which Cicero embraced, this opinion also of *acatalepsia* (I doubt) was not held sincerely: for that all those which excelled in copie of speech seem to have chosen that sect, as that which was fittest to give glory to their eloquence and variable discourses; being rather like progresses of pleasure, than journeys to an end. But assuredly many scattered in both Academies did hold it in subtility and integrity.[n] But here was their chief error; they charged the deceit upon the senses; which in my judgement (notwithstanding all their cavillations) are very sufficient to certify and

by touch: by testing. *satisfactory*: plausible.
Quae assensum: which procures assent but can do no work.
grossly: unskilfully. *variably*: unsystematically.
disable: belittle.
Tiberius: In Tacitus, *Annals* I. vii. 11.
variable: inconstant.
cavillations: making of quibbling objections.

report truth, though not always immediately, yet by comparison, by help of instrument,[n] and by producing and urging such things as are too subtile for the sense to some effect comprehensible by the sense, and other like assistance. But they ought to have charged the deceit upon the weakness of the intellectual powers, and upon the manner of collecting and concluding upon the reports of the senses. This I speak, not to disable the mind of man, but to stir it up to seek help: for no man, be he never so cunning or practised, can make a straight line or perfect circle[n] by steadiness of hand, which may be easily done by help of a ruler or compass.

5. *Experientia literata, and interpretatio naturae.* This part of invention, concerning the invention of sciences, I purpose (if God give me leave) hereafter to propound, having digested it into two parts; whereof the one I term *experientia literata*,[n] and the other *interpretatio naturae*:[n] the former being but a degree and rudiment of the latter. But I will not dwell too long, nor speak too great upon a promise.

6. The invention of speech[n] or argument is not properly an invention: for to invent is to discover that we know not, and not to recover or resummon that which we already know: and the use of this invention is no other but, out of the knowledge whereof our mind is already possessed, to draw forth or call before us that which may be pertinent to the purpose which we take into our consideration. So as to speak truly, it is no invention, but a remembrance or suggestion, with an application; which is the cause why the schools do place it after judgement, as subsequent and not precedent. Nevertheless, because we do account it a chase as well of deer in an inclosed park as in a forest at large, and that it hath already obtained the name, let it be called invention: so as it be perceived and discerned, that the scope and end of this invention is readiness and present use of our knowledge, and not addition or amplification thereof.

7. To procure this ready use of knowledge there are two courses, preparation and suggestion. The former of these seemeth scarcely a part of knowledge, consisting rather of diligence than of any artificial erudition. And herein Aristotle wittily, but hurtfully, doth deride the Sophists near his time, saying, 'They did as if one that professed the art of shoe-making should not teach how to make up a shoe, but only exhibit in a readiness a number of shoes of all fashions and sizes.' But yet a man might reply, that if a shoemaker

Experientia: Learned experience and interpretation of nature.
rudiment: an initial stage.
Aristotle: In *Sophistici Elenchi* ('Sophistical Refutations'), ii. 9.

should have no shoes in his shop, but only work as he is bespoken, he should be weakly customed. But our Saviour, speaking of divine knowledge, saith, 'That the kingdom of heaven is like a good householder, that bringeth forth both new and old store': and we see the ancient writers of rhetoric do give it in precept, that pleaders should have the places,[n] whereof they have most continual use, ready handled in all the variety that may be; as that, to speak for the literal interpretation of the law against equity,[n] and contrary; and to speak for presumptions and inferences against testimony, and contrary. And Cicero himself, being broken unto it by great experience, delivereth it plainly, that whatsoever a man shall have occasion to speak of (if he will take the pains), he may have it in effect premeditate and handled *in thesi*. So that when he cometh to a particular he shall have nothing to do, but to put to names, and times, and places, and such other circumstances of individuals. We see likewise the exact diligence of Demosthenes; who, in regard of the great force that the entrance and access into causes hath to make a good impression, had ready framed a number of prefaces for orations and speeches. All which authorities and precedents may overweigh Aristotle's opinion, that would have us change a rich wardrobe for a pair of shears.

8. But the nature of the collection of this provision or preparatory store, though it be common both to logic and rhetoric, yet having made an entry of it here, where it came first to be spoken of, I think fit to refer over the further handling of it to rhetoric.

9. The other part of invention, which I term suggestion, doth assign and direct us to certain marks, or places, which may excite our mind to return and produce such knowledge as it hath formerly collected, to the end we may make use thereof. Neither is this use (truly taken) only to furnish argument to dispute probably with others, but likewise to minister unto our judgement to conclude aright within ourselves. Neither may these places serve only to apprompt our invention, but also to direct our inquiry. For a faculty of wise interrogating is half a knowledge. For as Plato saith, 'Whosoever seeketh,

customed: patronized.
That the kingdom: Matthew 13:52.
broken into: taught.
in thesi: Cicero, *De Oratore* ii. 32–4. 'In propositions.'
put to: attach.
to rhetoric: At II. xviii. 6 f. below.
probably: plausibly.
apprompt: prompt, stimulate.*
Plato: *Meno*, 80.

knoweth that which he seeketh for in a general notion: else how shall he know it when he hath found it?' And therefore the larger your anticipation is, the more direct and compendious is your search. But the same places which will help us what to produce of that which we know already, will also help us, if a man of experience were before us, what questions to ask; or, if we have books and authors to instruct us, what points to search and revolve; so as I cannot report that this part of invention, which is that which the schools call topics, is deficient.

10. Nevertheless, topics are of two sorts, general and special. The general we have spoken to; but the particular hath been touched by some, but rejected generally as inartificial and variable. But leaving the humour which hath reigned too much in the schools (which is, to be vainly subtile in a few things which are within their command, and to reject the rest), I do receive particular topics, that is, places or directions of invention and inquiry in every particular knowledge, as things of great use, being mixtures of logic with the matter of sciences. For in these it holdeth, *ars inveniendi adolescit cum inventis*;[n] for as in going of a way, we do not only gain that part of the way which is passed, but we gain the better sight of that part of the way which remaineth: so every degree of proceeding in a science giveth a light to that which followeth; which light if we strengthen by drawing it forth into questions or places of inquiry, we do greatly advance our pursuit.

[OF JUDGEMENT: THE SYLLOGISM, INDUCTION, FALSE APPEARANCES]

XIV. 1. Now we pass unto the arts of judgement, which handle the natures of proofs and demonstrations; which as to induction hath a coincidence with invention. For in all inductions, whether in good or vicious form, the same action of the mind which inventeth, judgeth; all one as in the sense. But otherwise it is in proof by syllogism; for the proof being not immediate, but by mean,[n] the invention of the mean is one thing, and the judgement of the consequence is another; the one exciting only, the other examining. Therefore, for the real and exact form of judgement, we refer ourselves to that

inartificial: not in accordance with the principles of logic.
receive: admit.
coincidence: exact agreement.
otherwise it is: it is otherwise.
mean: the middle term of a syllogism.*
exciting: setting the process in motion.

which we have spoken of interpretation of nature.

2. For the other judgement by syllogism, as it is a thing most agreeable to the mind of man, so it hath been vehemently and excellently laboured. For the nature of man doth extremely covet to have somewhat in his understanding fixed and unmoveable, and as a rest and support of the mind. And therefore as Aristotle endeavoureth to prove, that in all motion there is some point quiescent; and as he elegantly expoundeth the ancient fable of Atlas (that stood fixed, and bare up the heaven from falling) to be meant of the poles or axle-tree of heaven, whereupon the conversion is accomplished: so assuredly men have a desire to have an Atlas or axle-tree within to keep them from fluctuation, which is like to a perpetual peril of falling. Therefore men did hasten to set down some principles about which the variety of their disputations might turn.

3. So then this art of judgement is but the reduction of propositions to principles in a middle term. The principles to be agreed by all and exempted from argument; the middle term to be elected at the liberty of every man's invention; the reduction to be of two kinds, direct and inverted; the one when the proposition is reduced to the principle, which they term a probation ostensive; the other, when the contradictory of the proposition is reduced to the contradictory of the principle, which is that which they call *per incommodum*, or pressing an absurdity; the number of middle terms to be as the proposition standeth degrees more or less removed from the principle.

4. But this art hath two several methods of doctrine, the one by way of direction, the other by way of caution; the former frameth and setteth down a true form of consequence, by the variations and deflections from which errors and inconsequences may be exactly judged. Toward the composition and structure of which form, it is incident to handle the parts thereof, which are propositions, and the parts of propositions, which are simple words. And this is that part of logic which is comprehended in the Analytics.

5. The second method of doctrine was introduced for expedite use and assurance sake; discovering the more subtile forms of sophisms and illaqueations with their redargutions, which is that which is termed *elenches*.[n] For although in the more gross sorts of fallacies it happeneth (as Seneca maketh the comparison well) as in

Aristotle: In *De Motu Animalium* ii. 3.
Atlas: Homer, *Odyssey* i. 52–4; Virgil, *Aeneid* iv. 481–2.
conversion: rotation. *expedite*: unencumbered, speedy.
illaqueations: sophisms, intanglings in arguments.*
redargutions: refutations.
Seneca: *Epistles to Lucilius* xlv. 8.

juggling feats, which, though we know not how they are done, yet we know well it is not as it seemeth to be; yet the more subtile sort of them doth not only put a man besides his answer, but doth many times abuse his judgement.

6. This part concerning *elenches* is excellently handled by Aristotle in precept, but more excellently by Plato in example; not only in the persons of the Sophists, but even in Socrates himself, who, professing to affirm nothing, but to infirm that which was affirmed by another, hath exactly expressed all the forms of objection, fallace, and redargution. And although we have said that the use of this doctrine is for redargution, yet it is manifest the degenerate and corrupt use is for caption and contradiction, which passeth for a great faculty, and no doubt is of very great advantage: though the difference be good which was made between orators and sophisters, that the one is as the greyhound, which hath his advantage in the race, and the other as the hare, which hath her advantage in the turn, so as it is the advantage of the weaker creature.

7. But yet further, this doctrine of *elenches* hath a more ample latitude and extent than is perceived; namely, unto divers parts of knowledge; whereof some are laboured and other omitted. For first, I conceive (though it may seem at first somewhat strange) that that part which is variably referred, sometimes to logic, sometimes to metaphysic, touching the common adjuncts of essences,[n] is but an *elenche*. For the great sophism of all sophisms being equivocation or ambiguity of words and phrase, specially of such words as are most general and intervene in every inquiry, it seemeth to me that the true and fruitful use (leaving vain subtilities and speculations) of the inquiry of majority, minority, priority, posteriority, identity, diversity, possibility, act, totality, parts, existence, privation, and the like, are but wise cautions against ambiguities of speech. So again the distribution of things into certain tribes, which we call categories[n] or predicaments, are but cautions against the confusion of definitions and divisions.

8. Secondly, there is a seducement that worketh by the strength of the impression, and not by the subtilty of the illaqueation; not so much perplexing the reason, as overruling it by power of the imagination. But this part I think more proper to handle when I shall speak of rhetoric.

9. But lastly, there is yet a much more important and profound

infirm: render questionable.
fallace: fallacy, sophistical argument.
caption: cavilling.*

kind of fallacies in the mind of man, which I find not observed or inquired at all, and think good to place here, as that which of all others appertaineth most to rectify judgement: the force whereof is such, as it doth not dazzle or snare the understanding in some particulars, but doth more generally and inwardly infect and corrupt the state thereof. For the mind of man is far from the nature of a clear and equal glass, wherein the beams of things should reflect according to their true incidence; nay, it is rather like an enchanted glass, full of superstition and imposture, if it be not delivered and reduced. For this purpose, let us consider the false appearances[n] that are imposed upon us by the general nature of the mind, beholding them in an example or two; as first, in that instance which is the root of all superstition, namely, that to the nature of the mind of all men it is consonant for the affirmative or active to affect more than the negative or privative.[n] So that a few times hitting or presence, countervails oft-times failing or absence; as was well answered by Diagoras to him that showed him in Neptune's temple the great number of pictures of such as had scaped shipwreck, and had paid their vows to Neptune, saying, 'Advise now, you that think it folly to invocate Neptune in tempest.' 'Yea, but (saith Diagoras) where are they painted that are drowned?' Let us behold it in another instance, namely, that the spirit of man, being of an equal and uniform substance, doth usually suppose and feign in nature a greater equality and uniformity than is in truth. Hence it cometh, that the mathematicians cannot satisfy themselves except they reduce the motions of the celestial bodies to perfect circles, rejecting spiral lines, and labouring to be discharged of eccentrics.[n] Hence it cometh, that whereas there are many things in nature, as it were *monodica, sui juris*;[n] yet the cogitations of man do feign unto them relatives, parallels, and conjugates, whereas no such thing is; as they have feigned an element of fire,[n] to keep square with earth, water, and air, and the like. Nay, it is not credible, till it be opened, what a number of fictions and fantasies the similitude of human actions and arts, together with the making of man *communis mensura*,[n] have brought into natural philosophy; not much better than the heresy of the Anthropomorphites,[n] bred in the cells of gross and solitary monks, and the opinion of Epicurus, answerable to the same in heathenism, who supposed the gods to be of human shape. And therefore Velleius the Epicurean needed not to have asked, why God should have adorned the

equal: smooth-surfaced.*
Diagoras: In Cicero, *On the Nature of the Gods* iii. 37.
conjugates: things connected or related in idea with others.*

heavens with stars, as if he had been an *aedilis*,[n] one that should have set forth some magnificent shows or plays. For if that great work-master had been of an human disposition, he would have cast the stars into some pleasant and beautiful works and orders, like the frets in the roofs of houses; whereas one can scarce find a posture in square, or triangle, or straight line, amongst such an infinite number; so differing an harmony there is between the spirit of man and the spirit of nature.

10. Let us consider again[n] the false appearances imposed upon us by every man's own individual nature and custom, in that feigned supposition that Plato maketh of the cave: for certainly if a child were continued in a grot or cave under the earth until maturity of age, and came suddenly abroad, he would have strange and absurd imaginations. So in like manner, although our persons live in the view of heaven, yet our spirits are included in the caves of our own complexions and customs, which minister unto us infinite errors and vain opinions, if they be not recalled to examination. But hereof we have given many examples in one of the errors, or peccant humours, which we ran briefly over in our first book.

11. *Elenchi magni, sive de idolis animi humani nativis et adventitiis.* And lastly, let us consider the false appearances that are imposed upon us by words,[n] which are framed and applied according to the conceit and capacities of the vulgar sort: and although we think we govern our words, and prescribe it well *loquendum ut vulgus sentiendum ut sapientes*; yet certain it is that words, as a Tartar's bow, do shoot back upon the understanding of the wisest, and mightily entangle and pervert the judgement. So as it is almost necessary, in all controversies and disputations, to imitate the wisdom of the mathematicians, in setting down in the very beginning the definitions of our words and terms, that others may know how we accept and understand them, and whether they concur with us or no. For it cometh to pass, for want of this, that we are sure to end there where we ought to have begun, which is, in questions and differences about words. To conclude therefore, it must be confessed that it is not possible to divorce ourselves from these fallacies and

posture: shape. *Plato*: *Republic* vii. *included*: shut up.
complexions: constitutions, derived from the combination of 'humours'.
first book: In I. v. 6.
Elenchi: Important refutations, or concerning the idols of the human spirit, innate and imposed.
loquendum: To speak as the common people do, to think as wise men do. Cf. Aristotle, *Topics* II. ii. 1.
almost: for the most part.

false appearances, because they are inseparable from our nature and condition of life; so yet nevertheless the caution of them (for all *elenches*, as was said, are but cautions) doth extremely import the true conduct of human judgement. The particular *elenches* or cautions against these three false appearances, I find altogether deficient.

12. *De analogia demonstrationum.* There remaineth one part of judgement of great excellency, which to mine understanding is so slightly touched, as I may report that also deficient; which is the application of the differing kinds of proofs of the differing kinds of subjects. For there being but four kinds of demonstrations, that is, by the immediate consent of the mind or sense, by induction, by syllogism, and by congruity, which is that which Aristotle calleth demonstration in orb or circle, and not *a notioribus*,[n] every of these hath certain subjects in the matter of sciences, in which respectively they have chiefest use; and certain others, from which respectively they ought to be excluded; and the rigour and curiosity in requiring the more severe proofs in some things, and chiefly the facility in contenting ourselves with the more remiss proofs in others, hath been amongst the greatest causes of detriment and hindrance to knowledge. The distributions and assignations of demonstrations, according to the analogy of sciences, I note as deficient.

[THE PRESERVATION OF KNOWLEDGE: WRITING AND MEMORY]

XV. 1. The custody or retaining of knowledge is either in writing or memory; whereof writing hath two parts, the nature of the character, and the order of the entry. For the art of characters, or other visible notes of words or things, it hath nearest conjugation with grammar;[n] and therefore I refer it to the due place. For the disposition and collocation of that knowledge which we preserve in writing, it consisteth in a good digest of common-places;[n] wherein I am not ignorant of the prejudice imputed to the use of common-place books, as causing a retardation of reading, and some sloth or relaxation of memory. But because it is but a counterfeit thing in knowledges to be forward and pregnant, except a man be deep and full, I hold the entry of common-places to be a matter of great use and essence

import: affect.
De analogia: Concerning the analogy of proofs.
collocation: arrangement.
pregnant: teeming with ideas. *essence*: importance.*

in studying, as that which assureth copie of invention, and contracteth judgement to a strength. But this is true, that of the methods of common-places that I have seen, there is none of any sufficient worth: all of them carrying merely the face of a school, and not of a world; and referring to vulgar matters and pedantical divisions, without all life or respect to action.

2. For the other principal part of the custody of knowledge, which is memory,[n] I find that faculty in my judgement weakly inquired of. An art there is extant of it; but it seemeth to me that there are better precepts than that art, and better practices of that art than those received. It is certain the art (as it is) may be raised to points of ostentation prodigious: but in use (as it is now managed) it is barren, not burdensome, nor dangerous to natural memory, as is imagined, but barren, that is, not dexterous to be applied to the serious use of business and occasions. And therefore I make no more estimation of repeating a great number of names or words upon once hearing, or the pouring forth of a number of verses or rhymes *ex tempore*, or the making of a satirical simile of everything, or the turning of everything to a jest, or the falsifying or contradicting of everything by cavil, or the like (whereof in the faculties of the mind there is great copie, and such as by device and practice may be exalted to an extreme degree of wonder), than I do of the tricks of tumblers, funambuloes, baladines; the one being the same in the mind that the other is in the body, matters of strangeness without worthiness.

3. This art of memory is but built upon two intentions; the one prenotion, the other emblem. Prenotion dischargeth the indefinite seeking of that we would remember, and directeth us to seek in a narrow compass, that is, somewhat that hath congruity with our place of memory. Emblem reduceth conceits intellectual to images sensible, which strike the memory more; out of which axioms may be drawn much better practique than that in use; and besides which axioms, there are divers moe touching help of memory, not inferior to them. But I did in the beginning distinguish, not to report those things deficient, which are but only ill managed.

funambuloes: rope-walkers.*
baladines: theatrical dancers.
prenotion: preconceived idea.*
emblem: an abstract use of the word, which refers to a picture expressing a moral or allegory.
practique: practice.
distinguish: decide.*

[TRANSMISSION OF KNOWLEDGE: SPEECH OR WRITING]

XVI. 1. There remaineth the fourth kind of rational knowledge, which is transitive, concerning the expressing or transferring our knowledge to others; which I will term by the general name of tradition or delivery. Tradition hath three parts; the first concerning the organ of tradition; the second concerning the method of tradition; and the third concerning the illustration of tradition.

2. For the organ of tradition, it is either speech or writing: for Aristotle saith well, 'Words are the images of cogitations, and letters are the images of words.' But yet it is not of necessity that cogitations be expressed by the medium of words. For whatsoever is capable of sufficient differences, and those perceptible by the sense, is in nature competent to express cogitations. And therefore we see in the commerce of barbarous people, that understand not one another's language, and in the practice of divers that are dumb and deaf, that men's minds are expressed in gestures, though not exactly, yet to serve the turn. And we understand further, that it is the use of China, and the kingdoms of the High Levant, to write in characters real,[n] which express neither letters nor words in gross, but things or notions; insomuch as countries and provinces, which understand not one another's language, can nevertheless read one another's writings, because the characters are accepted more generally than the languages do extend; and therefore they have a vast multitude of characters, as many (I suppose) as radical words.

3. *De notis rerum.* These notes of cogitations are of two sorts; the one when the note hath some similitude or congruity with the notion: the other *ad placitum*, having force only by contract or acceptation. Of the former sort are hieroglyphics[n] and gestures. For as to hieroglyphics (things of ancient use, and embraced chiefly by the Egyptians, one of the most ancient nations), they are but as continued impreses and emblems. And as for gestures, they are as transitory hierogylphics, and are to hieroglyphics as words spoken are to words written, in that they abide not; but they have evermore, as well as the other, an affinity with the things signified. As Periander,

transitive: concerned with transferring.
Aristotle: De Interpretatione i. 1.
De notis rerum: Concerning the notes of things.
ad placitum: arbitrary.
impreses: crests or heraldic devices, including the motto. Emblems are pictorial devices conveying a moral meaning.
Periander: In Aristotle, *Politics* iii. 8.

being consulted with how to preserve a tyranny newly usurped, bid the messenger attend and report what he saw him do; and went into his garden and topped all the highest flowers: signifying, that it consisted in the cutting off and keeping low of the nobility and grandees. *Ad placitum*, are the characters real before mentioned, and words: although some have been willing by curious inquiry, or rather by apt feigning, to have derived imposition of names from reason[n] and intendment; a speculation elegant, and, by reason it searcheth into antiquity, reverent; but sparingly mixed with truth, and of small fruit. This portion of knowledge, touching the notes of things, and cogitations in general, I find not inquired, but deficient. And although it may seem of no great use, considering that words and writings by letters do far excel all the other ways; yet because this part concerneth as it were the mint of knowledge (for words are the tokens current and accepted for conceits, as moneys are for values, and that it is fit men be not ignorant that moneys may be of another kind than gold and silver), I thought good to propound it to better inquiry.

4. Concerning speech and words, the consideration of them hath produced the science of grammar. For man still striveth to reintegrate himself in those benedictions, from which by his fault he hath been deprived; and as he hath striven against the first general curse by the invention of all other arts, so hath he sought to come forth of the second general curse (which was the confusion of tongues) by the art of grammar; whereof the use in a mother tongue is small, in a foreign tongue more; but most in such foreign tongues as have ceased to be vulgar tongues, and are turned only to learned tongues. The duty of it is of two natures: the one popular, which is for the speedy and perfect attaining languages, as well for intercourse of speech as for understanding of authors; the other philosophical, examining the power and nature of words, as they are the footsteps and prints of reason: which kind of analogy between words and reason is handled *sparsim*, brokenly though not entirely; and therefore I cannot report it deficient, though I think it very worthy to be reduced into a science by itself.

5. Unto grammar also belongeth, as an appendix, the consideration of the accidents of words; which are measure, sound, and eleva-

intendment: intention.
first general curse: Genesis 3:16–19. 'In the sweat of thy brow shalt thou eat bread.'
second general curse: Genesis 11:6–8. The confusion of tongues.
sparsim: dispersedly.
accidents: accessory qualities.

tion or accent, and the sweetness and harshness of them; whence hath issued some curious observations in rhetoric, but chiefly poesy, as we consider it, in respect of the verse and not of the argument. Wherein though men in learned tongues do tie themselves to the ancient measures,[n] yet in modern languages it seemeth to me as free to make new measures of verses as of dances: for a dance is a measured pace, as a verse is a measured speech. In these things the sense is better judge than the art;

> Coenae fercula nostrae
> Mallem convivis quam placuisse cocis.

And of the servile expressing antiquity in an unlike and an unfit subject, it is well said, *Quod tempore antiquum videtur, id incongruitate est maxime novum.*

6. For ciphers,[n] they are commonly in letters, or alphabets, but may be in words. The kinds of ciphers (besides the simple ciphers, with changes, and intermixtures of nulls and non-significants) are many, according to the nature or rule of the infolding, wheel-ciphers, key-ciphers, doubles, &c. But the virtues of them, whereby they are to be preferred, are three; that they be not laborious to write and read; that they be impossible to decipher; and, in some cases, that they be without suspicion. The highest degree whereof is to write *omnia per omnia*; which is undoubtedly possible, with a proportion quintuple at most of the writing infolding to the writing infolded, and no other restraint whatsoever. This art of ciphering hath for relative an art of deciphering, by supposition unprofitable, but, as things are, of great use. For suppose that ciphers were well managed, there be multitudes of them which exclude the decipherer. But in regard of the rawness and unskilfulness of the hands through which they pass, the greatest matters are many times carried in the weakest ciphers.

7. In the enumeration of these private and retired arts, it may be thought I seek to make a great muster-roll of sciences, naming them for show and ostentation, and to little other purpose. But let those which are skilful in them judge whether I bring them in only for appearance, or whether in that which I speak of them (though in

Coenae: Martial, *Epigrams* ix. 83.

> The dinner is for eating and my wish is
> That guests and not that cooks should like the dishes.

Quod tempore: There is nothing more new than an old thing that has ceased to fit.

nulls and non-significants: Bacon invents the words.

omnia per omnia: anything under cover of anything.

few words) there be not some seed of proficience. And this must be remembered, that as there be many of great account in their countries and provinces, which, when they come up to the seat of the estate, are but of mean rank and scarcely regarded; so these arts, being here placed with the principal and supreme sciences, seem petty things; yet to such as have chosen them to spend their labours and studies in them, they seem great matters.

[TRANSMISSION OF KNOWLEDGE: METHODS MAGISTRAL AND OF PROBATION]

XVII. 1. For the method of tradition, I see it hath moved a controversy[n] in our time. But as in civil business, if there be a meeting, and men fall at words, there is commonly an end of the matter for that time, and no proceeding at all; so in learning, where there is much controversy, there is many times little inquiry. For this part of knowledge of method seemeth to me so weakly inquired as I shall report it deficient.

2. Method hath been placed and that not amiss, in logic, as a part of judgement.[n] For as the doctrine of syllogisms comprehendeth the rules of judgement upon that which is invented, so the doctrine of method containeth the rules of judgement upon that which is to be delivered; for judgement precedeth delivery, as it followeth invention. Neither is the method or the nature of the tradition material only to the use of knowledge, but likewise to the progression of knowledge: for since the labour and life of one man cannot attain to perfection of knowledge, the wisdom of the tradition is that which inspireth the felicity of continuance and proceeding. And therefore the most real diversity of method is of method referred to use, and method referred to progression: whereof the one may be termed magistral,[n] and the other of probation.

3. The latter whereof seemeth to be *via deserta et interclusa.* For as knowledges are now delivered, there is a kind of contract of error between the deliverer and the receiver. For he that delivereth knowledge, desireth to deliver it in such form as may be best believed, and not as may be best examined; and he that receiveth knowledge, desireth rather present satisfaction, than expectant inquiry; and so rather not to doubt, than not to err: glory making the author not to lay open his weakness, and sloth making the disciple not to know his strength.

via deserta: Cicero, *Pro Caelio* xviii. 42. 'A way deserted and shut off.'

4. *De methodo sincera, sive ad filios scientiarum.* But knowledge that is delivered as a thread to be spun on,[n] ought to be delivered and intimated, if it were possible, in the same method wherein it was invented: and so is it possible of knowledge induced. But in this same anticipated and prevented knowledge, no man knoweth how he came to the knowledge which he hath obtained. But yet nevertheless, *secundum majus et minus*, a man may revisit and descend unto the foundations of his knowledge and consent; and so transplant it into another, as it grew in his own mind. For it is in knowledges as it is in plants: if you mean to use the plant, it is no matter for the roots; but if you mean to remove it to grow, then it is more assured to rest upon roots than slips: so the delivery of knowledges (as it is now used) is as of fair bodies of trees without the roots; good for the carpenter, but not for the planter. But if you will have sciences grow, it is less matter for the shaft or body of the tree, so you look well to the taking up of the roots. Of which kind of delivery the method of the mathematics, in that subject, hath some shadow: but generally I see it neither put in ure nor put in inquisition, and therefore note it for deficient.

5. Another diversity of method there is, which hath some affinity with the former, used in some cases by the discretion of the ancients, but disgraced since by the impostures of many vain persons, who have made it as a false light for their counterfeit merchandises; and that is, enigmatical and disclosed.[n] The pretence whereof is, to remove the vulgar capacities from being admitted to the secrets of knowledges, and to reserve them to selected auditors, or wits of such sharpness as can pierce the veil.

6. Another diversity of method, whereof the consequence is great, is the delivery of knowledge in aphorisms, or in methods;[n] wherein we may observe that it hath been too much taken into custom, out of a few axioms or observations upon any subject, to make a solemn and formal art, filling it with some discourses, and illustrating it with examples, and digesting it into a sensible method. But the writing in aphorisms hath many excellent virtues, whereto the writing in method doth not approach.

7. For first, it trieth the writer, whether he be superficial or solid: for aphorisms, except they should be ridiculous, cannot be made but

De methodo: Of the true method of the sciences, or the method of delivery to posterity.
induced: arrived at by induction.
prevented: i.e. knowledge which we already possess.
secundum: to a greater or less degree.
assured: secure. *slips*: cuttings.

of the pith and heart of sciences; for discourse of illustration is cut off; recitals of examples are cut off; discourse of connexion and order is cut off; descriptions of practice are cut off. So there remaineth nothing to fill the aphorisms but some good quantity of observation: and therefore no man can suffice, nor in reason will attempt, to write aphorisms, but he that is sound and grounded. But in methods,

> Tantum series juncturaque pollet,
> Tantum de medio sumptis accedit honoris.

as a man shall make a great show of an art, which, if it were disjointed, would come to little. Secondly, methods are more fit to win consent or belief, but less fit to point to action; for they carry a kind of demonstration in orb or circle, one part illuminating another, and therefore satisfy. But particulars being dispersed[n] do best agree with dispersed directions. And lastly, aphorisms, representing a knowledge broken, do invite men to inquire further; whereas methods, carrying the show of a total, do secure men, as if they were at furthest.

8. Another diversity of method, which is likewise of great weight, is the handling of knowledge by assertions and their proofs, or by questions[n] and their determinations. The latter kind whereof, if it be immoderately followed, is as prejudicial to the proceeding of learning, as it is to the proceeding of an army to go about to besiege every little fort or hold. For if the field be kept, and the sum of the enterprise pursued, those smaller things will come in of themselves: indeed a man would not leave some important piece enemy at his back. In like manner, the use of confutation in the delivery of sciences ought to be very sparing; and to serve to remove strong preoccupations and prejudgements, and not to minister and excite disputations and doubts.

9. Another diversity of methods is, according to the subject or matter which is handled. For there is a great difference in delivery of the mathematics, which are the most abstracted of knowledges, and policy, which is the most immersed. And howsoever contention hath been moved, touching an uniformity of method[n] in multiformity of matter, yet we see how that opinion, besides the weakness of it, hath been of ill desert towards learning, as that which taketh the

suffice: have the ability.
Tantum: Horace, *Art of Poetry*, ll. 242–3.

> Such grace can order and connection give;
> Such beauties common subjects may receive.
> (Francis)

way to reduce learning to certain empty and barren generalities; being but the very husks and shells of sciences, all the kernel being forced out and expulsed with the torture and press of the method. And therefore as I did allow well of particular topics for invention, so I do allow likewise of particular methods of tradition.

10. Another diversity of judgement[n] in the delivery and teaching of knowledge is, according unto the light and presuppositions of that which is delivered. For that knowledge which is new, and foreign from opinions received, is to be delivered in another form than that that is agreeable and familiar; and therefore Aristotle, when he thinks to tax Democritus, doth in truth commend him, where he saith, 'If we shall indeed dispute, and not follow after similitudes, &c.'[n] For those whose conceits are seated in popular opinions, need only but to prove or dispute; but those whose conceits are beyond popular opinions, have a double labour; the one to make themselves conceived, and the other to prove and demonstrate. So that it is of necessity with them to have recourse to similitudes and translations to express themselves. And therefore in the infancy of learning, and in rude times, when those conceits which are now trivial were then new, the world was full of parables and similitudes; for else would men either have passed over without mark, or else rejected for paradoxes that which was offered, before they had understood or judged. So in divine learning, we see how frequent parables and tropes are: for it is a rule, that whatsoever science is not consonant to presuppositions, must pray in aid of similitudes.

11. *De prudentia traditionis.* There be also other diversities of methods vulgar and received: as that of resolution or analysis, of constitution or systasis, of concealment or cryptic &c.,[n] which I do allow well of, though I have stood upon those which are least handled and observed. All which I have remembered to this purpose, because I would erect and constitute one general inquiry (which seems to me deficient) touching the wisdom of tradition.

12. But unto this part of knowledge, concerning method, doth further belong not only the architecture of the whole frame of a work, but also the several beams and columns thereof; not as to their stuff, but as to their quantity and figure. And therefore method considereth not only the disposition of the argument or subject, but likewise the propositions: not as to their truth or matter, but as to

trivial: commonplace.
pray in aid of: call in to one's assistance.
De prudentia: The wisdom of tradition.
vulgar: familiar.

their limitation and manner. For herein Ramus[n] merited better a great deal in reviving the good rules of propositions, Καθόλου πρῶτον, κατὰ παντός &c., than he did in introducing the canker of epitomes; and yet (as it is the condition of human things that, according to the ancient fables, 'the most precious things have the most pernicious keepers') it was so, that the attempt of the one made him fall upon the other. For he had need be well conducted that should design to make axioms convertible, if he make them not withal circular, and non-promovent,[n] or incurring into themselves; but yet the intention was excellent.

13. *De productione axiomatum.* The other considerations of method, concerning propositions, are chiefly touching the utmost propositions, which limit the dimensions of sciences: for every knowledge may be fitly said, besides the profundity (which is the truth and substance of it, that makes it solid), to have a longitude and a latitude; accounting the latitude towards other sciences, and the longitude towards action; that is, from the greatest generality to the most particular precept. The one giveth rule how far one knowledge ought to intermeddle within the province of another, which is the rule they call Καθαυτό;[n] the other giveth rule unto what degree of particularity a knowledge should descend: which latter I find passed over in silence, being in my judgement the more material. For certainly there must be somewhat left to practice; but how much is worthy the inquiry. We see remote and superficial generalities do but offer knowledge to scorn of practical men; and are no more aiding to practice, than an Ortelius'[n] universal map is to direct the way between London and York. The better sort of rules have been not unfitly compared to glasses of steel unpolished, where you may see the images of things, but first they must be filed: so the rules will help, if they be laboured and polished by practice. But how crystalline they may be made at the first, and how far forth they may be polished aforehand is the question; the inquiry whereof seemeth to me deficient.

14. There hath been also laboured and put in practice a method, which is not a lawful method, but a method of imposture; which is, to deliver knowledges in such manner, as men may speedily come to make a show of learning who have it not. Such was the travail of Raymundus Lullius,[n] in making that art which bears his name: not unlike to some books of typocosmy,[n] which have been made since;

fables: Referring to the dragons which kept the garden of the Hesperides and the Golden Fleece.
De productione: The production of axioms.

being nothing but a mass of words of all arts, to give men countenance, that those which use the terms might be thought to understand the art; which collections are much like a fripper's or broker's shop, that hath ends of everything, but nothing of worth.

[TRANSMISSION OF KNOWLEDGE: RHETORIC (WITH EXAMPLES OF DEFICIENCIES)]

XVIII. 1. Now we descend to that part which concerneth the illustration of tradition, comprehended in that science which we call rhetoric,[n] or art of eloquence; a science excellent, and excellently well laboured. For although in true value it is inferior to wisdom, as it is said by God to Moses, when he disabled himself for want of this faculty, 'Aaron shall be thy speaker, and thou shalt be to him as God'; yet with people it is the more mighty: for so Salomon saith, *Sapiens corde appellabitur prudens, sed dulcis eloquio majora reperiet*; signifying that profoundness of wisdom will help a man to a name or admiration, but that it is eloquence that prevaileth in an active life. And as to the labouring of it, the emulation of Aristotle with the rhetoricians of his time, and the experience of Cicero, hath made them in their works of rhetorics exceed themselves. Again, the excellency of examples of eloquence in the orations of Demosthenes and Cicero, added to the perfection of the precepts of eloquence, hath doubled the progression in this art; and therefore the deficiences which I shall note will rather be in some collections, which may as handmaids attend the art, than in the rules or use of the art itself.

2. Notwithstanding, to stir the earth a little about the roots of this science, as we have done of the rest; the duty and office of rhetoric is to apply reason to imagination for the better moving of the will.[n] For we see reason is disturbed in the administration thereof by three means; by illaqueation or sophism, which pertains to logic; by imagination or impression, which pertains to rhetoric; and by passion or affection, which pertains to morality. And as in negotiation with others, men are wrought by cunning, by importunity, and by vehe-

fripper's shop: an old clothes shop.
disabled: disparaged, depreciated.
Aaron: cf. Exodus 4:16.
Sapiens: Proverbs 16:21. 'The wise in heart shall be called prudent but he that is eloquent shall attain greater things.'
illaqueation: sophism.
morality: moral philosophy.
wrought: influenced.

mency; so in this negotiation within ourselves, men are undermined by inconsequences, solicited and importuned by impressions or observations, and transported by passions. Neither is the nature of man so unfortunately built, as that those powers and arts should have force to disturb reason, and not to establish and advance it. For the end of logic is to teach a form of argument to secure reason, and not to entrap it. The end of morality is to procure the affections to obey reason, and not to invade it. The end of rhetoric is to fill the imagination to second reason, and not to oppress it: for these abuses of arts come in but *ex obliquo*, for caution.

3. And therefore it was great injustice in Plato, though springing out of a just hatred to the rhetoricians of his time, to esteem of rhetoric but as a voluptuary art, resembling it to cookery, that did mar wholesome meats, and help unwholesome by variety of sauces to the pleasure of the taste. For we see that speech is much more conversant in adorning that which is good, than in colouring that which is evil; for there is no man but speaketh more honestly than he can do or think: and it was excellently noted by Thucydides in Cleon, that because he used to hold on the bad side in causes of estate, therefore he was ever inveighing against eloquence and good speech; knowing that no man can speak fair of courses sordid and base. And therefore as Plato said elegantly, 'That virtue, if she could be seen, would move great love and affection';[n] so seeing that she cannot be showed to the sense by corporal shape, the next degree is to show her to the imagination in lively representation: for to show her to reason only in subtility of argument was a thing ever derided in Chrysippus[n] and many of the Stoics, who thought to thrust virtue upon men by sharp disputations and conclusions, which have no sympathy with the will of man.

4. Again, if the affections in themselves were pliant and obedient to reason, it were true there should be no great use of persuasions and insinuations[n] to the will, more than of naked proposition and proofs; but in regard of the continual mutinies and seditions of the affections,

Video meliora, proboque,
Deteriora sequor,

inconsequences: arguments that lack logical sequence.*
ex obliquo: obliquely. *Plato*: *Gorgias* 462.
esteem of: reckon. *resembling*: comparing.
Thucydides: In his *History of the Peloponnesian War* iii. 42.
Video: Ovid, *Metamorphoses* vii. 20. 'I see the better course and approve it, but I follow the worse.'

reason would become captive and servile, if eloquence of persuasions did not practise and win the imagination from the affections' part, and contract a confederacy between the reason and imagination against the affections; for the affections themselves carry ever an appetite to good, as reason doth. The difference is, that the affection beholdeth merely the present; reason beholdeth the future and sum of time. And therefore the present filling the imagination more, reason is commonly vanquished; but after that force of eloquence and persuasion hath made things future and remote appear as present, then upon the revolt of the imagination reason prevaileth.

5. *De prudentia sermonis privati.* We conclude therefore that rhetoric can be no more charged with the colouring of the worse part, than logic with sophistry, or morality with vice. For we know the doctrines of contraries are the same, though the use be opposite. It appeareth also that logic differeth from rhetoric, not only as the fist from the palm,[n] the one close, the other at large; but much more in this, that logic handleth reason exact and in truth, and rhetoric handleth it as it is planted in popular opinions and manners. And therefore Aristotle doth wisely place rhetoric as between logic on the one side, and moral or civil knowledge on the other, as participating of both: for the proofs and demonstrations of logic are toward all men indifferent and the same; but the proofs and persuasions of rhetoric ought to differ according to the auditors:

Orpheus in sylvis, inter delphinas Arion.

Which application, in perfection of idea, ought to extend so far, that if a man should speak of the same thing to several persons, he should speak to them all respectively and several ways: though this politic part of eloquence in private speech it is easy for the greatest orators to want: whilst, by the observing their well-graced forms of speech, they leese the volubility of application: and therefore it shall not be amiss to recommend this to better inquiry, not being curious whether we place it here, or in that part which concerneth policy.

6. *Colores boni et mali, simplicis et comparati.* Now therefore will I descend to the deficiences, which (as I said) are but attendances: and first, I do not find the wisdom and diligence of Aristotle well

practise: conspire.
De prudentia: The wisdom of private discourse.
Aristotle: In the *Rhetoric* I. ii. 7.
Orpheus: Virgil, *Eclogues* viii. 56. 'An Orpheus in the woods, an Arion among the dolphins.'
leese: lose.
Colores boni: The colours of good and evil, simple and comparative.

pursued, who began to make a collection of the popular signs and colours of good and evil,[n] both simple and comparative, which are as the sophisms of rhetoric (as I touched before). For example:

Sophisma.
Quod laudatur, bonum: quod vituperatur, malum.
Redargutio.
Laudat venales qui vult extrudere merces.

Malum est, malum est (inquit emptor); sed cum recesserit, tum gloriabitur![n] The defects in the labour of Aristotle are three: one, that there be but a few of many; another, that their elenches are not annexed; and the third, that he conceiveth but a part of the use of them: for their use is not only in probation, but much more in impression. For many forms are equal in signification which are differing in impression; as the difference is great in the piercing of that which is sharp and that which is flat, though the strength of the percussion be the same. For there is no man but will be a little more raised by hearing it said, *Your enemies will be glad of this,*

Hoc Ithacus velit, et magno mercentur Atridae,

than by hearing it said only, *This is evil for you.*

7. Secondly, I do resume also that which I mentioned before, touching provision or preparatory store for the furniture of speech and readiness of invention, which appeareth to be of two sorts; the one in resemblance to a shop of pieces unmade up, the other to a shop of things ready made up; both to be applied to that which is frequent and most in request. The former of these I will call *antitheta*, and the latter *formulae*.

8. *Antitheta rerum. Antitheta*[n] are theses argued *pro et contra*; wherein men may be more large and laborious: but (in such as are able to do it) to avoid prolixity of entry, I wish the seeds of the several arguments to be cast up into some brief and acute sentences, not to be cited, but to be as skeins or bottoms of thread, to be unwinded at large when they come to be used; supplying authorities and examples by reference.

probation: proving. *impression*: affecting, moving.
Hoc Ithacus: Virgil, *Aeneid* ii. 104. 'This would the prince of Ithaca wish, and the sons of Atreus give large sums to purchase.'
before: at II. xiii. 7 above.
furniture of speech: materials for the embellishing of speech.
Antitheta: The antitheses of things.
acute: penetrating.
bottoms: skeins or balls.

Pro verbis legis.

Non est interpretatio, sed divinatio, quae recedit a litera:
Cum receditur a litera, judex transit in legislatorem.

Pro sententia legis.

Ex omnibus verbis est eliciendus sensus qui interpretatur singula.[n]

9. *Formulae*[n] are but decent and apt passages or conveyances of speech, which may serve indifferently for differing subjects; as of preface, conclusion, digression, transition, excusation, &c. For as in buildings there is great pleasure and use in the well casting of the staircases, entries, doors, windows, and the like; so in speech, the conveyances and passages are of special ornament and effect.

A conclusion in a deliberative.

So may we redeem the faults passed, and prevent the inconveniences future.

[TRANSMISSION OF KNOWLEDGE: EDITING, ANNOTATION, COMMENTARIES, ADVICE TO EDUCATORS]

XIX. 1. There remain two appendices touching the tradition of knowledge, the one critical, the other pedantical. For all knowledge is either delivered by teachers, or attained by men's proper endeavours: and therefore as the principal part of tradition of knowledge concerneth chiefly writing of books, so the relative part thereof concerneth reading of books; whereunto appertain incidently these considerations. The first is concerning the true correction and edition of authors; wherein nevertheless rash diligence hath done great prejudice. For these critics have often presumed that that which they understand not is false set down: as the priest that, where he found it written of S. Paul *Demissus est per sportam*, mended his book, and made it *Demissus est per portam*;[n] because *sporta* was an hard word, and out of his reading: and surely their errors, though they be not so palpable and ridiculous, yet are of the same kind. And therefore, as it hath been wisely noted, the most corrected copies are commonly the least correct.

The second is concerning the exposition and explication of authors, which resteth in annotations and commentaries: wherein it is over usual to blanch the obscure places and discourse upon the plain.

decent: appropriate. *proper*: own. *blanch*: leave unnoticed.*

The third is concerning the times, which in many cases give great light to true interpretations.

The fourth is concerning some brief censure and judgement of the authors; that men thereby may make some election unto themselves what books to read.

And the fifth is concerning the syntax and disposition of studies; that men may know in what order or pursuit to read.

2. For pedantical[n] knowledge, it containeth that difference of tradition which is proper for youth; whereunto appertain divers considerations of great fruit.

As first, the timing and seasoning of knowledges, as with what to initiate them, and from what for a time to refrain them.

Secondly, the consideration where to begin with the easiest, and so proceed to the more difficult; and in what courses to press the more difficult, and then to turn them to the more easy: for it is one method to practise swimming with bladders, and another to practise dancing with heavy shoes.

A third is the application of learning according unto the propriety of the wits; for there is no defect in the faculties intellectual, but seemeth to have a proper cure contained in some studies: as, for example, if a child be bird-witted, that is, hath not the faculty of attention, the mathematics giveth a remedy thereunto; for in them, if the wit be caught away but a moment, one is new to begin. And as sciences have a propriety towards faculties for cure and help, so faculties or powers have a sympathy towards sciences for excellency or speedy profiting: and therefore it is an inquiry of great wisdom, what kinds of wits and natures are most apt and proper for what sciences.

Fourthly, the ordering of exercises is matter of great consequence to hurt or help: for, as is well observed by Cicero, men in exercising their faculties, if they be not well advised, do exercise their faults and get ill habits as well as good; so as there is a great judgement to be had in the continuance and intermission of exercises. It were too long to particularize a number of other considerations of this nature, things but of mean appearance, but of singular efficacy. For as the wronging or cherishing of seeds or young plants is that that is most important to their thriving, and as it was noted that the first six kings being in truth as tutors of the state of Rome in the infancy

syntax: systematic arrangement.*
new: i.e. one must begin afresh. *Cicero*: In *De Oratore* i. 33.
it was noted: By Machiavelli, *Discourses on Livy* i. 19, who says *four* kings.
tutors: guardians.

thereof was the principal cause of the immense greatness of that state which followed, so the culture and manurance of minds in youth hath such a forcible (though unseen) operation, as hardly any length of time or contention of labour can countervail it afterwards. And it is not amiss to observe also how small and mean faculties gotten by education, yet when they fall into great men or great matters, do work great and important effects: whereof we see a notable example in Tacitus[n] of two stage players, Percennius and Vibulenus, who by their faculty of playing put the Pannonian armies into an extreme tumult and combustion. For there arising a mutiny amongst them upon the death of Augustus Caesar, Blaesus the lieutenant had committed some of the mutiners, which were suddenly rescued; whereupon Vibulenus got to be heard speak, which he did in this manner: 'These poor innocent wretches appointed to cruel death, you have restored to behold the light; but who shall restore my brother to me, or life unto my brother, that was sent hither in message from the legions of Germany, to treat of the common cause? and he hath murdered him this last night by some of his fencers and ruffians, that he hath about him for his executioners upon soldiers. Answer, Blaesus, what is done with his body? The mortalest enemies do not deny burial. When I have performed my last duties to the corpse with kisses, with tears, command me to be slain besides him; so that these my fellows, for our good meaning and our true hearts to the legions, may have leave to bury us.' With which speech he put the army into an infinite fury and uproar: whereas truth was he had no brother, neither was there any such matter; but he played it merely as if he had been upon the stage.

3. But to return: we are now come to a period of rational knowledges; wherein if I have made the divisions other than those that are received, yet would I not be thought to disallow all those divisions which I do not use. For there is a double necessity imposed upon me of altering the divisions. The one, because it differeth in end and purpose, to sort together those things which are next in nature, and those things which are next in use. For if a secretary of estate should sort his papers, it is like in his study or general cabinet he would sort together things of a nature, as treaties, instructions, &c. But in his boxes or particular cabinet he would sort together those that he were like to use together, though of several natures. So in this general cabinet of knowledge it was necessary for me to follow the divisions of the nature of things; whereas if myself

manurance: cultivation, training. *contention*: exertion.
suddenly: speedily. *particular*: private. *several*: different.

had been to handle any particular knowledge, I would have respected the divisions fittest for use. The other, because the bringing in of the deficiences did by consequence alter the partitions of the rest. For let the knowledge extant (for demonstration sake) be fifteen. Let the knowledge with the deficiences be twenty; the parts of fifteen are not the parts of twenty; for the parts of fifteen are three and five; the parts of twenty are two, four, five, and ten. So as these things are without contradiction, and could not otherwise be.

[OF ETHICS IN GENERAL]

XX. 1. We proceed now to that knowledge which considereth of the appetite and will[n] of man: whereof Salomon saith, *Ante omnia, fili, custodi cor tuum; nam inde procedunt actiones vitae.* In the handling of this science, those which have written seem to me to have done as if a man, that professed to teach to write, did only exhibit fair copies of alphabets and letters joined, without giving any precepts or directions for the carriage of the hand and framing of the letters. So have they made good and fair exemplars and copies, carrying the draughts and portraitures of good, virtue, duty, felicity; propounding them well described as the true objects and scopes of man's will and desires. But how to attain these excellent marks, and how to frame and subdue the will of man to become true and conformable to these pursuits, they pass it over altogether, or slightly and unprofitably. For it is not the disputing, that moral virtues are in the mind of man by habit and not by nature; or the distinguishing, that generous spirits are won by doctrines and persuasions, and the vulgar sort by reward and punishment;[n] and the like scattered glances and touches, that can excuse the absence of this part.

2. The reason of this omission I suppose to be that hidden rock whereupon both this and many other barks of knowledge have been cast away; which is, that men have despised to be conversant in ordinary and common matters, the judicious direction whereof nevertheless is the wisest doctrine (for life consisteth not in novelties nor subtilities), but contrariwise they have compounded sciences chiefly of a certain resplendent or lustrous mass of matter, chosen to give glory either to the subtility of disputations, or to the eloquence of discourses. But Seneca giveth an excellent check to elo-

without: beyond.
Ante omnia: Proverbs 4:23. 'Above all, my son, keep thy heart, for out of it proceed the actions of life.'

quence. *Nocet illis eloquentia, quibus non rerum cupiditatem facit, sed sui.* Doctrine should be such as should make men in love with the lesson, and not with the teacher; being directed to the auditor's benefit, and not to the author's commendation. And therefore those are of the right kind which may be concluded as Demosthenes concludes his counsel, *Quae si feceritis, non oratorem duntaxat in praesentia laudabitis, sed vosmetipsos etiam non ita multo post statu rerum vestrarum meliore.*

3. Neither needed men of so excellent parts to have despaired of a fortune, which the poet Virgil promised himself, and indeed obtained, who got as much glory of eloquence, wit, and learning in the expressing of the observations of husbandry, as of the heroical acts of Aeneas:

> Nec sum animi dubius, verbis ea vincere magnum
> Quam sit, et angustis his addere rebus honorem.

And surely, if the purpose be in good earnest, not to write at leisure that which men may read at leisure, but really to instruct and suborn action and active life, these Georgics of the mind, concerning the husbandry and tillage thereof, are no less worthy than the heroical descriptions of virtue, duty and felicity. Wherefore the main and primitive division of moral knowledge seemeth to be into the exemplar or platform of good, and the regiment or culture of the mind: the one describing the nature of good, the other prescribing rules how to subdue, apply, and accommodate the will of man thereunto.

4. The doctrine touching the platform or nature of good considereth it either simple or compared; either the kinds of good, or the degrees of good; in the latter whereof those infinite disputations which were touching the supreme degree thereof, which they term felicity, beatitude, or the highest good, the doctrines concerning which were as the heathen divinity,[n] are by the Christian faith discharged. And as Aristotle saith, 'That young men may be happy, but not otherwise but by hope'; so we must all acknowledge our minor-

Nocet illis: Seneca, *Epistles to Lucilius* lii. 14. 'Eloquence does harm to those who love it more than matter.'

Quae si: A Latin translation of Demosthenes, *Second Olynthiac* 31. 'If you do what I advise, you will not only praise the orator at the time, but soon afterwards yourselves also, by reason of the improved state of affairs.'

Nec sum: Virgil, *Georgics* iii. 289:

> How hard the task, alas, full well I know,
> With charm of words to grace a theme so low.

suborn: equip.

platform: model.

Aristotle: In *Ethics* i. 10, and *Rhetoric* II. xii. 8.

ity, and embrace the felicity which is by hope of the future world.

5. Freed therefore and delivered from this doctrine of the philosopher's heaven, whereby they feigned an higher elevation of man's nature than was (for we see in what height of style Seneca writeth, *Vere magnum, habere fragilitatem hominis, securitatem Dei*), we may with more sobriety and truth receive the rest of their inquiries and labours. Wherein for the nature of good positive or simple, they have set it down excellently in describing the forms of virtue and duty, with their situations and postures; in distributing them into their kinds, parts, provinces, actions, and administrations, and the like: nay further, they have commended them to man's nature and spirit with great quickness of argument and beauty of persuasions; yea, and fortified and entrenched them (as much as discourse can do) against corrupt and popular opinions. Again, for the degrees and comparative nature of good, they have also excellently handled it in their triplicity of good, in the comparisons between a contemplative and an active life, in the distinction between virtue with reluctation and virtue secured, in their encounters between honesty and profit, in their balancing of virtue with virtue, and the like; so as this part deserveth to be reported for excellently laboured.[n]

6. Notwithstanding, if before they had comen to the popular and received notions of virtue and vice, pleasure and pain, and the rest, they had stayed a little longer upon the inquiry concerning the roots of good and evil, and the strings of those roots, they had given, in my opinion, a great light to that which followed; and specially if they had consulted with nature, they had made their doctrines less prolix and more profound: which being by them in part omitted and in part handled with much confusion, we will endeavour to resume and open in a more clear manner.

7. There is formed in every thing a double nature of good: the one, as every thing is a total or substantive in itself; the other, as it is a part or member of a greater body: whereof the latter is in degree the greater and the worthier, because it tendeth to the conservation of a more general form. Therefore we see the iron in particular sympathy moveth to the loadstone; but yet if it exceed a certain quantity, it forsaketh the affection to the loadstone, and like a good patriot moveth to the earth, which is the region and country of massy

Vere magnum: Seneca, *Epistles to Lucilius* liii. 12. 'Look at that for an achievement, to have all the frailty of a human being and all the freedom from care of a god.' Cf. essay, 'Of Adversity'.
postures: relationships.*
reluctation: unwillingness.

bodies: so may we go forward, and see that water and massy bodies move to the centre of the earth; but rather than to suffer a divulsion in the continuance of nature, they will move upwards from the centre of the earth, forsaking their duty to the earth in regard of their duty to the world. This double nature of good, and the comparative thereof, is much more engraven upon man, if he degenerate not: unto whom the conservation of duty to the public ought to be much more precious than the conservation of life and being: according to that memorable speech of Pompeius Magnus, when being in commission of purveyance for a famine at Rome, and being dissuaded with great vehemency and instance by his friends about him, that he should not hazard himself to sea in an extremity of weather, he said only to them, *Necesse est ut eam, non ut vivam.* But it may be truly affirmed that there was never any philosophy, religion, or other discipline, which did so plainly and highly exalt the good which is communicative, and depress the good which is private and particular, as the Holy Faith; well declaring that it was the same God that gave the Christian law to men, who gave those laws of nature to inanimate creatures that we spake of before; for we read that the elected saints of God have wished themselves anathematized and razed out of the book of life,[n] in an ecstasy of charity and infinite feeling of communion.

8. This being set down and strongly planted, doth judge and determine most of the controversies wherein moral philosophy is conversant. For first, it decideth the question touching the preferment of the contemplative or active life,[n] and decideth it against Aristotle. For all the reasons which he bringeth for the contemplative are private, and respecting the pleasure and dignity of a man's self (in which respects no question the contemplative life hath the pre-eminence), not much unlike to that comparison, which Pythagoras made for the gracing and magnifying of philosophy and contemplation: who being asked what he was, answered, 'That if Hiero were ever at the Olympian games, he knew the manner, that some came to try their fortune for the prizes, and some came as merchants to utter their commodities, and some came to make good cheer and meet their friends, and some came to look on; and that he was one of them that came to look on.[n] But men must know, that in this theatre of man's life it is reserved only for God and angels to be lookers on.[n]

divulsion: a tearing asunder.
instance: urgent entreaty.
Necesse est: In Plutarch's 'Life' of Pompey, 50. 'It is necessary that I go, not that I live.'

Neither could the like question ever have been received in the church, notwithstanding their *Pretiosa in oculis Domini mors sanctorum ejus*, by which place they would exalt their civil death and regular professions, but upon this defence, that the monastical life is not simple contemplative, but performeth the duty either of incessant prayers and supplications, which hath been truly esteemed as an office in the church, or else of writing or taking instructions for writing concerning the law of God, as Moses did when he abode so long in the mount. And so we see Henoch the seventh from Adam, who was the first contemplative and walked with God, yet did also endow the church with prophecy, which Saint Jude citeth. But for contemplation which should be finished in itself, without casting beams upon society, assuredly divinity knoweth it not.

9. It decideth also the controversies between Zeno and Socrates,[n] and their schools and successions, on the one side, who placed felicity in virtue simply or attended, the actions and exercises whereof do chiefly embrace and concern society; and on the other side, the Cyrenaics and Epicureans, who placed it in pleasure, and made virtue (as it is used in some comedies of errors, wherein the mistress and the maid change habits) to be but as a servant, without which pleasure cannot be served and attended; and the reformed school of the Epicureans, which placed it in serenity of mind and freedom from perturbation; as if they would have deposed Jupiter again, and restored Saturn[n] and the first age, when there was no summer nor winter, spring nor autumn, but all after one air and season; and Herillus,[n] which placed felicity in extinguishment of the disputes of the mind, making no fixed nature of good and evil, esteeming things according to the clearness of the desires, or the reluctation; which opinion was revived in the heresy of the Anabaptists,[n] measuring things according to the motions of the spirit, and the constancy or wavering of belief: all which are manifest to tend to private repose and contentment, and not to point of society.

10. It censureth also the philosophy of Epictetus,[n] which presupposeth that felicity must be placed in those things which are in our power, lest we be liable to fortune and disturbance: as if it were not a thing much more happy to fail in good and virtuous ends for the public, than to obtain all that we can wish to ourselves in our proper fortune; as Consalvo[n] said to his soldiers, showing them Naples, and protesting he had rather die one foot forwards, than to

Pretiosa in Oculis: Psalms 116:15. 'Precious in the sight of the Lord is the death of his saints.'
Henoch: Enoch, in Genesis 5:24, and Jude 14.

have his life secured for long by one foot of retreat. Whereunto the wisdom of that heavenly leader hath signed, who hath affirmed that 'a good conscience is a continual feast'; showing plainly that the conscience of good intentions, howsoever succeeding, is a more continual joy to nature, than all the provision which can be made for security and repose.

11. It censureth likewise that abuse of philosophy, which grew general about the time of Epictetus, in converting it into an occupation or profession;[n] as if the purpose had been, not to resist and extinguish perturbations, but to fly and avoid the causes of them, and to shape a particular kind and course of life to that end; introducing such an health of mind, as was that health of body of which Aristotle speaketh of Herodicus, who did nothing all his life long but intend his health: whereas if men refer themselves to duties of society, as that health of body is best, which is ablest to endure all alterations and extremities; so likewise that health of mind is most proper, which can go through the greatest temptations and perturbations. So as Diogenes'[n] opinion is to be accepted, who commended not them which abstained, but them which sustained, and could refrain their mind *in praecipitio*, and could give unto the mind (as is used in horsemanship) the shortest stop or turn.

12. Lastly, it censureth the tenderness and want of application in some of the most ancient and reverend philosophers and philosophical men, that did retire too easily from civil business, for avoiding of indignities and perturbations: whereas the resolution of men truly moral ought to be such as the same Consalvo said the honour of a soldier should be, *e tela crassiore*, and not so fine as that every thing should catch in it and endanger it.

[OF PRIVATE AND PUBLIC GOOD]

XXI. 1. To resume private or particular good, it falleth into the division of good active and passive: for this difference of good (not unlike to that which amongst the Romans was expressed in the familiar or household terms of *promus* and *condus*)[n] is formed also in all things, and is best disclosed in the two several appetites in

signed: attested.
a good conscience: Proverbs 15:15.
Herodicus: Aristotle, *Rhetoric* I. v. 10.
intend: occupied himself with.
tenderness: sensitiveness.
application: assiduous attention.*
e tela crassiore: of a stouter web.

creatures; the one to preserve or continue themselves, and the other to dilate or multiply themselves; whereof the latter seemeth to be the worthier: for in nature the heavens, which are the more worthy, are the agent; and the earth, which is the less worthy, is the patient. In the pleasures of living creatures, that of generation is greater than that of food. In divine doctrine, *beatius est dare quam accipere.* And in life, there is no man's spirit so soft, but esteemeth the effecting of somewhat that he hath fixed in his desire, more than sensuality; which priority of the active good, is much upheld by the consideration of our estate to be mortal and exposed to fortune. For if we mought have a perpetuity and certainty in our pleasures, the state of them would advance their price. But when we see it is but *magni aestimamus mori tardius,* and *ne glorieris de crastino, nescis partum diei,*[n] it maketh us to desire to have somewhat secured and exempted from time, which are only our deeds and works: as it is said, *Opera eorum sequuntur eos.* The preeminence likewise of this active good is upheld by the affection which is natural in man towards variety and proceeding; which in the pleasures of the sense, which is the principal part of passive good, can have no great latitude. *Cogita quamdiu eadem feceris; cibus, somnus, ludus; per hunc circulum curritur; mori velle non tantum fortis, aut miser, aut prudens, sed etiam fastidiosus potest.* But in enterprises, pursuits, and purposes of life, there is much variety; whereof men are sensible with pleasure in their inceptions, progressions, recoils, reintegrations, approaches and attainings to their ends. So as it was well said, *Vita sine proposito languida et vaga est.* Neither hath this active good any identity with the good of society, though in some case it hath an incidence into it. For although it do many times bring forth acts of beneficence, yet it is with a respect private to a man's own power, glory, amplification, continuance; as appeareth plainly, when it findeth a contrary subject. For that gigantine[n] state of mind which possesseth the troublers of the world, such as was Lucius Sulla and infinite other in smaller model, who would have all men happy or

beatius est: Acts 20:35. 'It is more blessed to give than to receive.'
state: their stable condition.*
Opera eorum: Revelation 14:13. 'Their works follow them.'
proceeding: progress.
Cogita quamdiu: Seneca, *Epistles to Lucilius* lxxvii. 'Think how often you do the same thing over and over. Food, sleep, play come round in a perpetual circle; a man might wish to die, not only from fortitude or misery or wisdom, but merely from weariness of life.'
Vita sine: Seneca, *Epistles to Lucilius* xcv. 46. 'Life without a purpose is unsettled and languid.'
incidence into: coincidence with.

unhappy as they were their friends or enemies, and would give form to the world, according to their own humours (which is the true theomachy), pretendeth and aspireth to active good, though it recedeth furthest from good of society, which we have determined to be the greater.

2. To resume passive good, it receiveth a subdivision of conservative and perfective. For let us take a brief review of that which we have said: we have spoken first of the good of society, the intention whereof embraceth the form of human nature, whereof we are members and portions, and not our own proper and individual form: we have spoken of active good, and supposed it as a part of private and particular good. And rightly, for there is impressed upon all things a triple desire or appetite proceeding from love to themselves; one of preserving and continuing their form; another of advancing and perfecting their form; and a third of multiplying and extending their form upon other things: whereof the multiplying, or signature of it upon other things, is that which we handled by the name of active good. So as there remaineth the conserving of it, and perfecting or raising of it; which latter is the highest degree of passive good. For to preserve in state is the less, to preserve with advancement is the greater. So in man,

Igneus est ollis vigor, et caelestis origo.[n]

His approach or assumption to divine or angelical nature is the perfection of his form; the error or false imitation of which good is that which is the tempest of human life; while man, upon the instinct of an advancement formal and essential, is carried to seek an advancement local. For as those which are sick, and find no remedy, do tumble up and down and change place, as if by a remove local they could obtain a remove internal; so is it with men in ambition, when failing of the mean to exalt their nature, they are in a perpetual estuation to exalt their place. So then passive good is, as was said, either conservative or perfective.

3. To resume the good of conservation or comfort, which consisteth in the fruition of that which is agreeable to our natures; it seemeth to be the most pure and natural of pleasures, but yet the softest and the lowest. And this also receiveth a difference, which

theomachy: striving against God.
perfective: capable of being perfected or improved.
signature: the action of impressing.†
in state: in a sound condition.
estuation: agitation of mind.
softest: most unmanly.

hath neither been well judged of, nor well inquired: for the good of fruition or contentment is placed either in the sincereness of the fruition, or in the quickness and vigour of it; the one superinduced by equality, the other by vicissitude; the one having less mixture of evil, the other more impression of good. Whether of these is the greater good is a question controverted; but whether man's nature may not be capable of both is a question not inquired.

4. The former question being debated between Socrates and a sophist, Socrates placing felicity in an equal and constant peace of mind, and the sophist in much desiring and much enjoying, they fell from argument to ill words: the sophist saying that Socrates' felicity was the felicity of a block or stone; and Socrates saying that the sophist's felicity was the felicity of one that had the itch, who did nothing but itch and scratch. And both these opinions do not want their supports. For the opinion of Socrates is much upheld by the general consent even of the Epicures themselves, that virtue beareth a great part in felicity; and if so, certain it is, that virtue hath more use in clearing perturbations than in compassing desires. The sophist's opinion is much favoured by the assertion we last spake of, that good of advancement is greater than good of simple preservation; because every obtaining a desire hath a show of advancement, as motion though in a circle hath a show of progression.

5. But the second question, decided the true way, maketh the former superfluous. For can it be doubted, but that there are some who take more pleasure in enjoying pleasures than some other, and yet, nevertheless, are less troubled with the loss or leaving of them? So as this same, *Non uti ut non appetas, non appetere ut non metuas, sunt animi pusilli et diffidentis.*[n] And it seemeth to me, that most of the doctrines of the philosophers are more fearful and cautionary than the nature of things requireth. So have they increased the fear of death in offering to cure it. For when they would have a man's whole life to be but a discipline or preparation to die,[n] they must needs make men think that it is a terrible enemy, against whom there is no end of preparing. Better saith the poet:

> Qui finem vitae extremum inter munera ponat
> Naturae.

So have they sought to make men's minds too uniform and harmonical, by not breaking them sufficiently to contrary motions: the

Socrates: In Plato, *Gorgias* 462, 494.
Qui finem: Juvenal, *Satires* x. 357. 'He who would reckon the close of life among nature's gifts.'

reason whereof I suppose to be, because they themselves were men dedicated to a private, free and unapplied course of life. For as we see, upon the lute or like instrument, a ground, though it be sweet and have show of many changes, yet breaketh not the hand to such strange and hard stops and passages, as a set song or voluntary; much after the same manner was the diversity between a philosophical and a civil life. And therefore men are to imitate the wisdom of jewellers; who, if there be a grain, or a cloud, or an ice which may be ground forth without taking too much of the stone, they help it; but if it should lessen and abate the stone too much, they will not meddle with it: so ought men so to procure serenity as they destroy not magnanimity.

6. Having therefore deduced the good of man which is private and particular, as far as seemeth fit, we will now return to that good of man which respecteth and beholdeth society, which we may term duty; because the term of duty is more proper to a mind well framed and disposed towards others, as the term of virtue is applied to a mind well formed and composed in itself: though neither can a man understand virtue without some relation to society, nor duty without an inward disposition. This part may seem at first to pertain to science civil and politic: but not if it be well observed. For it concerneth the regiment and government of every man over himself, and not over others. And as in architecture the direction of framing the posts, beams, and other parts of building, is not the same with the manner of joining them and erecting the building; and in mechanicals, the direction how to frame an instrument or engine, is not the same with the manner of setting it on work and employing it; and yet nevertheless in expressing of the one you incidently express the aptness towards the other; so the doctrine of conjugation of men in society differeth from that of their conformity thereunto.

7. This part of duty is subdivided into two parts: the common duty of every man, as a man or member of a state; the other, the respective or special duty of every man, in his profession, vocation, and place. The first of these is extant and well laboured, as hath been said. The second likewise I may report rather dispersed than deficient; which manner of dispersed writing in this kind of argument I acknowledge to be best. For who can take upon him to write

ground: the plain-song or melody.
ice: a flaw, resembling a flake of ice, in a jewel.*
magnanimity: high-mindedness, nobility of feeling.
mechanicals: the science which relates to the construction of machines.*
conjugation: combination.*

of the proper duty, virtue, challenge, and right of every several vocation, profession, and place? For although sometimes a looker on may see more than a gamester, and there be a proverb more arrogant than sound, 'That the vale best discovereth the hill'; yet there is small doubt but that men can write best and most really and materially in their own professions; and that the writing of speculative men of active matter for the most part doth seem to men of experience, as Phormio's argument[n] of the wars seemed to Hannibal, to be but dreams and dotage. Only there is one vice which accompanieth them that write in their own professions, that they magnify them in excess. But generally it were to be wished (as that which would make learning indeed solid and fruitful) that active men would or could become writers.

8. In which kind I cannot but mention, *honoris causa*, your Majesty's excellent book[n] touching the duty of a king: a work richly compounded of divinity, morality, and policy, with great aspersion of all other arts; and being in mine opinion one of the most sound and healthful writings that I have read; not distempered in the heat of invention, nor in the coldness of negligence; not sick of dizziness, as those are who leese themselves in their order, nor of convulsions, as those which cramp in matters impertinent; not savouring of perfumes and paintings, as those do who seek to please the reader more than nature beareth; and chiefly well disposed in the spirits thereof, being agreeable to truth and apt for action; and far removed from that natural infirmity, whereunto I noted those that write in their own professions to be subject, which is, that they exalt it above measure. For your Majesty hath truly described, not a king of Assyria or Persia in their extern glory, but a Moses or a David, pastors of their people. Neither can I ever leese out of my remembrance what I heard your Majesty in the same sacred spirit of government deliver in a great cause of judicature, which was, 'That kings ruled by their laws, as God did by the laws of nature; and ought as rarely to put in use their supreme prerogative, as God doth his power of working miracles.' And yet notwithstanding, in your book of a free monarchy,[n] you do well give men to understand, than you know the plenitude of the power and right of a king, as well as the circle of his office and duty. Thus have I presumed to

challenge: claim.
gamester: one who plays a game.
materially: to the point.*
honoris causa: in the cause of honour.
aspersion: intermixture. *leese*: lose
cramp in: crush in.* *extern*: external.

allege this excellent writing of your Majesty, as a prime or eminent example of tractates concerning special and respective duties: wherein I should have said as much, if it had been written a thousand years since. Neither am I moved with certain courtly decencies, which esteem it flattery to praise in presence. No, it is flattery to praise in absence; that is, when either the virtue is absent, or the occasion is absent; and so the praise is not natural, but forced, either in truth or in time. But let Cicero be read in his oration *pro Marcello*,[n] which is nothing but an excellent table of Caesar's virtue, and made to his face; besides the example of many other excellent persons, wiser a great deal than such observers; and we will never doubt, upon a full occasion, to give just praises to present or absent.

9. *De cautelis et malis artibus.* But to return: there belongeth further to the handling of this part, touching the duties of professions and vocations, a relative or opposite, touching the frauds, cautels, impostures, and vices of every profession, which hath been likewise handled: but how? rather in a satire and cynically, than seriously and wisely: for men have rather sought by wit to deride and traduce much of that which is good in professions, than with judgement to discover and sever that which is corrupt. For, as Salomon saith, he that cometh to seek after knowledge with a mind to scorn and censure, shall be sure to find matter for his humour, but no matter for his instruction: *Quaerenti derisori scientiam ipsa se abscondit sed studioso fit obviam.* But the managing of this argument with integrity and truth, which I note as deficient, seemeth to me to be one of the best fortifications for honesty and virtue that can be planted. For, as the fable goeth of the basilisk,[n] that if he see you first, you die for it; but if you see him first, he dieth: so is it with deceits and evil arts; which, if they be first espied they leese their life; but if they prevent, they endanger. So that we are much beholden to Machiavel[n] and others, that write what men do, and not what they ought to do. For it is not possible to join serpentine wisdom with the columbine innocency, except men know exactly all the conditions of the serpent; his baseness and going upon his belly, his volubility and lubricity, his envy and sting, and the

table: picture. *De cautelis*: Of fraudulent and evil arts.
cautels: deceits.
Quaerenti: Proverbs 14:6. 'A scorner seeks wisdom and does not find it; to one of understanding, knowledge comes easily.'
prevent: are anticipated.
volubility: capacity for moving in a gliding manner.*
lubricity: slipperiness. *envy*: hostility.

rest; that is, all forms and natures of evil. For without this, virtue lieth open and unfenced. Nay, an honest man can do no good upon those that are wicked, to reclaim them, without the help of the knowledge of evil. For men of corrupted minds presuppose that honesty groweth out of simplicity of manners, and believing of preachers, schoolmasters, and men's exterior language. So as, except you can make them perceive that you know the utmost reaches of their own corrupt opinions, they despise all morality. *Non recipit stultus verba prudentiae, nisi ea dixeris quae versantur in corde ejus.*

10. Unto this part, touching respective duty, doth also appertain the duties between huband and wife, parent and child, master and servant. So likewise the laws of friendship and gratitude, the civil bond of companies, colleges, and politic bodies, of neighbourhood, and all other proportionate duties; not as they are parts of government and society, but as to the framing of the mind of particular persons.

11. The knowledge concerning good respecting society doth handle it also, not simply alone, but comparatively; whereunto belongeth the weighing of duties between person and person, case and case, particular and public. As we see in the proceeding of Lucius Brutus against his own sons, which was so much extolled; yet what was said?

Infelix, utcunque ferent ea facta minores.[n]

So the case was doubtful, and had opinion on both sides. Again, we see when M. Brutus and Cassius invited to a supper certain whose opinions they meant to feel, whether they were fit to be made their associates, and cast forth the question touching the killing of a tyrant being an usurper, they were divided in opinion; some holding that servitude was the extreme of evils, and others that tyranny was better than a civil war: and a number of the like cases there are of comparative duty. Amongst which that of all others is the most frequent, where the question is of a great deal of good to ensue of a small injustice. Which Jason of Thessalia determined against the truth: *Aliqua sunt injuste facienda, ut multa juste fieri possint.* But the reply is good, *Auctorem praesentis justitiae habes, sponsorem futurae non habes.* Men must pursue things which are just in present,

Non recipit: Proverbs 18:2. 'A fool will not listen to the words of the wise, unless you first tell him what is in his own heart.'
Brutus and Cassius: In Plutarch's 'Life' of Brutus, xii. 2.
Aliqua sunt: Plutarch, *Precepts of Statecraft* 818; *Advice about keeping well* 135. 'Some things must be done unjustly, that many may be done justly.' 'Present justice is in your power, for that which is to come you have no security.'

and leave the future to the divine Providence. So then we pass on from this general part touching the exemplar and description of good.

[OF MORAL CULTURE]

XXII. 1. *De cultura animi.* Now therefore that we have spoken of this fruit of life, it remaineth to speak of the husbandry that belongeth thereunto; without which part the former seemeth to be no better than a fair image, or statua, which is beautiful to contemplate, but is without life and motion; whereunto Aristotle himself subscribeth in these words: *Necesse est scilicet de virtute dicere, et quid sit, et ex quibus gignatur. Inutile enim fere fuerit virtutem quidem nosse, acquirendae autem ejus modos et vias ignorare. Non enim de virtute tantum, qua specie sit, quaerendum est, sed et quomodo sui copiam faciat: utrumque enim volumus, et rem ipsam nosse, et ejus compotes fieri: hoc autem ex voto non succedet, nisi sciamus et ex quibus et quomodo.*[n] In such full words and with such iteration doth he inculcate this part. So saith Cicero in great commendation of Cato the second, that he had applied himself to philosophy, *Non ita disputandi causa, sed ita vivendi.* And although the neglect of our times, wherein few men do hold any consultations touching the reformation of their life (as Seneca excellently saith, *De partibus vitae quisque deliberat, de summa nemo*), may make this part seem superfluous; yet I must conclude with that aphorism of Hippocrates, *Qui gravi morbo correpti dolores non sentiunt, iis mens aegrotat.* They need medicine, not only to assuage the disease, but to awake the sense. And if it be said, that the cure of men's minds belongeth to sacred divinity, it is most true: but yet moral philosophy may be preferred unto her as a wise servant and humble handmaid. For as the Psalm saith, 'That the eyes of the handmaid look perpetually towards the mistress,' and yet no doubt many things are left to the discretion of the handmaid, to discern of the mistress' will; so ought moral philosophy to give a constant attention to the doctrines of divinity,

De cultura animi: The culture of the mind.
Non ita: Cicero, *Pro Murena* xxx. 62, 'Not for the sake of arguing like a philosopher, but that he might live like one.'
De partibus: Seneca, *Epistles to Lucilius* lxxi. 2. 'Everyone thinks about the parts of his life, no man thinks about the whole.'
Qui gravi: Hippocrates, *Aphorisms* ii. 6. 'Those who are sick without feeling pain, are sick in their minds.'
preferred unto: recommended to.
That the eyes: Psalms 123:2.

and yet so as it may yield of herself (within due limits) many sound and profitable directions.

2. This part therefore, because of the excellency thereof, I cannot the find exceeding strange that it is not reduced to written inquiry: but rather, because it consisteth of much matter, wherein both speech and action is often conversant; and such wherein the common talk of men (which is rare, but yet cometh sometimes to pass) is wiser than their books. It is reasonable therefore that we propound it in the more particularity, both for the worthiness, and because we may acquit ourselves for reporting it deficient; which seemeth almost incredible, and is otherwise conceived and presupposed by those themselves that have written. We will therefore enumerate some heads or points[n] thereof, that it may appear the better what it is, and whether it be extant.

3. First therefore in this, as in all things which are practical, we ought to cast up our account, what is in our power, and what not; for the one may be dealt with by way of alteration, but the other by way of application only. The husbandman cannot command, neither the nature of the earth, nor the seasons of the weather; no more can the physician the constitution of the patient, nor the variety of accidents. So in the culture and cure of the mind of man, two things are without our command; points of nature, and points of fortune. For to the basis of the one, and the conditions of the other, our work is limited and tied. In these things therefore it is left unto us to proceed by application:

Vincenda est omnis fortuna ferendo:

and so likewise,

Vincenda est omnis Natura ferendo.

But when that we speak of suffering, we do not speak of a dull and neglected suffering, but of a wise and industrious suffering, which draweth and contriveth use and advantage out of that which seemeth adverse and contrary; which is that properly which we call accommodating or applying. Now the wisdom of application resteth principally in the exact and distinct knowledge of the precedent state or disposition, unto which we do apply: for we cannot fit a garment, except we first take measure of the body.

4. So then the first article of this knowledge is, to set down sound

accidents: symptoms.
Vincenda: Adapted from Virgil, *Aeneid* v. 710. 'All fortune may be overcome by suffering', and 'all nature may be overcome by suffering'.

and true distributions and descriptions of the several characters and tempers of men's natures and dispositions; specially having regard to those differences which are most radical in being the fountains and causes of the rest, or most frequent in concurrence or commixture; wherein it is not the handling of a few of them in passage, the better to describe the mediocrities of virtues, that can satisfy this intention. For if it deserve to be considered, that there are minds which are proportioned to great matters, and others to small (which Aristotle handleth or ought to have handled by the name of magnanimity), doth it not deserve as well to be considered, that there are minds proportioned to intend many matters, and others to few? So that some can divide themselves: others can perchance do exactly well, but it must be but in few things at once: and so there cometh to be a narrowness of mind, as well as a pusillanimity. And again, that some minds are proportioned to that which may be dispatched at once, or within a short return of time; others to that which begins afar off, and is to be won with length of pursuit:

Jam tum tenditque fovetque.

So that there may be fitly said to be a longanimity, which is commonly also ascribed to God as a magnanimity. So further deserved it to be considered by Aristotle, 'That there is a disposition in conversation (supposing it in things which do in no sort touch or concern a man's self) to soothe and please; and a disposition contrary to contradict and cross': and deserveth it not much better to be considered, 'That there is a disposition, not in conversation or talk, but in matter of more serious nature (and supposing it still in things merely indifferent), to take pleasure in the good of another: and a disposition contrariwise, to take distaste at the good of another?' which is that properly which we call good nature or ill nature, benignity or malignity: and therefore I cannot sufficiently marvel that this part of knowledge, touching the several characters of natures and dispositions, should be omitted both in morality and policy; considering it is of so great ministry and suppeditation to them both. A man shall find in the traditions of astrology[n] some pretty and apt divisions of men's natures, according to the predominances of the planets; lovers of quiet, lovers of action, lovers of victory, lovers

Aristotle: *Ethics* iv. 3. *pusillanimity*: littleness of mind.
Jam tum: Virgil, *Aeneid* i. 18. 'He begins to attend and nurse his project while it is yet in the cradle.'
longanimity: long-suffering. *Aristotle*: *Ethics* iv. 6.
ministry: service. *suppeditation*: usefulness.*

of honour, lovers of pleasure, lovers of arts, lovers of change, and so forth. A man shall find in the wisest sort of these relations which the Italians make touching conclaves,[n] the natures of the several cardinals handsomely and lively painted forth. A man shall meet with in every day's conference the denominations of sensitive,[n] dry, formal, real, humorous, certain, *huomo di prima impressione, huomo di ultima impressione*, and the like: and yet nevertheless this kind of observations wandereth in words, but is not fixed in inquiry. For the distinctions are found (many of them), but we conclude no precepts upon them: wherein our fault is the greater; because both history, poesy, and daily experience are as goodly fields where these observations grow; whereof we make a few posies to hold in our hands, but no man bringeth them to the confectionary, that receipts mought be made of them for use of life.

5. Of much like kind are those impressions of nature, which are imposed upon the mind by the sex, by the age, by the region, by health and sickness, by beauty and deformity, and the like, which are inherent and not extern; and again, those which are caused by extern fortune; as sovereignty, nobility, obscure birth, riches, want, magistracy, privateness, prosperity, adversity, constant fortune, variable fortune, rising *per saltum, per gradus*, and the like. And therefore we see that Plautus maketh it a wonder to see an old man beneficent, *benignitas hujus ut adolescentuli est.* Saint Paul concludeth that severity of discipline was to be used to the Cretans, *increpa eos dure*, upon the disposition of their country, *Cretenses semper mendaces, malae bestiae, ventres pigri.*[n] Sallust noteth that it is usual with kings to desire contradictories: *Sed plerumque regiae voluntates, ut vehementes sunt, sic mobiles, saepeque ipsae sibi adversae.*[n] Tacitus observeth how rarely raising of the fortune mendeth the disposition: *solus Vespasianus mutatus in melius.* Pindarus maketh an observation, that great and sudden fortune for the most part defeateth men *qui magnam felicitatem concoquere non possunt.*[n] So the Psalm showeth it is more easy to keep a measure in the enjoying of fortune, than in the increase of fortune: *Divitiae si affluant,*

confectionary: place where confections are made.*
region: climate.
magistracy: holding the office of magistrate.
per saltum: by leaps, by degrees.
benignitas: Plautus, *Miles Gloriosus* III. i. 40. 'His generosity is like that of a young man.'
solus Vespasianus: Tacitus, *History* i. 50. 'Vespasian alone was changed for the better.'
defeateth: ruins.
Divitiae: *Psalms* 62:10. 'Though riches increase, set not your heart upon them.'

nolite cor apponere. These observations and the like I deny not but are touched a little by Aristotle[n] as in passage in his Rhetorics, and are handled in some scattered discourses: but they were never incorporate into moral philosophy, to which they do essentially appertain; as the knowledge of the diversity of grounds and moulds doth to agriculture, and the knowledge of the diversity of complexions and constitutions doth to the physician; except we mean to follow the indiscretion of empirics, which minister the same medicines to all patients.

6. Another article of this knowledge is the inquiry touching the affections; for as in medicining of the body, it is in order first to know the divers complexions and constitutions; secondly, the diseases; and lastly, the cures: so in medicining of the mind, after knowledge of the divers characters of men's natures, it followeth in order to know the diseases and infirmities of the mind, which are no other than the perturbations and distempers of the affections. For as the ancient politiques[n] in popular estates were wont to compare the people to the sea, and the orators to the winds; because as the sea would of itself be calm and quiet, if the winds did not move and trouble it; so the people would be peaceable and tractable, if the seditious orators did not set them in working and agitation: so it may be fitly said, that the mind in the nature thereof would be temperate and stayed, if the affections, as winds, did not put it into tumult and perturbation. And here again I find strange, as before, that Aristotle should have written divers volumes of Ethics,[n] and never handled the affections, which is the principal subject thereof; and yet in his Rhetorics, where they are considered but collaterally and in a second degree (as they may be moved by speech), he findeth place for them, and handleth them well for the quantity; but where their true place is, he pretermitteth them. For it is not his disputations about pleasure and pain that can satisfy this inquiry, no more than he that should generally handle the nature of light can be said to handle the nature of colours; for pleasure and pain are to the particular affections, as light is to particular colours. Better travails, I suppose, had the Stoics[n] taken in this argument, as far as I can gather by that which we have at second hand. But yet it is like it was after their manner, rather in subtilty of definitions (which in a subject of this nature are but curiosities), than in active and ample

affections: passions.
it is in order first: i.e. the order is, first . . .
pretermitteth: does not mention.
pleasure and pain: Aristotle, *Ethics* ii. 3 f.

descriptions and observations. So likewise I find some particular writings of an elegant nature, touching some of the affections; as of anger, of comfort upon adverse accidents, of tenderness of countenance, and other.[n] But the poets and writers of histories are the best doctors of this knowledge; where we may find painted forth with great life, how affections are kindled and incited; and how pacified and refrained; and how again contained from act and further degree; how they disclose themselves; how they work; how they vary; how they gather and fortify; how they are enwrapped one within another; and how they do fight and encounter one with another; and other the like particularities. Amongst the which this last is of special use in moral and civil matters; how, I say, to set affection against affection, and to master one by another; even as we use to hunt beast with beast, and fly bird with bird, which otherwise percase we could not so easily recover: upon which foundation is erected that excellent use of *praemium* and *poena*,[n] whereby civil states consist: employing the predominant affections of fear and hope, for the suppressing and bridling the rest. For as in the government of states it is sometimes necessary to bridle one faction with another, so it is in the government within.

7. Now come we to those points which are within our own command, and have force and operation upon the mind, to affect the will and appetite, and to alter manners: wherein they ought to have handled custom, exercise, habit, education, example, imitation, emulation, company, friends, praise, reproof, exhortation, fame, laws, books, studies:[n] these as they have determinate use in moralities, from these the mind suffereth; and of these are such receipts and regiments compounded and described, as may serve to recover or preserve the health and good estate of the mind, as far as pertaineth to human medicine: of which number we will insist upon some one or two, as an example of the rest, because it were too long to prosecute all; and therefore we do resume custom and habit to speak of.

8. The opinion of Aristotle seemeth to me a negligent opinion, that of those things which consist by nature, nothing can be changed by custom; using for example, that if a stone be thrown ten thousand times up, it will not learn to ascend; and that by often seeing or hearing, we do not learn to see or hear the better. For though this principle be true in things wherein nature is peremptory (the reason

contained: restrained.*
fortify: gain strength.*
fly: hunt.
percase: perhaps.
consist: stand firm.
moralities: points of ethics.*
Aristotle: In *Ethics* ii. 1.
consist: exist.
peremptory: positively fixed.

whereof we cannot now stand to discuss), yet it is otherwise in things wherein nature admitteth a latitude. For he mought see that a strait glove will come more easily on with use; and that a wand will by use bend otherwise than it grew; and that by use of the voice we speak louder and stronger; and that by use of enduring heat or cold, we endure it the better, and the like: which latter sort have a nearer resemblance unto that subject of manners he handleth, than those instances which he allegeth. But allowing his conclusion, that virtues and vices consist in habit, he ought so much the more to have taught the manner of superinducing that habit: for there be many precepts of the wise ordering the exercises of the mind, as there is of ordering the exercises of the body; whereof we will recite a few.

9. The first shall be, that we beware we take not at the first, either too high a strain, or too weak: for if too high, in a diffident nature you discourage, in a confident nature you breed an opinion of facility, and so a sloth; and in all natures you breed a further expectation than can hold out, and so an insatisfaction in the end: if too weak, of the other side, you may not look to perform and overcome any great task.

10. Another precept is, to practise all things chiefly at two several times, the one when the mind is best disposed, the other when it is worst disposed; that by the one you may gain a great step, by the other you may work out the knots and stonds of the mind, and make the middle times the more easy and pleasant.

11. Another precept is, that which Aristotle mentioneth by the way, which is to bear ever towards the contrary extreme of that whereunto we are by nature inclined; like unto the rowing against the stream, or making a wand straight by bending him contrary to his natural crookedness.

12. Another precept is, that the mind is brought to anything better, and with more sweetness and happiness, if that whereunto you pretend be not first in the intention, but *tanquam aliud agendo*, because of the natural hatred of the mind against necessity and constraint. Many other axioms there are touching the managing of exercise and custom; which being so conducted, doth prove indeed another nature; but being governed by chance, doth commonly prove

mought: might. *strait*: tight-fitting.
insatisfaction: dissatisfaction.
stonds: impediments. A word used only by Bacon.
Aristotle: In *Ethics* ii. 9. *pretend*: aim, aspire.
tanquam: while you are attending to something else.

but an ape of nature, and bringeth forth that which is lame and counterfeit.

13. So if we should handle books and studies, and what influence and operation they have upon manners, are there not divers precepts of great caution and direction appertaining thereunto? Did not one of the fathers in great indignation call poesy *vinum daemonum*,[n] because it increaseth temptations, perturbations, and vain opinions? Is not the opinion of Aristotle worthy to be regarded, wherein he saith, That young men are no fit auditors of moral philosophy, because they are not settled from the boiling heat of their affections, nor attempered with time and experience?[n] And doth it not hereof come, that those excellent books and discourses of the ancient writers (whereby they have persuaded unto virtue most effectually, by representing her in state and majesty, and popular opinions against virtue in their parasites' coats fit to be scorned and derided), are of so little effect towards honesty of life, because they are not read and revolved by men in their mature and settled years, but confined almost to boys and beginners? But is it not true also, that much less young men are fit auditors of matters of policy, till they have been throughly seasoned in religion and morality; lest their judgements be corrupted, and made apt to think that there are no true differences of things, but according to utility and fortune, as the verse describes it, *Prosperum et felix scelus virtus vocatur*; and again, *Ille crucem pretium sceleris tulit, hic diadema*: which the poets do speak satirically, and in indignation on virtue's behalf; but books of policy do speak it seriously and positively; for so it pleaseth Machiavel to say, 'That if Caesar had been overthrown, he would have been more odious than ever was Catiline'; as if there had been no difference, but in fortune, between a very fury of lust and blood, and the most excellent spirit (his ambition reserved) of the world? Again, is there not a caution likewise to be given of the doctrines of moralities themselves (some kinds of them), lest they make men too precise, arrogant, incompatible; as Cicero saith of Cato, *In Marco Catone haec bona quae videmus divina et egregia, ipsius*

attempered: well-balanced.
Prosperum: Seneca, *Hercules Furens* 251. 'Successful crime is called virtue.'
Ille crucem: Juvenal, *Satires* xiii. 105. 'For the same crime one man is hanged, another is crowned.'
Machiavel: In *Discourses on Livy* i. 10.
precise: strict in the observance of rule.
In Marco: Cicero, *Pro Murena* xxix. 61. 'The divine and excellent qualities we see in Marcus Cato are his own; but the defects we sometimes find are derived not from his nature but from his teachers.'

scitote esse propria; quae nonnunquam requirimus, ea sunt omnia non a natura, sed a magistro? Many other axioms and advices there are touching those proprieties and effects, which studies do infuse and instil into manners. And so likewise is there touching the use of all those other points, of company, fame, laws, and the rest, which we recited in the beginning in the doctrine of morality.

14. But there is a kind of culture of the mind that seemeth yet more accurate and elaborate than the rest, and is built upon this ground; that the minds of all men are at some times in a state more perfect, and at other times in a state more depraved. The purpose therefore of this practice is to fix and cherish the good hours of the mind, and to obliterate and take forth the evil. The fixing of the good hath been practised by two means, vows or constant resolutions, and observances or exercises; which are not to be regarded so much in themselves, as because they keep the mind in continual obedience. The obliteration of the evil hath been practised by two means, some kind of redemption or expiation of that which is past, and an inception or account *de novo* for the time to come. But this part seemeth sacred and religious, and justly; for all good moral philosophy (as was said) is but an handmaid to religion.

15. Wherefore we will conclude with that last point, which is of all other means the most compendious and summary, and again, the most noble and effectual to the reducing of the mind unto virtue and good estate; which is, the electing and propounding unto a man's self good and virtuous ends of his life, such as may be in a reasonable sort within his compass to attain. For if these two things be supposed, that a man set before him honest and good ends, and again, that he be resolute, constant, and true unto them; it will follow that he shall mould himself into all virtue at once. And this is indeed like the work of nature; whereas the other course is like the work of the hand. For as when a carver makes an image, he shapes only that part whereupon he worketh; as if he be upon the face, that part which shall be the body is but a rude stone still, till such times as he comes to it. But contrariwise when nature makes a flower or living creature, she formeth rudiments of all the parts at one time. So in obtaining virtue by habit, while a man practiseth temperance, he doth not profit much to fortitude, nor the like: but when he dedicateth and applieth himself to good ends, look, what virtue soever the pursuit and passage towards those ends doth commend unto him, he is invested of a precedent disposition to conform him-

accurate: worked out with care.* *de novo*: anew.

self thereunto. Which state of mind Aristotle doth excellently express himself, that it ought not to be called virtuous, but divine: his words are these: *Immanitati autem consentaneum est opponere eam, quae supra humanitatem est, heroicam sive divinam virtutem*; and a little after, *Nam ut ferae neque vitium neque virtus est, sic neque Dei: sed hic quidem status altius quiddam virtute est, ille aliud quiddam a vitio.* And therefore we may see what celsitude of honour Plinius Secundus attributeth to Trajan in his funeral oration; where he said, 'That men needed to make no other prayers to the gods, but that they would continue as good lords to them as Trajan had been'; as if he had not been only an imitation of divine nature, but a pattern of it. But these be heathen and profane passages, having but a shadow of that divine state of mind, which religion and the holy faith doth conduct men unto, by imprinting upon their souls charity, which is excellently called the bond of perfection, because it comprehendeth and fasteneth all virtues together. And as it is elegantly said by Menander of vain love, which is but a false imitation of divine love, *Amor melior Sophista laevo ad humanam vitam*, that love teacheth a man to carry himself better than the sophist or preceptor, which he calleth left-handed, because, with all his rules and preceptions, he cannot form a man so dexteriously, nor with that facility to prize himself and govern himself, as love can do: so certainly, if a man's mind be truly inflamed with charity, it doth work him suddenly into greater perfection than all the doctrine of morality can do, which is but a sophist in comparison of the other. Nay further, as Xenophon observed truly, that all other affections, though they raise the mind, yet they do it by distorting and uncomeliness of ecstasies or excesses; but only love doth exalt the mind, and nevertheless at the same instant doth settle and compose it: so in all other excellencies, though they advance nature, yet they are subject to excess. Only charity

Immanitati: Aristotle, *Ethics* vii. 1. 'To brutal vice we may naturally oppose that divine or heroical virtue which is above humanity.'
Nam ut: ibid. 'For as beasts are incapable of virtue or vice, so likewise is the Deity, for this latter state is something higher than virtue, as the former is something other than vice.'
celsitude: eminence.
That men needed: Pliny, *Panegyric* 74. It was not a funeral oration.
charity: See Bacon's *Meditationes Sacrae*, 'Of the Miracles of our Saviour'. Charity for Bacon meant meeting people's needs.
bond of perfection: Colossians 3:14.
Amor melior: Not Menander, but a fragment of Anaxandrides, 'Love is a better teacher for human life than a left-handed Sophist.'
preceptions: precepts.†
dexteriously: dexterously.
Xenophon: In his *Symposium* i. 10.

admitteth no excess. For so we see, aspiring to be like God in power, the angels transgressed and fell; *Ascendam, et ero similis altissimo*: by aspiring to be like God in knowledge, man transgressed and fell; *Eritis sicut Dii, scientes bonum et malum*: but by aspiring to a similitude of God in goodness or love, neither man nor angel ever transgressed, or shall transgress. For unto that imitation we are called: *Diligite inimicos vestros, benefacite eis qui oderunt vos, et orate pro persequentibus et calumniantibus vos, ut sitis filii Patris vestri qui in coelis est, qui solem suum oriri facit super bonos et malos, et pluit super justos et injustos.* So in the first platform of the divine nature itself, the heathen religion speaketh thus, *Optimus Maximus*: and the sacred scriptures thus, *Misericordia ejus super omnia opera ejus.*

16. Wherefore I do conclude this part of moral knowledge, concerning the culture and regiment of the mind; wherein if any man, considering the parts thereof which I have enumerated, do judge that my labour is but to collect into an art or science that which hath been pretermitted by others, as matter of common sense and experience, he judgeth well. But as Philocrates sported with Demosthenes, 'You may not marvel (Athenians) that Demosthenes and I do differ; for he drinketh water, and I drink wine'; and like as we read of an ancient parable of the two gates of sleep,

> Sunt geminae somni portae: quarum altera fertur
> Cornea, qua veris facilis datur exitus umbris:
> Altera candenti perfecta nitens elephanto,
> Sed falsa ad coelum mittunt insomnia manes:

so if we put on sobriety and attention, we shall find it a sure maxim in knowledge, that the more pleasant liquor (*of wine*) is the more

Ascendam: Isaiah 14:14. 'I will ascend and be like the highest.'
Eritis: Genesis 3:5. 'You shall be as gods, knowing good and evil.'
Diligite: Matthew 5:44 and Luke 6:27, 28. 'Love your enemies, do good to them that hate you, and pray for them that despitefully use you and persecute you; that you may be the children of your father which is in Heaven, who makes his sun to rise on the evil and on the good, and sendeth rain on the just and on the unjust.'
platform: model. *Optimus*: Best and Greatest.
Misericordia: Psalms 145:9. 'For his mercy is over all his works.'
pretermitted: omitted.
You may not marvel: Demosthenes, *De Falsa Legatione* 46.
Sunt geminae: Virgil, *Aeneid* vi. 893–6.

> Two gates the entrance of Sleep's house adorn;
> Of ivory one, the other simple horn;
> Through horn a crowd of real visions stream,
> Through ivory portals pass delusive dreams.

vaporous, and the braver gate (*of ivory*) sendeth forth the falser dreams.

17. But we have now concluded that general part of human philosophy, which contemplateth man segregate, and as he consisteth of body and spirit. Wherein we may further note, that there seemeth to be a relation or conformity between the good of the mind and the good of the body. For as we divided the good of the body into health, beauty, strength, and pleasure; so the good of the mind, inquired in rational and moral knowledges, tendeth to this, to make the mind sound, and without perturbation; beautiful, and graced with decency; and strong and agile for all duties of life. These three, as in the body, so in the mind, seldom meet, and commonly sever. For it is easy to observe, that many have strength of wit and courage, but have neither health from perturbations, nor any beauty or decency in their doings: some again have an elegancy and fineness of carriage, which have neither soundness of honesty, nor substance of sufficiency: and some again have honest and reformed minds, that can neither become themselves nor manage business: and sometimes two of them meet, and rarely all three. As for pleasure, we have likewise determined that the mind ought not to be reduced to stupid, but to retain pleasure; confined rather in the subject of it, than in the strength and vigour of it.

[OF CIVIL KNOWLEDGE—IN CONVERSATION, NEGOTIATION OR BUSINESS (WITH RULES FOR SELF-ADVANCEMENT), AND GOVERNMENT (INCLUDING LAWS)]

XXIII. 1. Civil knowledge is conversant about a subject which of all others is most immersed in matter, and hardliest reduced to axiom. Nevertheless, as Cato the Censor said, 'That the Romans were like sheep, for that a man were better drive a flock of them, than one of them; for in a flock, if you could get but some few go right, the rest would follow': so in that respect moral philosophy is more difficile than policy. Again, moral philosophy propoundeth to itself the framing of internal goodness; but civil knowledge requireth only an external goodness; for that as to society sufficeth. And therefore it cometh oft to pass that there be evil times in good governments: for so we find in the holy story, when the kings were good,

braver: finer. *decency*: comeliness.
sufficiency: ability, capacity. *hardliest*: with most difficulty.
Cato: In Plutarch's 'Life' of Cato, 8.
difficile: hard to construct.

yet it is added, *Sed adhuc populus non direxerat cor suum ad Dominum Deum patrum suorum.* Again, states, as great engines, move slowly, and are not so soon put out of frame: for as in Egypt the seven good years sustained the seven bad, so governments for a time well grounded, do bear out errors following; but the resolution of particular persons is more suddenly subverted. These respects do somewhat qualify the extreme difficulty of civil knowledge.

2. This knowledge hath three parts, according to the three summary actions of society; which are conversation, negotiation, and government. For man seeketh in society comfort, use, and protection:[n] and they be three wisdoms of divers natures, which do often sever: wisdom of the behaviour, wisdom of business, and wisdom of state.

3. The wisdom of conversation ought not to be over much affected, but much less depised; for it hath not only an honour in itself, but an influence also into business and government. The poet saith, *Nec vultu destrue verba tuo*: a man may destroy the force of his words with his countenance: so may he of his deeds, saith Cicero, recommending to his brother affability and easy access; *Nil interest habere ostium apertum, vultum clausum*; it is nothing won to admit men with an open door, and to receive them with a shut and reserved countenance. So we see Atticus, before the first interview between Caesar and Cicero, the war depending, did seriously advise Cicero touching the composing and ordering of his countenance and gesture. And if the government of the countenance be of such effect, much more is that of the speech, and other carriage appertaining to conversation; the true model whereof seemeth to me well expressed by Livy, though not meant for this purpose: *Ne aut arrogans videar, aut obnoxius; quorum alterum est alienae libertatis obliti, alterum*

Sed adhuc: 2 Chronicles 20:33. 'Howbeit the people had not yet set their hearts aright to the Lord God of their fathers.'
out of frame: out of order.
seven good years: Genesis 41.
comfort, use, and protection: i.e. comfort against solitude, assistance in business, and protection against injuries. 'Conversation' covers social intercourse, behaviour in everyday life.
Nec vultu: Ovid, *The Art of Love* ii. 312. 'Let not your looks your word betray.'
Nil interest: From a work wrongly ascribed to Cicero's brother, Quintus Cicero, *De Petitione Consulatus* xi. 44. 'It is of no use to have an open door and a shut countenance.'
Caesar and Cicero: in *Letters to Atticus* xi. 24.
depending: being imminent.*
Ne aut: Livy, *History* xxiii. 12. 'That I may not appear to be either arrogant or servile; the arrogant man is unmindful of others' liberty, the servile of his own.'

suae: the sum of behaviour is to retain a man's own dignity, without intruding upon the liberty of others. On the other side, if behaviour and outward carriage be intended too much, first it may pass into affectation, and then *Quid deformius quam scenam in vitam transferre*, to act a man's life? But although it proceed not to that extreme, yet it consumeth time, and employeth the mind too much. And therefore as we use to advise young students from company keeping, by saying, *Amici fures temporis*: so certainly the intending of the discretion of behaviour is a great thief of meditation. Again, such as are accomplished in that form of urbanity please themselves in it, and seldom aspire to higher virtue; whereas those that have defect in it do seek comeliness by reputation; for where reputation is, almost everything becometh; but where that is not, it must be supplied by *puntos* and compliments. Again, there is no greater impediment of action than an over-curious observance of decency, and the guide of decency, which is time and season. For as Salomon saith, *Qui respicit ad ventos, non seminat; et qui respicit ad nubes, non metet*: a man must make his opportunity, as oft as find it. To conclude, behaviour seemeth to me as a garment of the mind, and to have the conditions of a garment. For it ought to be made in fashion; it ought not to be too curious; it ought to be shaped so as to set forth any good making of the mind and hide any deformity; and above all, it ought not to be too strait or restrained for exercise or motion. But this part of civil knowledge hath been elegantly handled,[n] and therefore I cannot report it for deficient.

4. *De negotiis gerendis.* The wisdom touching negotiation or business hath not been hitherto collected into writing, to the great derogation of learning, and the professors of learning. For from this root springeth chiefly that note or opinion, which by us is expressed in adage to this effect, that there is no great concurrence between learning and wisdom. For of the three wisdoms which we have set down to pertain to civil life, for wisdom of behaviour, it is by learned men

sum: ultimate end.
Quid deformius: From Bacon's Antitheta on Ceremonies in the *De Augmentis* vi. 3. 'What is more unseemly than to carry the manners of the stage into real life?'
Amici: Friends are the thieves of time.
intending of: occupying oneself with.
puntos: minute observances. *decency*: propriety.
Qui respicit: Ecclesiastes 11:4. 'He that considereth the wind shall not sow, and he that looketh to the clouds shall not reap.' Cf. the Essay 'Of Ceremonies' with this passage.
De negotiis: Of the conduct of affairs.
negotiation: transaction of business.

for the most part despised, as an inferior to virtue and an enemy to meditation; for wisdom of government, they acquit themselves well when they are called to it, but that happeneth to few; but for the wisdom of business, wherein man's life is most conversant, there be no books of it, except some few scattered advertisements, that have no proportion to the magnitude of this subject. For if books were written of this as the other, I doubt not but learned men with mean experience, would far excel men of long experience without learning, and outshoot them in their own bow.

5. Neither needeth it at all to be doubted, that this knowledge should be so variable as it falleth not under precept; for it is much less infinite than science of government, which we see is laboured and in some part reduced. Of this wisdom it seemeth some of the ancient Romans in the saddest and wisest times were professors; for Cicero reporteth, that it was then in use for senators that had name and opinion for general wise men, as Coruncanius, Curius, Laelius, and many others, to walk at certain hours in the Place, and to give audience to those that would use their advice; and that the particular citizens would resort unto them, and consult with them of the marriage of a daughter, or of the employing of a son, or of a purchase or bargain, or of an accusation, and every other occasion incident to man's life. So as there is a wisdom of counsel and advice even in private causes, arising out of an universal insight into the affairs of the world; which is used indeed upon particular cases propounded, but is gathered by general observation of cases of like nature. For so we see in the book which Q. Cicero writeth to his brother, *De petitione consulatus* (being the only book of business that I know written by the ancients), although it concerned a particular action then on foot, yet the substance thereof consisteth of many wise and politic axioms, which contain not a temporary, but a perpetual direction in the case of popular elections. But chiefly we may see in those aphorisms which have place amongst divine writings, composed by Salomon the king, of whom the scriptures testify that his heart was as the sands of the sea, encompassing the world and all worldly matters, we see, I say, not a few profound and excellent cautions, precepts, positions, extending to much variety of occasions;

advertisements: notices. *mean*: moderate.
saddest: most serious.
Cicero: In *De Oratore* III. xxxiii. 133–5.
De Petitione: A work on the art of canvassing, wrongly attributed to Cicero's brother.
sands of the sea: 1 Kings 4:29.
positions: maxims.

whereupon we will stay a while, offering to consideration some number of examples.

6. *Sed et cunctis sermonibus qui dicuntur ne accommodes aurem tuam, ne forte audias servum tuum maledicentem tibi.* Here is commended the provident stay of inquiry of that which we would be loth to find: as it was judged great wisdom in Pompeius Magnus that he burned Sertorius' papers unperused.

Vir sapiens, si cum stulto contenderit, sive irascatur, sive rideat, non inveniet requiem. Here is described the great disadvantage which a wise man hath in undertaking a lighter person than himself; which is such an engagement as, whether a man turn the matter to jest, or turn it to heat, or howsoever he change copy, he can no ways quit himself well of it.

Qui delicate a pueritia nutrit servum suum, postea sentiet eum contumacem. Here is signified, that if a man begin too high a pitch in his favours, it doth commonly end in unkindness and unthankfulness.

Vidisti virum velocem in opere suo? coram regibus stabit, nec erit inter ignobiles. Here is observed, that of all virtues for rising to honour, quickness of despatch is the best; for superiors many times love not to have those they employ too deep or too sufficient, but ready and diligent.

Vidi cunctos viventes qui ambulant sub sole, cum adolescente secundo qui consurgit pro eo. Here is expressed that which was noted by Sulla first, and after him by Tiberius; *Plures adorant solem orientem quam occidentem vel meridianum.*

Si spiritus potestatem habentis ascenderit super te, locum tuum ne dimiseris; quia curatio faciet cessare peccata maxima. Here

Sed et cunctis: Ecclesiastes 7:21. 'Do not pay attention to everything men say, or you may hear your servant disparage you.' The story of Pompey is in Plutarch's 'Life' of Pompey 20, and of Sertorius 27.

Vir sapiens: Proverbs 29:9. 'If a wise man goes to law with a fool, he will meet abuse or derision, but get no remedy.'

undertaking a lighter person: contending with a less grave person.

change copy: change style.

Qui delicate: Proverbs 29:21. 'Pamper a slave from boyhood, and in the end he will prove ungrateful.'

Vidisti virum: Proverbs 22:29. 'If you see a man that is quick in his business, he will stand before kings, not among common men.'

Vidi cunctos: Ecclesiastes 4:15. 'I have studied all life here under the sun, and I saw his place taken by yet another young man.'

Plures adorant: Tacitus, *Annals* vi. 46. 'There be more that worship the rising sun than the sun setting or at midday.'

Si spiritus: Ecclesiastes 10:14. 'If your ruler breaks out in anger against you, do not resign your post; submission will make amends for great mistakes.'

caution is given, that upon displeasure, retiring is of all courses the unfittest; for a man leaveth things at worst, and depriveth himself of means to make them better.

Erat civitas parva, et pauci in ea viri: venit contra eam rex magnus, et vallavit eam, instruxitque munitiones per gyrum, et perfecta est obsidio; inventusque est in ea vir pauper et sapiens, et liberavit eam per sapientiam suam; et nullus deinceps recordatus est hominis illius pauperis. Here the corruption of states is set forth, that esteem not virtue or merit longer than they have use of it.

Mollis responsio frangit iram. Here is noted that silence or rough answer exasperateth; but an answer present and temperate pacifieth.

Iter pigrorum quasi sepes spinarum. Here is lively represented how laborious sloth proveth in the end: for when things are deferred till the last instant, and nothing prepared beforehand, every step findeth a brier or impediment, which catcheth or stoppeth.

Melior est finis orationis quam principium. Here is taxed the vanity of formal speakers, that study more about prefaces and inducements, than upon the conclusions and issues of speech.

Qui cognoscit in judicio faciem, non bene facit; iste et pro buccella panis deseret veritatem. Here is noted, that a judge were better be a briber than a respecter of persons; for a corrupt judge offendeth not so lightly as a facile.

Vir pauper calumnians pauperes similis est imbri vehementi, in quo paratur fames. Here is expressed the extremity of necessitous extortions, figured in the ancient fable of the full and the hungry horseleech.

Fons turbatus pede, et vena corrupta, est justus cadens coram impio. Here is noted, that one judicial and exemplar iniquity in the

Erat civitas: Ecclesiastes 9:14, 15. 'There was a small town with few inhabitants, and a great king came against it, and besieged it and raised great siege works against it. There was in it a poor, wise man, and he by his wisdom saved the city. But no one remembered that poor wise man.'
Mollis responsio: Proverbs 15:1. 'A soft answer turns away anger.'
present: immediate.
Iter pigrorum: Proverbs 15:19. 'The path of the sluggard is a tangle of weeds.'
Melior est: Ecclesiastes 7:8. 'Better is the end of a speech than the beginning.'
Qui cognoscit: Proverbs 28:21. 'He who shows favour in judgment does not do well; even for a piece of bread will that man depart from the truth.'
briber: one who takes bribes.
lightly: easily. *facile*: easily swayed.
Vir pauper: Proverbs 28:3. 'A poor man that oppresses the poor is like driving rain, which produces a famine.' The fable of the horse leech is similar to Aesop's fable of the fox and the fleas, told in Aristotle's *Rhetoric* ii. 20.
Fons turbatus: Proverbs 25:26. 'Like a muddied spring or a tainted well is a righteous man who gives way to a wicked one.'
exemplar: conspicuous.

face of the world doth trouble the fountains of justice more than many particular injuries passed over by connivance.

Qui subtrahit aliquid a patre et a matre, et dicit hoc non esse peccatum, particeps est homicidii. Here is noted, that whereas men in wronging their best friends use to extenuate their fault, as if they mought presume or be bold upon them, it doth contrariwise indeed aggravate their fault, and turneth it from injury to impiety.

Noli esse amicus homini iracundo, nec ambulato cum homine furioso. Here caution is given, that in the election of our friends we do principally avoid those which are impatient, as those that will espouse us to many factions and quarrels.

Qui conturbat domum suam, possidebit ventum. Here is noted, that in domestical separations and breaches men do promise to themselves quieting of their mind and contentment; but still they are deceived of their expectation, and it turneth to wind.

Filius sapiens laetificat patrem: filius vero stultus moestitia est matri suae. Here is distinguished, that fathers have most comfort of the good proof of their sons; but mothers have most discomfort of their ill proof, because women have little discerning of virtue, but of fortune.

Qui celat delictum, quaerit amicitiam; sed qui altero sermone repetit, separat foederatos. Here caution is given, that reconcilement is better managed by an amnesty, and passing over that which is past, than by apologies and excusations.

In omni opere bono erit abundantia; ubi autem verba sunt plurima, ibi frequenter egestas. Here is noted, that words and discourse aboundeth most where there is idleness and want.

Primus in sua causa justus; sed venit altera pars, et inquiret in eum. Here is observed, that in all causes the first tale possesseth

Qui subtrahit: Proverbs 28:24. 'He who robs his father and mother and says he has done no wrong, is a murderer.'

Noli esse: Proverbs 22:24. 'Never make friends with an angry man, nor keep company with a bad-tempered one.'

Qui conturbat: Proverbs 11:29. 'He who brings trouble on his own family inherits the wind.'

domestical: domestic.

Filius sapiens: Proverbs 10:1. 'A wise son brings joy to his father, a foolish son is his mother's bane.'

good proof: turning out well.

Qui celat: Proverbs 17:9. 'He who conceals another's offence seeks his good will, but he who harps on something breaks up friendship.'

In omni opere: Proverbs 14:23. 'In every good work there is abundance; but where there are many words there is commonly penury.'

Primus in sua: Proverbs 18:17. 'In a law suit the first speaker seems right, until another steps forward and cross-questions him.'

possesseth: influences opinion.

much; in sort, that the prejudice thereby wrought will be hardly removed, except some abuse or falsity in the information be detected.

Verba bilinguis quasi simplicia, et ipsa perveniunt ad interiora ventris. Here is distinguished, that flattery and insinuation, which seemeth set and artificial, sinketh not far; but that entereth deep which hath show of nature, liberty, and simplicity.

Qui erudit derisorem, ipse sibi injuriam facit; et qui arguit impium, sibi maculam generat. Here caution is given how we tender reprehension to arrogant and scornful natures, whose manner is to esteem it for contumely, and accordingly to return it.

Da sapienti occasionem, et addetur ei sapientia. Here is distinguished the wisdom brought into habit, and that which is but verbal and swimming only in conceit; for the one upon the occasion presented is quickened and redoubled, the other is amazed and confused.

Quomodo in aquis resplendent vultus prospicientium, sic corda hominum manifesta sunt prudentibus. Here the mind of a wise man is compared to a glass, wherein the images of all diversity of natures and customs are represented; from which representation proceedeth that application,

Qui sapit, innumeris moribus aptus erit.

7. Thus have I stayed somewhat longer upon these sentences politic of Salomon than is agreeable to the proportion of an example; led with a desire to give authority to this part of knowledge, which I noted as deficient, by so excellent a precedent; and have also attended them with brief observations, such as to my understanding offer no violence to the sense, though I know they may be applied to a more divine use: but it is allowed, even in divinity, that some interpretations, yea, and some writings, have more of the eagle[n] than others; but taking them as instructions for life, they mought have received large discourse, if I would have broken them and illustrated them by deducements and examples.

Verba bilinguis: Proverbs 18:8. 'The words of the double-tongued man seem artless, and they go down into the innermost parts of the belly.'
Qui erudit: Proverbs 9:7. 'He that instructs a scorner gets to himself shame, and he that rebukes the wicked gets himself a blot.'
Da sapienti: Proverbs 9:9. 'Give opportunity to a wise man and he will increase his wisdom.'
Quomodo in aquis: Proverbs 27:19. 'As the face is reflected in the water, so is the heart of man manifest to the wise.'
Qui sapit: Ovid, *Art of Love* i. 760. 'The wise man will know how to adapt himself to all sorts of characters.'
deducements: deductions.*

8. Neither was this in use only with the Hebrews, but it is generally to be found in the wisdom of the more ancient times; that as men found out any observation that they thought was good for life, they would gather it and express it in parable or aphorism or fable. But for fables, they were vicegerents and supplies where examples failed: now that the times abound with history, the aim is better when the mark is alive. And therefore the form of writing which of all others is fittest for this variable argument of negotiation and occasions is that which Machiavel[n] chose wisely and aptly for government; namely, discourse upon histories or examples. For knowledge drawn freshly and in our view out of particulars, knoweth the way best to particulars again. And it hath much greater life for practice when the discourse attendeth upon the example, than when the example attendeth upon the discourse. For this is no point of order, as it seemeth at first, but of substance. For when the example is the ground, being set down in an history at large, it is set down with all circumstances, which may sometimes control the discourse thereupon made, and sometimes supply it, as a very pattern for action; whereas the examples alleged for the discourse's sake are cited succinctly, and without particularity, and carry a servile aspect towards the discourse which they are brought in to make good.

9. But this difference is not amiss to be remembered, that as history of times is the best ground for discourse of government, such as Machiavel handleth, so histories of lives is the most proper for discourse of business, because it is more conversant in private actions. Nay, there is a ground of discourse for this purpose fitter than them both, which is discourse upon letters, such as are wise and weighty, as many are of Cicero *ad Atticum*,[n] and others. For letters have a great and more particular representation of business than either chronicles or lives. Thus have we spoken both of the matter and form of this part of civil knowledge, touching negotiation, which we note to be deficient.

10. But yet there is another part of this part, which differeth as much from that whereof we have spoken as *sapere* and *sibi sapere*, the one moving as it were to the circumference, the other to the centre. For there is a wisdom of counsel, and again there is a wisdom of pressing a man's own fortune; and they do sometimes meet, and often sever. For many are wise in their own ways that are weak for

vicegerents: substitutes.
sapere: to be wise; *sibi sapere*: to be wise for oneself.
pressing: pursuing eagerly.
sever: are separated.

government or counsels; like ants, which is a wise creature for itself, but very hurtful for the garden. This wisdom the Romans did take much knowledge of: *Nam pol sapiens* (saith the comical poet) *fingit fortunam sibi*; and it grew to an adage, *Faber quisque fortunae propriae*; and Livy attributed it to Cato the first, *In hoc viro tanta vis animi et ingenii inerat, ut quocunque loco natus esset sibi ipse fortunam facturus videretur.*

11. This conceit or position, if it be too much declared and professed, hath been thought a thing impolitic and unlucky, as was observed in Timotheus the Athenian, who, having done many great services to the estate in his government, and giving an account thereof to the people as the manner was, did conclude every particular with this clause, 'And in this fortune had no part'. And it came so to pass, that he never prospered in any thing he took in hand afterward. For this is too high and too arrogant, savouring of that which Ezekiel saith of Pharaoh, *Dicis, Fluvius est meus et ego feci memet ipsum*: or of that which another prophet speaketh, that men offer sacrifices to their nets and snares; and that which the poet expresseth,

> Dextra mihi Deus, et telum quod missile libro,
> Nunc adsint!

For these confidences were ever unhallowed, and unblessed: and therefore those that were great politiques indeed ever ascribed their successes to their felicity, and not to their skill or virtue. For so Sulla surnamed himself Felix, not Magnus. So Caesar said to the master of the ship, *Caesarem portas et fortunam ejus.*

Nam pol: *Plautus, Trinummus* II. ii. 82. 'For a wise man fashions his fortune for himself.'

Faber quisque: Every man is the architect of his own fortune. A saying attributed to Appius Claudius (*fl.* 300 B.C.). Cf. the essays 'Of Wisdom for a Man's Self' and 'Of Fortune'.

In hoc viro: Livy, *History* xxxix. 40. 'Such was his force of mind and genius that wherever he had been born he would have made his fortune.'

position: maxim.

And in this: Plutarch's 'Life' of Sulla vi. 5.

Dicis, Fluvius: Ezekiel 29:3. 'My river is mine own, it was I who made it.'

nets and snares: Habakkuk 1:16.

Dextra mihi Deus: Virgil, *Aeneid* x. 773. 'Let this right hand, my god, and the hurtling weapon I poise, now be my aid.'

confidences: acts of over-boldness.

felicity: good fortune.*

Sulla: In Plutarch's 'Life' of Sulla vi. 5 he is called the 'Fortunate' not the 'Great'.

Caesarem: Plutarch's 'Life' of Caesar 38. 'You carry Caesar and his fortune.'

12. *Faber fortunae, sive de ambitu vitae.* But yet nevertheless these positions, *Faber quisque fortunae suae: Sapiens dominabitur astris: Invia virtuti nulla est via,* and the like, being taken and used as spurs to industry, and not as stirrups to insolency, rather for resolution than for the presumption or outward declaration, have been ever thought sound and good; and are no question imprinted in the greatest minds, who are so sensible of this opinion, as they can scarce contain it within. As we see in Augustus Caesar (who was rather diverse from his uncle than inferior in virtue), how when he died he desired his friends about him to give him a *plaudite*, as if he were conscient to himself that he had played his part well upon the stage. This part of knowledge we do report also as deficient: not but that it is practised too much, but it hath not been reduced to writing. And therefore lest it should seem to any that it is not comprehensible by axiom, it is requisite, as we did in the former, that we set down some heads or passages of it.

13. Wherein it may appear at the first a new and unwonted argument to teach men how to raise and make their fortune; a doctrine wherein every man perchance will be ready to yield himself a disciple, till he see the difficulty: for fortune layeth as heavy impositions as virtue; and it is as hard and severe a thing to be a true politique, as to be truly moral. But the handling hereof concerneth learning greatly, both in honour and in substance. In honour, because pragmatical men may not go away with an opinion that learning is like a lark, that can mount, and sing, and please herself, and nothing else; but may know that she holdeth as well of the hawk, that can soar aloft, and can also descend and strike upon the prey. In substance, because it is the perfect law of inquiry of truth, that nothing be in the globe of matter,[n] which should not be likewise in the globe of crystal, of form; that is, that there be not any thing in being and action, which should not be drawn and collected into contemplation and doctrine. Neither doth learning admire or esteem of this architecture of fortune, otherwise than as of an inferior work: for no man's fortune can be an end worthy of his being; and many times the worthiest men do abandon their fortune willingly for better respects: but nevertheless fortune as an organ of virtue and merit deserveth the consideration.

Faber fortunae: The architect of fortune, or how to get on in life.
Faber quisque: Ovid, *Metamorphoses* xiv. 113. 'Every man is the architect of his own fortune; a wise man shall rule the stars; there is no path that virtue cannot tread.'
plaudite: In Suetonius, *Octavius* 99. 'Applause.'
conscient: conscious.* *esteem of*: reckon.

14. First therefore the precept which I conceive to be most summary towards the prevailing in fortune, is to obtain that window which Momus[n] did require: who seeing in the frame of man's heart such angles and recesses, found fault there was not a window to look into them; that is, to procure good informations of particulars touching persons, their natures, their desires and ends, their customs and fashions, their helps and advantages, and whereby they chiefly stand: so again their weaknesses and disadvantages, and where they lie most open and obnoxious; their friends, factions, dependences; and again their opposites, enviers, competitors, their moods and times, *Sola viri molles aditus et tempora noras*; their principles, rules, and observations, and the like: and this not only of persons, but of actions; what are on foot from time to time, and how they are conducted, favoured, opposed, and how they import, and the like. For the knowledge of present actions is not only material in itself, but without it also the knowledge of persons is very erroneous: for men change with the actions; and whiles they are in pursuit they are one, and when they return to their nature they are another. These informations of particulars, touching persons and actions, are as the minor propositions in every active syllogism; for no excellency of observations (which are as the major propositions) can suffice to ground a conclusion, if there be error and mistaking in the minors.

15. That this knowledge is possible, Salomon is our surety, who saith, *Consilium in corde viri tanquam aqua profunda; sed vir prudens exhauriet illud*. And although the knowledge itself falleth not under precept, because it is of individuals, yet the instructions for the obtaining of it may.

16. We will begin therefore with this precept, according to the ancient opinion, that the sinews of wisdom are slowness of belief and distrust; that more trust be given to countenances and deeds than to words; and in words rather to sudden passages and surprised words than to set and purposed words. Neither let that be feared which is said, *Fronti nulla fides*, which is meant of a general outward behaviour, and not of the private and subtile motions and labours

obnoxious: exposed.
dependences: dependents.†
opposites: antagonists.
Sola viri: Virgil, *Aeneid* iv. 423. 'You alone know the favourable times for approaching him.'
Consilium: Proverbs 20:5. 'Counsel in another's heart is like deep water, but a discerning man will draw it up.'
sinews of wisdom: Cicero, *Letters to Atticus* i. 19.
Fronti nulla: Juvenal, *Satires* ii. 8. 'Trust not to a man's face.'

of the countenance and gesture; which, as Q. Cicero elegantly saith, is *Animi janua, the gate of the mind.* None more close than Tiberius, and yet Tacitus saith of Gallus, *Etenim vultu offensionem conjectaverat.* So again, noting the differing character and manner of his commending Germanicus and Drusus in the senate, he said, touching his fashion wherein he carried his speech of Germanicus, thus; *Magis in speciem adornatis verbis, quam ut penitus sentire crederetur*: but of Drusus thus; *Paucioribus sed intentior, et fida oratione*: and in another place, speaking of his character of speech, when he did any thing that was gracious and popular, he saith, that in other things he was *velut eluctantium verborum*; but then again, *solutius loquebatur quando subveniret.* So that there is no such artificer of dissimulation, nor no such commanded countenance (*vultus jussus*), that can sever from a feigned tale some of these fashions, either a more slight and careless fashion, or more set and formal, or more tedious and wandering, or coming from a man more drily and hardly.

17. Neither are deeds such assured pledges, as that they may be trusted without a judicious consideration of their magnitude and nature: *Fraus sibi in parvis fidem praestruit ut majore emolumento fallat*; and the Italian thinketh himself upon the point to be bought and sold, when he is better used than he was wont to be without manifest cause. For small favours, they do but lull men asleep, both as to caution and as to industry; and are, as Demosthenes calleth them, *Alimenta socordiae.* So again we see how false the nature of some deeds are, in that particular which Mutianus practised upon Antonius Primus, upon that hollow and unfaithful reconcilement which was made between them; whereupon Mutianus advanced many of the friends of Antonius, *Simul amicis ejus praefecturas et tribunatus largitur*: wherein, under pretence to strengthen him, he did desolate him, and won from him his dependences.

Cicero: Quintus Cicero, *De Petitione Consulatus* xi. 44.
Etenim: Tacitus, *Annals* i. 12. 'He guessed that he was offended by his looks.'
Magis in speciem: Tacitus, *Annals* i. 52. 'It was in words too laboured and specious to be taken for what he really felt. But of Drusus, he said less, but more earnestly and in a style of sincerity.'
velut: Tacitus, *Annals* iv. 31. 'Of a kind of struggling speech', but then again, 'he spoke with more freedom when he was speaking in a man's favour'.
Fraus sibi: Livy, *History* xxviii. 42. 'The treacherous man begins by being faithful in small things, that he may afterwards deceive with greater profit.'
Italian: Referring to the proverb, 'Chi mi fa più caresse chè non suole / O m'a ingannato, o ingannar mi vuole.'
Alimenta: Demosthenes, *Third Olynthiac* 33. 'Sops to feed sloth.'
Simul amicis: Tacitus, *Histories* iv. 39. 'At the same time he bestows tribuneships and governments on his friends.'
dependences: dependents†.

18. As for words, though they be like waters to physicians, full of flattery and uncertainty, yet they are not to be despised, specially with the advantage of passion and affection. For so we see Tiberius, upon a stinging and incensing speech of Agrippina, came a step forth of his dissimulation, when he said, 'You are hurt because you do not reign'; of which Tacitus saith, *Audita haec raram occulti pectoris vocem elicuere; correptamque Graeco versu admonuit, ideo laedi quia non regnaret.* And therefore the poet doth elegantly call passions tortures, that urge men to confess their secrets:

Vino tortus et ira.

And experience showeth, there are few men so true to themselves and so settled, but that, sometimes upon heat, sometimes upon bravery, sometimes upon kindness, sometimes upon trouble of mind and weakness, they open themselves; specially if they be put to it with a counter-dissimulation, according to the proverb of Spain, *Di mentira, y sacaras verdad*: 'Tell a lie and find a truth.'

19. As for the knowing of men which is at second hand from reports; men's weaknesses and faults are best known from their enemies, their virtues and abilities from their friends, their customs and times from their servants, their conceits and opinions from their familiar friends, with whom they discourse most. General fame is light, and the opinions conceived by superiors or equals are deceitful; for such men are more masked: *Verior fama e domesticis emanat.*

20. But the soundest disclosing and expounding of men is by their natures and ends, wherein the weakest sort of men are best interpreted by their natures, and the wisest by their ends. For it was both pleasantly and wisely said (though I think very untruly) by a nuncio of the pope, returning from a certain nation where he served as lidger; whose opinion being asked touching the appointment of one to go in his place, he wished that in any case they did not send one that was too wise; because no very wise man would ever imagine

waters to physicians: Doctors examined a patient's urine as a means of diagnosing his illness.
Audita haec: Tacitus, *Annals* iv. 52. 'These words drew from Tiberius the voice, so rarely heard, of his secret heart; he retorted upon her, in a Greek verse, that she was thus offended because she did not reign.'
Vino tortus: Horace, *Epistles* I. xviii. 38. 'Tortured with wine and wrath.'
fame: report.
Verior fama: Quintus Cicero, *De Petitione Consulatus* v. 17. 'The truest character comes from a man's own household.'
lidger: resident ambassador.

what they in that country were like to do. And certainly it is an error frequent for men to shoot over, and to suppose deeper ends, and more compass reaches than are: the Italian proverb being elegant, and for the most part true:

> Di danari, di senno, e di fede,
> C'è ne manco che non credi:

There is commonly less money, less wisdom, and less good faith than men do account upon.

21. But princes, upon a far other reason, are best interpreted by their natures, and private persons by their ends. For princes being at the top of human desires, they have for the most part no particular ends whereto they aspire, by distance from which a man mought take measure and scale of the rest of their actions and desires; which is one of the causes that maketh their hearts more inscrutable. Neither is it sufficient to inform ourselves in men's ends and natures of the variety of them only, but also of the predominancy, what humour reigneth most, and what end is principally sought. For so we see, when Tigellinus saw himself outstripped by Petronius Turpilianus in Nero's humours of pleasures, *metus ejus rimatur*, he wrought upon Nero's fears, whereby he brake the other's neck.

22. But to all this part of inquiry the most compendious way resteth in three things: the first, to have general acquaintance and inwardness with those which have general acquaintance and look most into the world; and specially according to the diversity of business, and the diversity of persons, to have privacy and conversation with some one friend at least which is perfect and well intelligenced in every several kind. The second is to keep a good mediocrity in liberty of speech and secrecy; in most things liberty: secrecy where it importeth; for liberty of speech inviteth and provoketh liberty to be used again, and so bringeth much to a man's knowledge; and secrecy on the other side induceth trust and inwardness. The last is the reducing of a man's self to this watchful and serene habit, as to make account and purpose, in every conference and action, as well to observe as to act. For as Epictetus would have a philosopher in every particular action to say to himself, *Et hoc volo, et etiam institutum servare*; so a politic man in everything should say to him-

shoot over: overshoot the mark.*
compass reaches: roundabout schemes.
account upon: reckon.
metus ejus: Tacitus, *Annals* xvi. 18. 'He wrought on [Nero's] fears.' It was not Petronius Turpilianus but Caius Petronius.
inwardness: close friendship. *importeth*: is of importance.

self, *Et hoc volo, ac etiam aliquid addiscere.* I have stayed the longer upon this precept of obtaining good information, because it is a main part by itself, which answereth to all the rest. But, above all things, caution must be taken that men have a good stay and hold of themselves, and that this much knowing do not draw on much meddling; for nothing is more unfortunate than light and rash intermeddling in many matters. So that this variety of knowledge tendeth in conclusion but only to this, to make a better and freer choice of those actions which may concern us, and to conduct them with the less error and the more dexterity.

23. The second precept concerning this knowledge is, for men to take good information touching their own person, and well to understand themselves: knowing that, as St. James saith, though men look oft in a glass, yet they do suddenly forget themselves; wherein as the divine glass is the word of God, so the politic glass is the state of the world, or times wherein we live, in the which we are to behold ourselves.

24. For men ought to take an unpartial view of their own abilities and virtues; and again of their wants and impediments; accounting these with the most, and those other with the least;[n] and from this view and examination to frame the considerations following.

25. First, to consider how the constitution of their nature sorteth with the general state of the times; which if they find agreeable and fit, then in all things to give themselves more scope and liberty; but if differing and dissonant, then in the whole course of their life to be more close retired, and reserved: as we see in Tiberius, who was never seen at a play, and came not into the Senate in twelve of his last years; whereas Augustus Caesar lived ever in men's eyes, which Tacitus observeth, *alia Tiberio morum via.*

26. Secondly, to consider how their nature sorteth with professions and courses of life, and accordingly to make election, if they be free; and, if engaged, to make the departure at the first opportunity: as we see was done by Duke Valentine,[n] that was designed by his father to a sacerdotal profession, but quitted it soon after in regard of his parts and inclination; being such, nevertheless, as a man cannot tell well whether they were worse for a prince or for a priest.

27. Thirdly, to consider how they sort with those whom they are

Et hoc volo: Epictetus, *Encheiridion* 4. 'I both wish to do this and also to keep to my rule.' And, 'I both wish to do this and also to learn something from it.'
but only: i.e. only.
St. James: James 1:23, 24.
sorteth: agrees with.
alia Tiberio: Tacitus, *Annals* i. 54. 'The habits of Tiberius were different.'

like to have competitors and concurrents; and to take that course wherein there is most solitude, and themselves like to be most eminent: as Caesar Julius did, who at first was an orator or pleader; but when he saw the excellency of Cicero, Hortensius, Catulus, and others, for eloquence, and saw there was no man of reputation for the wars but Pompeius, upon whom the state was forced to rely, he forsook his course begun toward a civil and popular greatness, and transferred his designs to a martial greatness.

28. Fourthly, in the choice of their friends and dependences, to proceed according to the composition of their own nature: as we may see in Caesar, all whose friends and followers were men active and effectual, but not solemn, or of reputation.[n]

29. Fifthly, to take special heed how they guide themselves by examples, in thinking they can do as they see others do; whereas perhaps their natures and carriages are far differing. In which error it seemeth Pompey was, of whom Cicero saith, that he was wont often to say, *Sulla potuit, ego non potero?* Wherein he was much abused, the natures and proceedings of himself and his example being the unlikest in the world; the one being fierce, violent, and pressing the fact; the other solemn, and full of majesty and circumstance, and therefore the less effectual.

But this precept touching the politic knowledge of ourselves hath many other branches, whereupon we cannot insist.

30. Next to the well understanding and discerning of a man's self, there followeth the well opening and revealing a man's self; wherein we see nothing more usual than for the more able man to make the less show. For there is a great advantage in the well setting forth of a man's virtues, fortunes, merits; and again, in the artificial covering of a man's weaknesses, defects, disgraces; staying upon the one, sliding from the other; cherishing the one by circumstances, gracing the other by exposition, and the like. Wherein we see what Tacitus saith of Mutianus, who was the greatest politique of his time, *Omnium quae dixerat feceratque arte quadam ostentator*: which requireth indeed some art, lest it turn tedious and arrogant; but yet so, as ostentation (though it be the first degree of vanity) seemeth to me rather a vice in manners than in policy: for as it is said,

concurrents: rivals.
Caesar: In Plutarch's 'Life' of Caesar, 3.
dependences: dependents.†
Sulla potuit: Cicero, *Letters to Atticus* ix. 10. 'Sulla could do this, why should not I?'
Omnium quae: Tacitus, *Histories* ii. 80. 'He had a certain art of setting forth to advantage everything he said or did.'

Audacter calumniare, semper aliquid haeret: so, except it be in a ridiculous degree of deformity, *Audacter te vendita, semper aliquid haeret.* For it will stick with the more ignorant and inferior sort of men, though men of wisdom and rank do smile at it and despise it; and yet the authority won with many doth countervail the disdain of a few. But if it be carried with decency and government, as with a natural, pleasant, and ingenious fashion; or at times when it is mixed with some peril and unsafety (as in military persons); or at times when others are most envied; or with easy and careless passage to it and from it, without dwelling too long, or being too serious; or with an equal freedom of taxing a man's self, as well as gracing himself; or by occasion of repelling or putting down others' injury or insolency; it doth greatly add to reputation: and surely not a few solid natures, that want this ventosity and cannot sail in the height of the winds, are not without some prejudice and disadvantage by their moderation.

31. But for these flourishes and enhancements of virtue, as they are not perchance unnecessary, so it is at least necessary that virtue be not disvalued and imbased under the just price; which is done in three manners: by offering and obtruding a man's self; wherein men think he is rewarded, when he is accepted; by doing too much, which will not give that which is well done leave to settle, and in the end induceth satiety; and by finding too soon the fruit of a man's virtue, in commendation, applause, honour, favour; wherein if a man be pleased with a little, let him hear what is truly said; *Cave ne insuetus rebus majoribus videaris, si haec te res parva sicuti magna delectat.*

32. But the covering of defects is of no less importance than the valuing of good parts; which may be done likewise in three manners, by caution, by colour, and by confidence. Caution is when men do ingeniously and discreetly avoid to be put into those things for which they are not proper: whereas contrariwise bold and unquiet spirits will thrust themselves into matters without difference, and so publish and proclaim all their wants. Colour is when men make a

Audacter: cf. Plutarch, *How to tell a Flatterer from a Friend* 65. 'Slander boldly, for some of it will stick.' And 'Boldly sound your own praises, for some of it will stick'. With this paragraph, cf. the essay 'Of Vain Glory'.
ingenious: ingenuous.
injury: intentionally offensive speech.
ventosity: vanity.
disvalued: depreciated.
Cave ne: *Rhetorica ad Herennium* iv. 4. 'If he take so much delight in a little thing, he will be thought unused to greater things.'
colour: giving a specious appearance.

way for themselves to have a construction made of their faults or wants, as proceeding from a better cause or intended for some other purpose. For of the one it is well said,

Saepe latet vitium proximitate boni,

and therefore whatsoever want a man hath, he must see that he pretend the virtue that shadoweth it; as if he be dull, he must affect gravity; if a coward, mildness; and so the rest. For the second, a man must frame some probable cause why he should not do his best, and why he should dissemble his abilities; and for that purpose must use to dissemble those abilities which are notorious in him, to give colour that his true wants are but industries and dissimulations. For confidence, it is the last but the surest remedy; namely, to depress and seem to despise whatsoever a man cannot attain; observing the good principle of the merchants, who endeavour to raise the price of their own commodities, and to beat down the price of others. But there is a confidence that passeth this other; which is to face out a man's own defects, in seeming to conceive that he is best in those things wherein he is failing; and, to help that again, to seem on the other side that he hath least opinion of himself in those things wherein he is best: like as we shall see it commonly in poets, that if they show their verses, and you except to any, they will say 'That that line cost them more labour than any of the rest'; and presently will seem to disable and suspect rather some other line, which they know well enough to be the best in the number. But above all, in this righting and helping of a man's self in his own carriage, he must take heed he show not himself dismantled and exposed to scorn and injury, by too much dulceness, goodness, and facility of nature; but show some sparkles of liberty, spirit, and edge. Which kind of fortified carriage, with a ready rescussing of a man's self from scorns, is sometimes of necessity imposed upon men by somewhat in their person or fortune; but it ever succeedeth with good felicity.

33. Another precept of this knowledge is by all possible endeavour to frame the mind to be pliant and obedient to occasion;

Saepe latet: Ovid, *Art of Love* ii. 662. 'Vice often hides itself in the neighbourhood of virtue.'
dismantled: literally, 'without his cloak'.
dulceness: sweetness.
rescussing: rescuing. A word not in the *OED*. *Rescous* is an old law term for 'rescue'.
person or fortune: i.e. in the case of deformed people, bastards, men branded with some mark of disgrace.

for nothing hindereth men's fortunes so much as this: *Idem manebat, neque idem decebat*, men are where they were, when occasions turn: and therefore to Cato, whom Livy maketh such an architect of fortune, he addeth that he had *versatile ingenium*. And thereof it cometh that these grave solemn wits, which must be like themselves and cannot make departures, have more dignity than felicity. But in some it is nature to be somewhat viscous and inwrapped, and not easy to turn. In some it is a conceit that is almost a nature, which is, that men can hardly make themselves believe that they ought to change their course, when they have found good by it in former experience. For Machiavel noted wisely, how Fabius Maximus would have been temporizing still, according to his old bias, when the nature of the war was altered and required hot pursuit. In some other it is want of point and penetration in their judgement, that they do not discern when things have a period, but come in too late after the occasion; as Demosthenes compareth the people of Athens to country fellows, when they play in a fence school, that if they have a blow, then they remove their weapon to that ward, and not before. In some other it is a lothness to leese labours passed, and a conceit that they can bring about occasions to their ply; and yet in the end, when they see no other remedy, then they come to it with disadvantage; as Tarquinius, that gave for the third part of Sibylla's books the treble price, when he mought at first have had all three for the simple. But from whatsoever root or cause this restiveness of mind proceedeth, it is a thing most prejudicial; and nothing is more politic than to make the wheels of our mind concentric and voluble with the wheels of fortune.

34. Another precept of this knowledge, which hath some affinity with that we last spake of, but with difference, is that which is well expressed, *Fatis accede deisque*, that men do not only turn with the occasions, but also run with the occasions, and not strain their credit or strength to over-hard or extreme points; but choose in

Idem manebat: Cicero, *Brutus* 95. 'He continued the same, when it no longer became him.'
versatile: Livy, *History* xxxix. 40. 'A versatile nature.'
make departures: set off on new courses.
Machiavel: In his *Discourses on Livy* iii. 9.
period: conclusion.
Demosthenes: *First Philippic* 46.
ply: inclination.
Tarquinius: In Aulus Gellius, *Noctes Atticae* i. 19.
simple: single.
voluble: capable of rotation on an axis.
Fatis: Lucan, *Pharsalia* viii. 485. 'Yield to destiny and the gods.'

their actions that which is most passable: for this will preserve men from foil, not occupy them too much about one matter, win opinion of moderation, please the most, and make a show of a perpetual felicity in all they undertake; which cannot but mightily increase reputation.

35. Another part of this knowledge seemeth to have some repugnancy with the former two, but not as I understand it; and it is that which Demosthenes uttereth in high terms; *Et quemadmodum receptum est, ut exercitum ducat imperator, sic et a cordatis viris res ipsae ducendae; ut quae ipsis videntur, ea gerantur, et non ipsi eventus persequi cogantur.* For if we observe we shall find two differing kinds of sufficiency in managing of business: some can make use of occasions aptly and dextrously, but plot little; some can urge and pursue their own plots well, but cannot accommodate nor take in; either of which is very unperfect without the other.

36. Another part of this knowledge is the observing a good mediocrity in the declaring, or not declaring a man's self: for although depth of secrecy, and making way (*qualis est via navis in mari,* which the French calleth *sourdes menées,* when men set things in work without opening themselves at all), be sometimes both prosperous and admirable; yet many times *dissimulatio errores parit, qui dissimulatorem ipsum illaqueant.* And therefore we see the greatest politiques have in a natural and free manner professed their desires, rather than been reserved and disguised in them. For so we see that Lucius Sulla made a kind of profession, 'that he wished all men happy or unhappy, as they stood his friends or enemies'. So Caesar, when he went first into Gaul, made no scruple to profess 'That he had rather be first in a village than second at Rome'. So again, as soon as he had begun the war, we see what Cicero saith of him, *Alter* (meaning of Caesar) *non recusat, sed quodammodo postulat, ut* (*ut est*) *sic appelletur tyrannus.* So we may see in a letter of Cicero

passable: able to be accomplished.
foil: defeat.
most: majority.
Et quemadmodum: Demosthenes, *First Philippic* 39. 'Wise men should lead events as a general leads an army; they should get that done which they think good to be done, not merely follow events.'
qualis est: Proverbs 30:19. 'The way of a ship out at sea.'
sourdes menées: underhand practices.
dissimulatio: 'Dissimulation breeds mistakes in which the dissembler himself is caught.'
Sulla: In Plutarch's 'Life' of Sulla 38.
Caesar: In Plutarch's 'Life' of Caesar xi. 2.
Alter non recusat: Cicero, *Letters to Atticus* x. 4. 'He does not refuse, but in a manner demands, to be called what he is—tyrant.'

to Atticus, that Augustus Caesar, in his very entrance into affairs, when he was a darling of the senate, yet in his harangues to the people would swear, *Ita parentis honores consequi liceat* (which was no less than the tyranny), save that, to help it, he would stretch forth his hand towards a statua of Caesar's that was erected in the place: and men laughed, and wondered, and said, Is it possible? or, Did you ever hear the like? and yet thought he meant no hurt; he did it so handsomely and ingenuously. And all these were prosperous: whereas Pompey, who tended to the same ends, but in a more dark and dissembling manner, as Tacitus saith of him, *Occultior non melior*, wherein Sallust concurreth, *Ore probo, animo inverecundo*, made it his design, by infinite secret engines, to cast the state into an absolute anarchy and confusion, that the state mought cast itself into his arms for necessity and protection, and so the sovereign power be put upon him, and he never seen in it: and when he had brought it (as he thought) to that point, when he was chosen consul alone, as never any was, yet he could make no great matter of it, because men understood him not; but was fain in the end to go the beaten track of getting arms into his hands, by colour of the doubt of Caesar's designs: so tedious, casual, and unfortunate are these deep dissimulations: whereof it seemeth Tacitus made this judgement, that they were a cunning of an inferior form in regard of true policy; attributing the one to Augustus, the other to Tiberius; where, speaking of Livia, he saith, *Et cum artibus mariti simulatione filii bene composita*: for surely the continual habit of dissimulation is but a weak and sluggish cunning, and not greatly politic.

37. Another precept of this architecture of fortune is to accustom our minds to judge of the proportion or value of things, as they conduce and are material to our particular ends: and that to do substantially, and not superficially. For we shall find the logical part (as I may term it) of some men's minds good, but the mathematical part erroneous; that is, they can well judge of consequences, but not of proportions and comparison, preferring things of show and

Ita parentis: Cicero, *Letters to Atticus* xvi. 15. 'By the hope of attaining his father's honours.'
tyranny: absolute power.
Occultior: Tacitus, *Histories* ii. 38. 'A more reserved but not a better character.'
Ore probo: Quoted by Suetonius, *On Grammarians* xv. 'Of honest tongue and shameless mind.'
engines: plots.
Et cum artibus: Tacitus, *Annals* v. 1. 'She was a match for the diplomacy of her husband and the dissimulation of her son.'
proportions: the relative value of things.

sense before things of substance and effect. So some fall in love with access to princes, others with popular fame and applause, supposing they are things of great purchase, when in many cases they are but matters of envy, peril, and impediment. So some measure things according to the labour and difficulty or assiduity which are spent about them; and think, if they be ever moving, that they must needs advance and proceed; as Caesar saith in a despising manner of Cato the second, when he describeth how laborious and indefatigable he was to no great purpose, *Haec omnia magno studio agebat.* So in most things men are ready to abuse themselves in thinking the greatest means to be best, when it should be the fittest.

38. As for the true marshalling of men's pursuits towards their fortune, as they are more or less material, I hold them to stand thus. First the amendment of their own minds. For the remove of the impediments of the mind will sooner clear the passages of fortune, than the obtaining fortune will remove the impediments of the mind. In the second place I set down wealth and means; which I know most men would have placed first, because of the general use which it beareth towards all variety of occasions. But that opinion I may condemn with like reason as Machiavel doth that other, that moneys were the sinews of the wars; whereas (saith he) the true sinews of the wars are the sinews of men's arms, that is, a valiant, populous, and military nation: and he voucheth aptly the authority of Solon, who, when Croesus showed him his treasury of gold, said to him, that if another came that had better iron, he would be master of his gold. In like manner it may be truly affirmed, that it is not moneys that are the sinews of fortune, but it is the sinews and steel of men's minds, wit, courage, audacity, resolution, temper, industry, and the like. In the third place I set down reputation, because of the peremptory tides and currents it hath; which, if they be not taken in their due time, are seldom recovered, it being extreme hard to play an after game of reputation. And lastly I place honour, which is more easily won by any of the other three, much more by all, than any of them can be purchased by honour. To conclude this precept, as there is order and priority in matter, so is there in time, the preposterous placing whereof is one of the commonest errors: while men fly to their ends when they should intend their beginnings, and do not

purchase: value.
Haec omnia: Caesar, *Civil War* i. 30. 'He did all these things with great energy.'
remove: removal.
Machiavel: In his *Discourses on Livy* ii. 20.
peremptory: fixed.
after game: a second game, played to reverse or improve the issue of the first.
preposterous: placing last that which should be first.

take things in order of time as they come on, but marshal them according to greatness and not according to instance; not observing the good precept, *Quod nunc instat agamus.*

39. Another precept of this knowledge is not to embrace any matters which do occupy too great a quantity of time, but to have that sounding in a man's ears, *Sed fugit interea fugit irreparabile tempus*: and that is the cause why those which take their course of rising by professions of burden, as lawyers, orators, painful divines, and the like, are not commonly so politic for their own fortune, otherwise than in their ordinary way, because they want time to learn particulars, to wait occasions, and to devise plots.

40. Another precept of this knowledge is to imitate nature which doth nothing in vain; which surely a man may do if he do well interlace his business, and bend not his mind too much upon that which he principally intendeth. For a man ought in every particular action so to carry the motions of his mind, and so to have one thing under another, as if he cannot have that he seeketh in the best degree, yet to have it in a second, or so in a third; and if he can have no part of that which he purposed, yet to turn the use of it to somewhat else; and if he cannot make anything of it for the present, yet to make it as a seed of somewhat in time to come; and if he can contrive no effect or substance from it, yet to win some good opinion by it, or the like. So that he should exact an account of himself of every action, to reap somewhat, and not to stand amazed and confused if he fail of that he chiefly meant: for nothing is more impolitic than to mind actions wholly one by one. For he that doth so leeseth infinite occasions which intervene, and are many times more proper and propitious for somewhat that he shall need afterwards, than for that which he urgeth for the present; and therefore men must be perfect in that rule, *Haec oportet facere, et illa non omittere.*

41. Another precept of this knowledge is, not to engage a man's self peremptorily in any thing, though it seem not liable to accident; but ever to have a window to fly out at, or a way to retire: following the wisdom in the ancient fable of the two frogs, which consulted when their plash was dry whither they should go; and the one

Quod nunc: Virgil, *Eclogues* ix. 66. 'Let us attend to present business.'
Sed fugit: Virgil, *Georgics* iii. 284. 'But meanwhile time, which can never be recalled, is flying.'
painful: diligent. *interlace*: vary by mixing.
Haec oportet: Matthew 23:23. 'These things ought you to do without leaving the others undone.'
plash: shallow pool.

moved to go down into a pit, because it was not likely the water would dry there; but the other answered, True, but if it do, how shall we get out again?

42. Another precept of this knowledge is that ancient precept of Bias, construed not to any point of perfidiousness, but only to caution and moderation, *Et ama tanquam inimicus futurus et odi tanquam amaturus*. For it utterly betrayeth all utility for men to embark themselves too far into unfortunate friendships, troublesome spleens, and childish and humorous envies or emulations.

43. But I continue this beyond the measure of an example; led, because I would not have such knowledges, which I note as deficient, to be thought things imaginative or in the air, or an observation or two much made of, but things of bulk and mass, whereof an end is hardlier made than a beginning. It must be likewise conceived, that in these points which I mention and set down, they are far from complete tractates of them, but only as small pieces for patterns. And lastly, no man I suppose will think that I mean fortunes are not obtained without all this ado; for I know they come tumbling into some men's laps; and a number obtain good fortunes by diligence in a plain way, little intermeddling, and keeping themselves from gross errors.

44. But as Cicero, when he setteth down an idea of a perfect orator, doth not mean that every pleader should be such; and so likewise, when a prince or a courtier hath been described by such as have handled those subjects, the mould hath used to be made according to the perfection of the art, and not according to common practice: so I understand it, that it ought to be done in the description of a politic man, I mean politic for his own fortune.

45. But it must be remembered all this while, that the precepts which we have set down are of that kind which may be counted and called *Bonae Artes*. As for evil arts, if a man would set down for himself that principle of Machiavel, 'That a man seek not to attain virtue itself, but the appearance only thereof; because the credit of virtue is a help, but the use of it is cumber': or that other of his principles, 'That he presuppose, that men are not fitly to be

moved: proposed.
Et ama: Attributed to Bias (*fl.* 550 B.C.), in Aristotle, *Rhetoric* ii. 13. 'Love as if you were sometime to hate, and hate as if you were sometime to love.'
spleens: fits of anger.
Cicero: Referring to Cicero's *The Orator*, and later to such works as Castiglione, *The Courtier*, Elyot, *The Governor*, Machiavelli, *The Prince*.
Bonae Artes: Good arts.
That a man: Quotations from chapters 18 and 17 of *The Prince*.
cumber: a hindrance.

wrought otherwise but by fear; and therefore that he seek to have every man obnoxious, low, and in strait' which the Italians call *seminar spine*, to sow thorns: or that other principle, contained in the verse which Cicero citeth *Cadant amici, dummodo inimici intercidant*, as the triumvirs, which sold every one to other the lives of their friends for the deaths of their enemies: or that other protestation of L. Catilina, to set on fire and trouble states, to the end to fish in droumy waters, and to unwrap their fortunes, *Ego si quid in fortunis meis excitatum sit incendium, id non aqua sed ruina restinguam*: or that other principle of Lysander, 'That children are to be deceived with comfits, and men with oaths: and the like evil and corrupt positions, whereof (as in all things) there are more in number than of the good: certainly with these dispensations from the laws of charity and integrity, the pressing of a man's fortune may be more hasty and compendious. But it is in life as it is in ways, the shortest way is commonly the foulest, and surely the fairer way is not much about.

46. But men, if they be in their own power, and do bear and sustain themselves, and be not carried away with a whirlwind or tempest of ambition, ought in the pursuit of their own fortune to set before their eyes not only that general map of the world, 'That all things are vanity and vexation of spirit,' but many other more particular cards and directions: chiefly that, that being without well-being is a curse, and the greater being the greater curse; and that all virtue is most rewarded, and all wickedness most punished in itself: according as the poet saith excellently:

> Quae vobis, quae digna, viri, pro laudibus istis
> Praemia posse rear solvi? pulcherrima primum
> Dii moresque dabunt vestri.

And so of the contrary. And secondly they ought to look up to the eternal providence and divine judgement, which often subverteth

obnoxious: in dread of punishment. *in strait*: in a difficulty.
Cadant amici: Cicero, *For King Deiotarus* ix. 25. 'Let friends fall, provided our enemies perish with them.' For the triumvirs, see Shakespeare's *Julius Caesar* IV. i.
droumy: turbid.* *unwrap*: disentangle.
Ego si quid: Cicero, *For Murena* xxv. 51. 'If fire breaks out among my possessions I will put it out not with water but with destruction.'
That children: Plutarch's 'Life' of Lysander 8.
That all things: Ecclesiastes 2:11. *cards*: maps.
Quae vobis: Virgil, *Aeneid* ix. 252–4.

> What recompense, O friends, can I hold out
> Worthy such deeds? The best is that ye have—
> God's blessing and your proper nobleness.

the wisdom of evil plots and imaginations, according to that scripture, 'He hath conceived mischief, and shall bring forth a vain thing.' And although men should refrain themselves from injury and evil arts, yet this incessant and Sabbathless pursuit of a man's fortune leaveth not tribute which we owe to God of our time; who (we see) demandeth a tenth of our substance, and a seventh, which is more strict, of our time: and it is to small purpose to have an erected face towards heaven, and a perpetual groveling spirit upon earth, eating dust as doth the serpent, *Atque affigit humo divinae particulam aurae*. And if any man flatter himself that he will employ his fortune well, though he should obtain it ill, as was said concerning Augustus Caesar, and after of Septimius Severus, 'That either they should never have been born, or else they should never have died', they did so much mischief in the pursuit and ascent of their greatness, and so much good when they were established; yet these compensations and satisfactions are good to be used, but never good to be purposed. And lastly, it is not amiss for men in their race toward their fortune, to cool themselves a little with that conceit which is elegantly expressed by the Emperor Charles the Fifth, in his instructions to the king his son, 'That fortune hath somewhat of the nature of a woman, that if she be too much wooed she is the farther off.' But this last is but a remedy for those whose tastes are corrupted: let men rather build upon that foundation which is as a cornerstone of divinity and philosophy, wherein they join close, namely that same *Primum quaerite*. For divinity saith, *Primum quaerite regnum Dei, et ista omnia adjicienter vobis*: and philosophy saith, *Primum quaerite bona animi; caetera aut aderunt, aut non oberunt*. And although the human foundation hath somewhat of the sands, as we see in M. Brutus, when he brake forth into that speech,

Te colui (Virtus) ut rem; ast tu nomen inane es;

He hath conceived: Job 15:35.
Atque affigit: Horace, *Satires* II. ii. 79. 'The particle of the divine spirit cleaves to the ground.'
That either: Aurelius Victor, *Epitome de Caesaribus* i. 28 and Spartianus, *Life of Severus* 18.
Charles the Fifth: King of Spain (1516) and Emperor of Germany (1519) until he abdicated in favour of his son, Philip II, in 1556.
Primum quaerite: Seek ye first. *Divinity*: Matthew 6:33, 'Seek ye first the kingdom of God and all those things shall be added unto you.' *Philosophy*: 'Seek ye first the good things of the mind, and the rest will either be supplied, or their loss will not be felt.'
sands: Matthew 7:24–7.
Te colui: Dio Cassius, *Roman History* xlvii. 49.

Virtue I worshipped and, as real, sought,
But found her empty and a thing of nought.

yet the divine foundation is upon the rock. But this may serve for a taste of that knowledge which I noted as deficient.

47. Concerning government, it is a part of knowledge secret and retired in both these respects in which things are deemed secret; for some things are secret because they are hard to know, and some because they are not fit to utter. We see all governments are obscure and invisible:

> Totamque infusa per artus
> Mens agitat molem, et magno se corpore miscet.

Such is the description of governments. We see the government of God over the world is hidden, insomuch as it seemeth to participate of much irregularity and confusion. The government of the soul in moving the body is inward and profound, and the passages thereof hardly to be reduced to demonstration. Again, the wisdom of antiquity (the shadows whereof are in the poets) in the description of torments and pains, next unto the crime of rebellion, which was the giants' offence, doth detest the offence of futility,[n] as in Sisyphus and Tantalus. But this was meant of particulars: nevertheless even unto the general rules and discourses of policy and government there is due a reverent and reserved handling.

48. But contrariwise in the governors towards the governed, all things ought as far as the frailty of man permitteth to be manifest and revealed. For so it is expressed in the scriptures touching the government of God, that this globe, which seemeth to us a dark and shady body, is in the view of God as crystal: *Et in conspectu sedis tanquam mare vitreum simile crystallo.* So unto princes and states, and specially towards wise senates and councils, the natures and dispositions of the people, their conditions and necessities, their factions and combinations, their animosities and discontents, ought to be, in regard of the variety of their intelligences, the wisdom of their observations, and the height of their station where they keep sentinel, in great part clear and transparent. Wherefore, considering that I write to a king that is a master of this science, and is so well assisted, I think it decent to pass over this part in silence, as willing to obtain the certificate which one of the ancient philosophers aspired unto; who being silent, when others contended to make demonstration of their abilities by speech, desired it mought be certified for his part,

Totamque infusa: Virgil, *Aeneid* vi. 726–7. 'The mind works in each member of the frame, and stirs the mighty whole.'
inward: secret.
Et in conspectu: Revelation 4:6. 'And before the throne there was a sea of glass, like unto crystal.'

'That there was one that knew how to hold his peace'.

49. *De prudentia legislatoria, sive, de fontibus juris.* Notwithstanding, for the more public part of government, which is laws, I think good to note only one deficience; which is, that all those which have written of laws, have written either as philosophers or as lawyers, and none as statesmen. As for the philosophers, they make imaginary laws for imaginary commonwealths, and their discourses are as the stars, which give little light because they are so high. For the lawyers, they write according to the states where they live what is received law, and not what ought to be law: for the wisdom of a lawmaker is one, and of a lawyer is another. For there are in nature certain fountains of justice, whence all civil laws are derived but as streams: and like as waters do take tinctures and tastes from the soils through which they run, so do civil laws vary according to the regions and governments where they are planted, though they proceed from the same fountains. Again, the wisdom of a lawmaker consisteth not only in a platform of justice, but in the application thereof; taking into consideration by what means laws may be made certain, and what are the causes and remedies of the doubtfulness and incertainty of law; by what means laws may be made apt and easy to be executed, and what are the impediments and remedies in the execution of laws; what influence laws touching private right of *meum* and *tuum* have into the public state, and how they may be made apt and agreeable; how laws are to be penned and delivered, whether in texts or in acts, brief or large, with preambles, or without; how they are to be pruned and reformed from time to time, and what is the best means to keep them from being too vast in volumes, or too full of multiplicity and crossness; how they are to be expounded, when upon causes emergent and judicially discussed, and when upon responses and conferences touching general points or questions; how they are to be pressed, rigorously or tenderly; how they are to be mitigated by equity and good conscience, and whether discretion and strict law are to be mingled in the same courts, or kept apart in several courts; again, how the practice, profession, and erudition of law is to be censured and governed; and many other points touching the administration, and (as I may term it) animation of laws. Upon which I insist the less, because I purpose (if God give me leave),

That there was one: Told of Zeno, in Plutarch, *Of Garrulity* 504.
De prudentia: The wisdom of a legislator, or the fountains of equity.
imaginary laws: Thinking of Plato in the *Laws*.
platform: model, pattern.
meum and tuum: mine and thine.
crossness: intricacy.*

having begun a work of this nature in aphorisms, to propound it hereafter, noting it in the mean time for deficient.

50. And for your Majesty's laws of England, I could say much of their dignity, and somewhat of their defect; but they cannot but excel the civil laws in fitness for the government: for the civil law was *non hos quaesitum munus in usus*; it was not made for the countries which it governeth. Hereof I cease to speak, because I will not intermingle matter of action with matter of general learning.

[CONCLUSION]

XXIV. Thus have I concluded this portion of learning touching civil knowledge; and with civil knowledge have concluded human philosophy; and with human philosophy, philosophy in general. And being now at some pause, looking back into that I have passed through, this writing seemeth to me (*si nunquam fallit imago*), as far as a man can judge of his own work, not much better than that noise or sound which musicians make while they are in tuning their instruments: which is nothing pleasant to hear, but yet is a cause why the music is sweeter afterwards. So have I been content to tune the instruments of the Muses, that they may play that have better hands. And surely, when I set before me the condition of these times, in which learning hath made her third visitation or circuit in all the qualities thereof; as the excellency and vivacity of the wits of this age; the noble helps and lights which we have by the travails of ancient writers; the art of printing, which communicateth books to men of all fortunes; the openness of the world by navigation, which hath disclosed multitudes of experiments, and a mass of natural history; the leisure wherewith these times abound, not employing men so generally in civil business, as the states of Grecia did, in respect of their popularity, and the state of Rome, in respect of the greatness of their monarchy; the present disposition of these times at this instant to peace; the consumption of all that ever can be said in controversies of religion, whch have so much diverted men from other sciences; the perfection of your Majesty's learning, which as a phoenix may call whole vollies

begun a work: A specimen of this, a treatise on Universal Justice, in aphorisms, is given in *De Augmentis* viii. 3. And see *Works* vii. 309 f. for his *Maxims of the Law*.
civil law: i.e. Roman law.
Non hos: Virgil, *Aeneid* iv. 647. A gift meant for different uses.
si nunquam: Virgil, *Eclogues* ii. 27. If the image does not deceive me.
third visitation: See note to p. 78.
popularity: democratic character.
vollies: flights (as of birds).

of wits to follow you; and the inseparable propriety of time, which is ever more and more to disclose truth; I cannot but be raised to this persuasion that this third period of time will far surpass that of the Grecian and Roman learning: only if men will know their own strength, and their own weakness both; and take, one from the other, light of invention, and not fire of contradiction; and esteem of the inquisition of truth as of an enterprise, and not as of a quality or ornament; and employ wit and magnificence to things of worth and excellency, and not to things vulgar and of popular estimation. As for my labours, if any man shall please himself or others in the reprehension of them, they shall make that ancient and patient request, *Verbera, sed audi*; let men reprehend them, so they observe and weigh them. For the appeal is lawful (though it may be it shall not be needful) from the first cogitations of men to their second, and from the nearer times to the times further off. Now let us come to that learning, which both the former times were not so blessed as to know, sacred and inspired divinity, the Sabbath and port of all men's labours and peregrinations.

[OF THEOLOGY]

XXV. 1. The prerogative of God extendeth as well to the reason as to the will of man; so that as we are to obey his law, though we find a reluctation in our will, so we are to believe his word, though we find a reluctation in our reason. For if we believe only that which is agreeable to our sense, we give consent to the matter, and not to the author; which is no more than we would do towards a suspected and discredited witness; but that faith which was accounted to Abraham for righteousness was of such a point as whereat Sarah laughed,[n] who therein was an image of natural reason.

2. Howbeit (if we will truly consider of it) more worthy it is to believe than to know as we now know. For in knowledge man's mind suffereth from sense; but in belief it suffereth from spirit, such one as it holdeth for more authorised than itself, and so suffereth from the worthier agent. Otherwise it is of the state of man glorified; for then faith shall cease, and we shall know as we are known.

3. Wherefore we conclude that sacred theology (which in our idiom we call divinity) is grounded only upon the word and oracle of

propriety: particular nature.
Verbera: Plutarch's 'Life' of Themistocles, 11. 'Strike, but hear.'
reluctation: unwillingness. *suffereth*: is subjected to.
Otherwise it is: it is otherwise. See I Corinthians 13:12.

God, and not upon the light of nature: for it is written, *Coeli enarrant gloriam Dei*; but it is not written, *Coeli enarrant voluntatem Dei*: but of that it is said, *Ad legem et testimonium: si non fecerint secundum verbum istud &c.* This holdeth not only in those points of faith which concern the great mysteries of the Deity, of the creation, of the redemption, but likewise those which concern the law moral truly interpreted: 'Love your enemies: do good to them that hate you: Be like to your heavenly Father, that suffereth his rain to fall upon the just and unjust.' To this it ought to be applauded, *Nec vox hominem sonat*: it is a voice beyond the light of nature. So we see the heathen poets, when they fall upon a libertine passion, do still expostulate with laws and moralities, as if they were opposite and malignant to nature; *Et quod natura remittit, invida jura negant.* So said Dendamis the Indian unto Alexander's messengers, that he had heard somewhat of Pythagoras, and some other of the wise men of Grecia, and that he held them for excellent men: but that they had a fault, which was that they had in too great reverence and veneration a thing they called law and manners. So it must be confessed, that a great part of the law moral is of that perfection, whereunto the light of nature cannot aspire: how then is it that man is said to have, by the light and law of nature, some notions and conceits of virtue and vice, justice and wrong, good and evil? Thus, because the light of nature is used in two several senses; the one, that which springeth from reason, sense, induction, argument, according to the laws of heaven and earth; the other, that which is imprinted upon the spirit of man by an inward instinct, according to the law of conscience, which is a sparkle of the purity of his first estate; in which latter sense only he is participant of some light and discerning touching the perfection of the moral law: but how? sufficient to check the vice, but not to inform the duty. So then the doctrine of religion, as well moral as mystical, is not to be attained but by inspiration and revelation from God.

4. The use notwithstanding of reason in spiritual things,[n] and the latitude thereof, is very great and general: for it is not for nothing

Coeli . . . Psalms 19:1. 'The heavens declare the glory of God'; but not 'the heavens declare the will of God'. *Ad Legem*: Isaiah 8:20. 'To the law and to the testimony: if men do not according to this word [it is because there is no light in them].'

Love your enemies: Matthew 5:44, 45.

Nec vox: Virgil, *Aeneid* i. 328. 'The words do not sound like those of men.'

fall upon: come across.†

Et quod: Ovid, *Metamorphoses* x. 330. 'What nature grants the envious laws deny.'

Dendamis: Plutarch's 'Life' of Alexander, 65.

that the apostle calleth religion 'our reasonable service of God'; insomuch as the very ceremonies and figures of the old law were full of reason and signification, much more than the ceremonies of idolatry and magic, that are full of non-significants and surd characters. But most specially the Christian faith, as in all things so in this, deserveth to be highly magnified; holding and preserving the golden mediocrity in this point between the law of the heathen and the law of Mahumet, which have embraced the two extremes. For the religion of the heathen had no constant belief or confession, but left all to the liberty of argument; and the religion of Mahumet on the other side interdicteth argument altogether: the one having the very face of error, and the other of imposture: whereas the Faith doth both admit and reject disputation with difference.[n]

5. The use of human reason in religion is of two sorts: the former, in the conception and apprehension of the mysteries of God to us revealed; the other, in the inferring and deriving of doctrine and direction thereupon. The former extendeth to the mysteries themselves; but how? by way of illustration, and not by way of argument. The latter consisteth indeed of probation and argument. In the former we see God vouchsafeth to descend to our capacity, in the expressing of his mysteries in sort as may be sensible unto us; and doth grift his revelations and holy doctrine upon the notions of our reason, and applieth his inspirations to open our understanding, as the form of the key to the ward of the lock. For the latter, there is allowed us an use of reason and argument, secondary and respective, although not original and absolute.[n] For after the articles and principles of religion are placed and exempted from examination of reason, it is then permitted unto us to make derivations and inferences from and according to the analogy of them, for our better direction. In nature this holdeth not; for both the principles are examinable by induction, though not by a medium or syllogism; and besides, those principles or first positions have no discordance with that reason which draweth down and deduceth the inferior positions. But yet it holdeth not in religion alone, but in many knowledges, both of greater and smaller nature, namely, wherein there are not only

our reasonable service: St. Paul in Romans 12:1.
non-significants: A word invented by Bacon, cf. p. 133 above.
surd: meaningless. Only Bacon uses the word in this sense.
mediocrity: mean.
interdicteth: prohibits.
grift: graft.
examinable: capable of being examined.*

posita but *placita*; for in such there can be no use of absolute reason. We see it familiarly in games of wit, as chess, or the like. The draughts and first laws of the game are positive, but how? merely *ad placitum*, and not examinable by reason; but then how to direct our play thereupon with best advantage to win the game, is artificial and rational. So in human laws there be many grounds and maxims which are *placita juris*, positive upon authority, and not upon reason, and therefore not to be disputed: but what is most just, not absolutely but relatively, and according to those maxims, that affordeth a long field of disputation. Such therefore is that secondary reason, which hath place in divinity, which is grounded upon the *placets* of God.

6. *De usu legitimo rationis humanae in divinis.* Here therefore I note this deficience, that there hath not been, to my understanding, sufficiently inquired and handled the true limits and use of reason in spiritual things, as a kind of divine dialectic: which for that it is not done, it seemeth to me a thing usual, by pretext of true conceiving that which is revealed, to search and mine into that which is not revealed; and by pretext of *enucleating* inferences and contradictories, to examine that which is positive. The one sort falling into the error of Nicodemus, demanding to have things made more sensible than it pleaseth God to reveal them, *Quomodo possit homo nasci cum sit senex?* The other sort into the error of the disciples, which were scandalized at a show of contradiction, *Quid est hoc quod dicit nobis? Modicum, et non videbitis me; et iterum, modicum, et videbitis me,* &c.

7. Upon this I have insisted the more, in regard of the great and blessed use thereof; for this point well laboured and defined of would in my judgement be an opiate to stay and bridle not only the vanity of curious speculations, wherewith the schools labour, but the fury of controversies, wherewith the church laboureth.[n] For it can-

postia: positive.
placita: arbitrary.
ad placitum: arbitrary.
placita juris: arbitrary rules of law.
placets: what is pleasing to.
De usu: The legitimate use of reason in divine subjects.
enucleating: making clear.
Quomodo: John 3:4. 'How can a man be born when he is old?' This is Nicodemus' question, when Christ says, 'except a man be born again, he cannot see the kingdom of god'.
Quid est: John 16:17. 'What is this that he saith unto us, a little while, and ye shall not see me; and again, a little while and ye shall see me.'
defined of: defined.

not but open men's eyes, to see that many controversies do merely pertain to that which is either not revealed or positive; and that many others do grow upon weak and obscure inferences or derivations: which latter sort, if men would revive the blessed style of that great doctor of the Gentiles, would be carried thus, *ego, non dominus*; and again, *secundum consilium meum*, in opinions and counsels, and not in positions and oppositions. But men are now over-ready to usurp the style, *non ego, sed dominus*; and not so only, but to bind it with the thunder and denunciation of curses and anathemas, to the terror of those which have not sufficiently learned out of Salomon, that 'The causeless curse shall not come.'

8. Divinity hath two principal parts; the matter informed or revealed, and the nature of the information or revelation: and with the latter we will begin, because it hath most coherence with that which we have now last handled. The nature of the information consisteth of three branches; the limits of the information, the sufficiency of the information, and the acquiring or obtaining the information. Unto the limits of the information belong these considerations; how far forth particular persons continue to be inspired; how far forth the Church is inspired; and how far forth reason may be used: the last point whereof I have noted as deficient. Unto the sufficiency of the information belong two considerations; what points of religion are fundamental, and what perfective, being matter of further building and perfection upon one and the same foundation; and again, how the gradations of light according to the dispensation of times are material to the sufficiency of belief.

9. *De gradibus unitatis in civitate Dei*. Here again I may rather give it in advice than note it as deficient, that the points fundamental, and the points of further perfection only, ought to be with piety and wisdom distinguished: a subject tending to much like end as that I noted before; for as that other were likely to abate the number of controversies, so this is like to abate the heat of many of them. We see Moses when he saw the Israelite and the Egyptian fight, he did not say, 'Why strive you?' but drew his sword and slew the Egyptian: but when he saw the two Israelites fight, he said, 'You are brethren, why strive you?' If the point of doctrine be an Egyptian, it must be slain by the sword of the spirit, and not recon-

ego, non dominus: St. Paul in I Corinthians 7:12. 'I, not the Lord.'
secundum: I Corinthians 7:40. 'This is my opinion.'
The causeless: Proverbs 26:2.
perfective: capable of being perfected.
De gradibus: On the degree of unity in the kingdom of God.
Moses: Exodus 2:11–13, Acts 7:24–6.

ciled; but if it be an Israelite, though in the wrong, then, 'Why strive you?' We see of the fundamental points, our Saviour penneth the league thus, 'He that is not with us is against us'; but of points not fundamental, thus, 'He that is not against us is with us.' So we see the coat of our Saviour was entire without seam, and so is the doctrine of the scriptures in itself; but the garment of the church was of divers colours and yet not divided. We see the chaff may and ought to be severed from the corn in the ear, but the tares may not be pulled up from the corn in the field. So as it is a thing of great use well to define what, and of what latitude those points are, which do make men merely aliens and disincorporate from the Church of God.[n]

10. For the obtaining of the information, it resteth upon the true and sound interpretation of the scriptures, which are the fountains of the water of life. The interpretations of the scriptures are of two sorts; methodical, and solute or at large. For this divine water, which excelleth so much that of Jacob's well, is drawn forth much in the same kind as natural water useth to be out of wells and fountains; either it is first forced up into a cistern, and from thence fetched and derived for use; or else it is drawn and received in buckets and vessels immediately where it springeth. The former sort whereof, though it seem to be the more ready, yet in my judgement is more subject to corrupt. This is that method which hath exhibited unto us the scholastical divinity;[n] whereby divinity hath been reduced into an art, as into a cistern, and the streams of doctrine or positions fetched and derived from thence.

11. In this men have sought three things, a summary brevity, a compacted strength, and a complete perfection; whereof the two first they fail to find, and the last they ought not to seek. For as to brevity, we see in all summary methods, while men purpose to abridge, they give cause to dilate. For the sum or abridgement by contraction becometh obscure; the obscurity requireth exposition, and the exposition is deduced into large commentaries, or into common places and titles, which grow to be more vast than the original writings, whence the sum was at first extracted. So we see the volumes of the schoolmen are greater much than the first writings of the

He that is: Matthew 12:30, Luke 9:50.
without seam: John 19:23.
chaff . . . tares: cf. Psalms 1:4, 35:5, Matthew 13:29.
disincorporate: separated from.*
solute: free, discursive.*
Jacob's well: John 4:13–14.
to corrupt: to become corrupt.

fathers, whence the Master of the Sentences[n] made his sum or collection. So in like manner the volumes of the modern doctors of the civil law exceed those of the ancient jurisconsults, of which Tribonian[n] compiled the digest. So as this course of sums and commentaries is that which doth infallibly make the body of sciences more immense in quantity, and more base in substance.

12. And for strength, it is true that knowledges reduced into exact methods[n] have a show of strength, in that each part seemeth to support and sustain the other; but this is more satisfactory than substantial: like unto buildings which stand by architecture and compaction, which are more subject to ruin than those that are built more strong in their several parts, though less compacted. But it is plain that the more you recede from your grounds, the weaker do you conclude: and as in nature, the more you remove yourself from particulars, the greater peril of error you do incur: so much more in divinity, the more you recede from the scriptures by inferences and consequences, the more weak and dilute are your positions.

13. And as for perfection or completeness in divinity, it is not to be sought; which makes this course of artificial divinity the more suspect. For he that will reduce a knowledge into an art, will make it round and uniform: but in divinity many things must be left abrupt, and concluded with this: *O altitudo sapientiae et scientiae Dei! quam incomprehensibilia sunt judicia ejus, et non investigabiles viae ejus*. So again the apostle saith, *Ex parte scimus*: and to have the form of a total, where there is but matter for a part, cannot be without supplies by supposition and presumption. And therefore I conclude, that the true use of these sums and methods hath place in institutions or introductions preparatory unto knowledge: but in them, or by deducement from them, to handle the main body and substance of a knowledge, is in all sciences prejudicial, and in divinity dangerous.

14. As to the interpretation of the scriptures solute and at large, there have been divers kinds introduced and devised; some of them rather curious and unsafe than sober and warranted. Notwithstanding, thus much must be confessed, that the scriptures, being given by inspiration and not by human reason, do differ from all other books in the author: which by consequence doth draw on some

satisfactory: plausible.
dilute: enfeebled.*
O altitudo: Romans 11:33. 'O the depth of the wisdom and knowledge of God! How incomprehensible are his judgements, and his ways past finding out.'
Ex parte: 1 Corinthians 13:9. 'We know in part.'
deducement: deduction.

difference to be used by the expositor. For the inditer of them did know four things which no man attains to know; which are, the mysteries of the kingdom of glory, the perfection of the laws of nature, the secrets of the heart of man, and the future succession of all ages. For as to the first it is said, 'He that presseth into the light, shall be oppressed of the glory.' And again, 'No man shall see my face and live.' To the second, 'When he prepared the heavens I was present, when by law and compass he inclosed the deep.' To the third, 'Neither was it needful that any should bear witness to him of man, for he knew well what was in man.' And to the last, 'From the beginning are known to the Lord all his works.'

15. From the former two of these have been drawn certain senses and expositions of scriptures, which had need be contained within the bounds of sobriety; the one anagogical,[n] and the other philosophical. But as to the former, man is not to prevent his time: *Videmus nunc per speculum in aenigmate, tunc autem facie ad faciem*: wherein nevertheless there seemeth to be a liberty granted, as far forth as the polishing of this glass, or some moderate explication of this aenigma. But to press too far into it, cannot but cause a dissolution and overthrow of the spirit of man. For in the body there are three degrees of that we receive into it, aliment, medicine, and poison: whereof aliment is that which the nature of man can perfectly alter and overcome; medicine is that which is partly converted by nature, and partly converteth nature; and poison is that which worketh wholly upon nature, without that, that nature can in any part work upon it. So in the mind, whatsoever knowledge reason cannot at all work upon and convert is a mere intoxication, and endangereth a dissolution of the mind and understanding.

16. But for the latter, it hath been extremely set on foot of late time by the school of Paracelsus,[n] and some others, that have pretended to find the truth of all natural philosophy in the scriptures; scandalizing and traducing all other philosophy as heathenish and profane. But there is no such enmity between God's word and his works; neither do they give honour to the scriptures, as they suppose, but much imbase them. For to seek heaven and earth in the word of God, whereof it is said, 'Heaven and earth shall pass, but my word shall not pass', is to seek temporary things amongst eternal: and

He that: Proverbs 25:27 (trs. the Vulgate version); *No man*: Exodus 33:20; *When he*: Proverbs 8:27; *Neither was*: John 2:25; *From the*: Acts 15:18.
prevent: anticipate.
Videmus: I Corinthians 13:12. 'For now we see through a glass darkly, but then face to face.'
Heaven and earth: Mark 13:31.

as to seek divinity in philosophy is to seek the living amongst the dead,[n] so to seek philosophy in divinity is to seek the dead amongst the living: neither are the pots or lavers, whose place was in the outward part of the temple, to be sought in the holiest place of all, where the ark of the testimony was seated. And again, the scope or purpose of the spirit of God is not to express matters of nature in the scriptures, otherwise than in passage, and for application to man's capacity and to matters moral or divine. And it is a true rule, *Auctoris aliud agentis parva auctoritas.* For it were a strange conclusion, if a man should use a similitude for ornament or illustration sake, borrowed from nature or history according to vulgar conceit, as of a basilisk, an unicorn, a centaur, a Briareus, an hydra, or the like, that therefore he must needs be thought to affirm the matter thereof positively to be true. To conclude therefore these two interpretations, the one by reduction or aenigmatical, the other philosophical or physical, which have been received and pursued in imitation of the rabbins and cabalists, are to be confined with a *noli altum sapere, sed time.*

17. But the two latter points, known to God and unknown to man, touching the secrets of the heart and the successions of time, doth make a just and sound difference between the manner of the exposition of the scriptures and all other books. For it is an excellent observation which hath been made upon the answers of our Saviour Christ to many of the questions which were propounded to him, how that they are impertinent to the state of the question demanded; the reason whereof is, because not being like man, which knows man's thoughts by his words, but knowing man's thoughts immediately, he never answered their words, but their thoughts. Much in the like manner it is with the scriptures, which being written to the thoughts of men, and to the succession of all ages, with a foresight of all heresies, contradictions, differing estates of the church, yea and particularly of the elect, are not to be interpreted only according to the latitude of the proper sense of the place, and respectively towards that present occasion whereupon the words were uttered, or in precise congruity or contexture with the words before or after, or in contemplation of the principal scope of the place; but have in themselves, not only totally or collectively, but distributively in clauses

lavers: basins for washing.
Auctoris: 'What a man says incidentally about matters which are not in question has little authority.'
noli altum: 'Be not overwise, but fear.'
answers: e.g. in Luke 9:47–8.

and words, infinite springs and streams of doctrine to water the church in every part. And therefore as the literal sense is, as it were, the main stream or river; so the moral sense chiefly, and sometimes the allegorical or typical,[n] are they whereof the church hath most use: not that I wish men to be bold in allegories, or indulgent or light in allusions; but that I do much condemn that interpretation of the scripture which is only after the manner as men use to interpret a profane book.

18. *Emanationes scripturarum in doctrinas positivas.* In this part touching the exposition of the scriptures, I can report no deficience; but by way of remembrance this I will add. In perusing books of divinity, I find many books of controversies, and many of commonplaces and treatises, a mass of positive divinity, as it is made an art: a number of sermons and lectures, and many prolix commentaries upon the scriptures, with harmonies and concordances. But that form of writing in divinity which in my judgement is of all others most rich and precious, is positive divinity, collected upon particular texts of scriptures in brief observations; not dilated into commonplaces, not chasing after controversies, not reduced into method of art; a thing abounding in sermons, which will vanish, but defective in books which will remain, and a thing wherein this age excelleth. For I am persuaded, and I may speak it with an *absit invidia verbo*, and no ways in derogation of antiquity, but as in a good emulation between the vine and the olive, that if the choice and best of those observations upon texts of scriptures, which have been made dispersedly in sermons[n] within this your Majesty's island of Brittany by the space of these forty years and more (leaving out the largeness of exhortations and applications thereupon) had been set down in a continuance, it had been the best work in divinity which had been written since the Apostles' times.

19. The matter informed by divinity is of two kinds; matter of belief and truth of opinion, and matter of service and adoration; which is also judged and directed by the former: the one being as the internal soul of religion, and the other as the external body thereof. And therefore the heathen religion was not only a worship of idols,[n] but the whole religion was an idol in itself; for it had no soul, that is, no certainty of belief or confession: as a man may well think, considering the chief doctors of their church were the poets: and the reason was, because the heathen gods were no jealous gods, but were glad to be admitted into part, as they had

Emanationes: The emanations of the Scriptures in positive doctrines.
absit invidia: without wishing to offend.

reason. Neither did they respect the pureness of heart, so they mought have external honour and rites.

20. But out of these two do result and issue four main branches of divinity; faith, manners, liturgy, and government. Faith containeth the doctrine of the nature of God, of the attributes of God, and of the works of God. The nature of God consisteth of three persons in unity of Godhead. The attributes of God are either common to the Deity, or respective to the persons. The works of God summary are two, that of the creation and that of the redemption; and both these works, as in total they appertain to the unity of the Godhead, so in their parts they refer to the three persons: that of the creation, in the mass of the matter, to the Father; in the disposition of the form, to the Son; and in the continuance and conservation of the being, to the Holy Spirit. So that of the redemption, in the election and counsel, to the Father; in the whole act and consummation, to the Son; and in the application, to the Holy Spirit; for by the Holy Ghost was Christ conceived in flesh, and by the Holy Ghost are the elect regenerate in spirit. This work likewise we consider either effectually, in the elect; or privatively, in the reprobate; or according to appearance, in the visible church.

21. For manners, the doctrine thereof is contained in the law, which discloseth sin. The law itself is divided, according to the edition thereof, into the law of nature, the law moral, and the law positive; and according to the style, into negative and affirmative, prohibitions and commandments. Sin, in the matter and subject thereof, is divided according to the commandments; in the form thereof, it referreth to the three persons in Deity: sins of infirmity against the Father, whose more special attribute is power; sins of ignorance against the Son, whose attribute is wisdom; and sins of malice against the Holy Ghost, whose attribute is grace or love. In the motions of it, it either moveth to the right hand or to the left; either to blind devotion, or to profane and libertine transgression; either in imposing restraint where God granteth liberty, or in taking liberty where God imposeth restraint. In the degrees and progress of it, it divideth itself into thought, word, or act.[n] And in this part I commend much the deducing of the law of God to cases of conscience; for that I take indeed to be a breaking, and not exhibiting whole of the bread of life. But that which quickeneth both these doctrines of faith and manners, is the elevation and consent of the heart; where-

edition: promulgation.

unto appertain books of exhortation, holy meditation, Christian resolution, and the like.

22. For the liturgy or service, it consisteth of the reciprocal acts between God and man; which, on the part of God, are the preaching of the word, and the sacraments, which are seals to the covenant, or as the visible word; and on the part of man, invocation of the name of God; and under the law, sacrifices; which were as visible prayers or confessions: but now the adoration being *in spiritu et veritate*, there remaineth only *vituli labiorum*; although the use of holy vows of thankfulness and retribution may be accounted also as sealed petitions.

23. And for the government of the church, it consisteth of the patrimony of the church, the franchises of the church, and the offices and jurisdictions of the church, and the laws of the church directing the whole; all which have two considerations, the one in themselves, the other how they stand compatible and agreeable to the civil estate.

24. This matter of divinity is handled either in form of instruction of truth, or in form of confutation of falsehood. The declinations from religion, besides the privative, which is atheism and the branches thereof, are three; heresies, idolatry, and witchcraft: heresies, when we serve the true God with a false worship; idolatry, when we worship false gods, supposing them to be true; and witchcraft, when we adore false gods, knowing them to be wicked and false. For so your Majesty[n] doth excellently well observe, that witchcraft is the height of idolatry. And yet we see though these be true degrees, Samuel teacheth us that they are all of a nature, when there is once a receding from the word of God; for so he saith, *Quasi peccatum ariolandi est repugnare, et quasi scelus idololatriae nolle acquiescere*.

25. These things I have passed over so briefly because I can report no deficience concerning them: for I can find no space or ground that lieth vacant and unsown in the matter of divinity: so diligent have men been, either in sowing of good seed, or in sowing of tares.

Thus have I made as it were a small globe of the intellectual world,

in spiritu: John 4:23. 'In spirit and in truth.'
vituli: Hosea 14:2. 'Offerings of the lips.'
declinations: fallings away.
Quasi peccatum: 1 Samuel 15:23. 'Rebellion is as the sin of witchcraft, and stubbornness as the sin of idolatry.'

as truly and faithfully as I could discover; with a note and description of those parts which seem to me not constantly occupate, or not well converted by the labour of man. In which, if I have in any point receded from that which is commonly received, it hath been with a purpose of proceeding *in melius*, and not *in aliud*; a mind of amendment and proficience, and not of change and difference. For I could not be true and constant to the argument I handle, if I were not willing to go beyond others; but yet not more willing than to have others go beyond me again: which may the better appear by this, that I have propounded my opinions naked and unarmed, not seeking to preoccupate the liberty of men's judgements by confutations. For in anything which is well set down, I am in good hope, that if the first reading move an objection, the second reading will make an answer. And in those things wherein I have erred, I am sure I have not prejudiced the right by litigious arguments; which certainly have this contrary effect and operation, that they add authority to error, and destroy the authority of that which is well invented. For question is an honour and preferment to falsehood, as on the other side it is a repulse to truth. But the errors I claim and challenge to myself as mine own. The good, if any be, is due *tanquam adeps sacrificii*, to be incensed to the honour, first of the Divine Majesty, and next of your Majesty, to whom on earth I am most bounden.

in melius: for the better.
in aliud: to be different.
preoccupate: influence, usurp.
question: to call in question.
tanquam: Leviticus 3:16. 'As the fat of the sacrifice.'
incensed: to be offered as incense.*

New Atlantis

TO THE READER

This fable my Lord devised, to the end that he might exhibit therein a model or description of a college instituted for the interpreting of nature and the producing of great and marvellous works for the benefit of men, under the name of Salomon's House, or the College of the Six Days' Works. And even so far his Lordship hath proceeded, as to finish that part. Certainly the model is more vast and high than can possibly be imitated in all things; notwithstanding most things therein are within men's power to effect. His Lordship thought also in this present fable to have composed a frame of Laws, or of the best state or mould of a commonwealth; but foreseeing it would be a long work, his desire of collecting the Natural History diverted him, which he preferred many degrees before it.

This work of the *New Atlantis*[n] (as much as concerneth the English edition) his Lordship designed for this place;[n] in regard it hath so near affinity (in one part of it) with the preceding Natural History.

W. RAWLEY

New Atlantis

We sailed from Peru, (where we had continued by the space of one whole year,) for China and Japan, by the South Sea[1]; taking with us victuals for twelve months; and had good winds from the east, though soft and weak, for five months' space and more. But then the wind came about, and settled in the west for many days, so as we could make little or no way, and were sometimes in purpose to turn back. But then again there arose strong and great winds from the south, with a point east; which carried us up (for all that we could do) towards the north: by which time our victuals failed us, though we had made good spare of them. So that finding ourselves in the midst of the greatest wilderness of waters in the world, without victual, we gave ourselves for lost men, and prepared for death. Yet we did lift up our hearts and voices to God above, who 'showeth his wonders in the deep'; beseeching him of his mercy, that as in the beginning he discovered the face of the deep, and brought forth dry land, so he would now discover land to us, that we might not perish. And it came to pass that the next day about evening, we saw within a kenning before us, towards the north, as it were thick clouds, which did put us in some hope of land; knowing how that part of the South Sea was utterly unknown; and might have islands or continents, that hitherto were not come to light. Wherefore we bent our course thither, where we saw the appearance of land, all that night; and in the dawning of the next day, we might plainly discern that it was a land; flat to our sight, and full of boscage; which made it shew the more dark. And after an hour and a half's sailing, we entered into a good haven, being the port of a fair city; not great indeed, but well built, and that gave a pleasant view from the sea: and we thinking every minute long till we were on land, came close to the shore, and offered to land. But straightways we saw divers of the people, with bastons in their hands, as it were forbidding us to land; yet without any cries or fierceness, but only as warning us off by signs that they made. Whereupon being not a little discomforted, we were advising with ourselves what we should do. During which time there made forth to us a small boat, with about eight persons in it; whereof one of them had in his hand a tipstaff of a

good spare: frugal use.
showeth his wonders: Psalms, 107:24.
within a kenning: within sight, i.e. about 20 miles.
boscage: trees and shrubs.
divers: various persons. *bastons*: staves.
tipstaff: a staff carried as a badge by an official.

yellow cane, tipped at both ends with blue, who came aboard our ship, without any show of distrust at all. And when he saw one of our number present himself somewhat afore the rest, he drew forth a little scroll of parchment, (somewhat yellower than our parchment, and shining like the leaves of writing tables, but otherwise soft and flexible,) and delivered it to our foremost man. In which scroll were written in ancient Hebrew, and in ancient Greek, and in good Latin of the School, and in Spanish, these words; 'Land ye not, none of you; and provide to be gone from this coast within sixteen days, except you have further time given you. Meanwhile, if you want fresh water, or victual, or help for your sick, or that your ship needeth repair, write down your wants, and you shall have that which belongeth to mercy.' This scroll was signed with a stamp of cherubins' wings, not spread but hanging downwards, and by them a cross. This being delivered, the officer returned, and left only a servant with us to receive our answer. Consulting hereupon amongst ourselves, we were much perplexed. The denial of landing and hasty warning us away troubled us much; on the other side, to find that the people had languages and were so full of humanity, did comfort us not a little. And above all, the sign of the cross to that instrument was to us a great rejoicing, and as it were a certain presage of good. Our answer was in the Spanish tongue; 'That for our ship, it was well; for we had rather met with calms and contrary winds than any tempests. For our sick, they were many, and in very ill case; so that if they were not permitted to land, they ran danger of their lives.' Our wants we set down in particular; adding, 'that we had some little store of merchandise, which if it pleased them to deal for, it might supply our wants without being chargeable unto them'. We offered some reward in pistolets unto the servant, and a piece of crimson velvet to be presented to the officer; but the servant took them not, nor would scarce look upon them; and so left us, and went back in another little boat which was sent for him.

About three hours after we had dispatched our answer, there came towards us a person (as it seemed) of place. He had on him a gown with wide sleeves, of a kind of water chamolet, of an excellent azure colour, far more glossy than ours; his under apparel was green; and so was his hat, being in the form of a turban, daintily made, and

tables: thin sheets of ivory or wood, for making notes on.
School: correct classical Latin.
pistolets: Spanish gold coins.
of place: of high station.
water chamolet: a fabric made of goat's hair, and wavy like water.

not so huge as the Turkish turbans; and the locks of his hair came down below the brims of it. A reverend man was he to behold. He came in a boat, gilt in some part of it, with four persons more only in that boat; and was followed by another boat, wherein were some twenty. When he was come within a flight-shot of our ship, signs were made to us that we should send forth some to meet him upon the water; which we presently did in our ship-boat, sending the principal man amongst us save one, and four of our number with him. When we were come within six yards of their boat, they called to us to stay, and not to approach farther; which we did. And thereupon the man whom I before described stood up, and with a loud voice in Spanish, asked, 'Are ye Christians?' We answered, 'We were'; fearing the less, because of the cross we had seen in the subscription. At which answer the said person lifted up his right hand towards heaven, and drew it softly to his mouth, (which is the gesture they use when they thank God,) and then said: 'If ye will swear (all of you) by the merits of the Saviour that ye are no pirates, nor have shed blood lawfully nor unlawfully within forty days past, you may have licence to come on land.' We said, 'We were all ready to take that oath.' Whereupon one of those that were with him, being (as it seemed) a notary, made an entry of this act. Which done, another of the attendants of the great person, which was with him in the same boat, after his lord had spoken a little to him, said aloud; 'My lord would have you know, that it is not of pride or greatness that he cometh not aboard your ship; but for that in your answer you declare that you have many sick amongst you, he was warned by the Conservator of Health of the city that he should keep a distance.' We bowed ourselves towards him, and answered, 'We were his humble servants; and accounted for great honour and singular humanity towards us that which was already done; but hoped well that the nature of the sickness of our men was not infectious.' So he returned; and a while after came the notary to us aboard our ship; holding in his hand a fruit of that country, like an orange, but of colour between orange-tawney and scarlet, which cast a most excellent odour. He used it (as it seemeth) for a preservative against infection. He gave us our oath; 'By the name of Jesus and his merits': and after told us that the next day by six of the clock in the morning we should be sent to, and brought to the Strangers' House, (so he called it,) where we should be accommodated of things both for our whole and for our sick. So he left us; and when we offered him

flight-shot: a bowshot, i.e. 200 to 300 yards.
presently: at once. *for that*: because.

some pistolets, he smiling said, 'He must not be twice paid for one labour': meaning (as I take it) that he had salary sufficient of the state for his service. For (as I after learned) they call an officer that taketh rewards, *twice paid*.

The next morning early, there came to us the same officer that came to us at first with his cane, and told us, 'He came to conduct us to the Strangers' House; and that he had prevented the hour, because we might have the whole day before us for our business. 'For,' said he, 'if you will follow my advice, there shall first go with me some few of you, and see the place, and how it may be made convenient for you; and then you may send for your sick, and the rest of your number which ye will bring on land.' We thanked him, and said, 'That this care which he took of desolate strangers God would reward.' And so six of us went on land with him: and when we were on land, he went before us, and turned to us, and said, 'He was but our servant, and our guide.' He led us through three fair streets; and all the way we went there were gathered some people on both sides standing in a row; but in so civil a fashion, as if it had been not to wonder at us but to welcome us; and divers of them, as we passed by them, put their arms a little abroad; which is their gesture when they bid any welcome. The Strangers' House is a fair and spacious house, built of brick, of somewhat a bluer colour than our brick; and with handsome windows, some of glass, some of a kind of cambric oiled. He brought us first into a fair parlour above stairs, and then asked us, 'What number of persons we were? And how many sick?' We answered 'We were in all (sick and whole) one and fifty persons, whereof our sick were seventeen.' He desired us to have patience a little, and to stay till he came back to us; which was about an hour after; and then he led us to see the chambers which were provided for us, being in number nineteen: they having cast it (as it seemeth) that four of those chambers, which were better than the rest, might receive four of the principal men of our company, and lodge them alone by themselves; and the other fifteen chambers were to lodge us two and two together. The chambers were handsome and cheerful chambers, and furnished civilly. Then he led us to a long gallery, like a dorture, where he showed us all along the one side (for the other side was but wall and window) seventeen cells, very neat ones, having partitions of cedar wood. Which gallery and cells, being in all forty, (many more than we needed,) were instituted as an infirmary for sick persons. And he

prevented: come before. *cast*: arranged. *dorture*: dormitory.

told us withal, that as any of our sick waxed well, he might be removed from his cell to a chamber; for which purpose there were set forth ten spare chambers, besides the number we spake of before. This done, he brought us back to the parlour, and lifting up his cane a little, (as they do when they give any charge or command,) said to us, 'Ye are to know that the custom of the land requireth, that after this day and to-morrow, (which we give you for removing of your people from your ship,) you are to keep within doors for three days. But let it not trouble you, nor do not think yourselves restrained, but rather left to your rest and ease. You shall want nothing, and there are six of our people appointed to attend you, for any business you may have abroad.' We gave him thanks with all affection and respect, and said, 'God surely is manifested in this land.' We offered him also twenty pistolets; but he smiled, and only said; 'What? twice paid!' And so he left us. Soon after our dinner was served in; which was right good viands, both for bread and meat: better than any collegiate diet that I have known in Europe. We had also drink of three sorts, all wholesome and good; wine of the grape; a drink of grain, such as is with us our ale, but more clear; and a kind of cider made of a fruit of that country; a wonderful pleasing and refreshing drink. Besides, there were brought in to us great store of those scarlet oranges for our sick; which (they said) were an assured remedy for sickness taken at sea. There was given us also a box of small grey or whitish pills, which they wished our sick should take, one of the pills every night before sleep; which (they said) would hasten their recovery. The next day, after that our trouble of carriage and removing of our men and goods out of our ship was somewhat settled and quiet, I thought good to call our company together; and when they were assembled said unto them; 'My dear friends, let us know ourselves, and how it standeth with us. We are men cast on land, as Jonas was out of the whale's belly, when we were as buried in the deep: and now we are on land, we are but between death and life; for we are beyond both the old world and the new; and whether ever we shall see Europe, God only knoweth. It is a kind of miracle hath brought us hither: and it must be little less that shall bring us hence. Therefore in regard of our deliverance past, and our danger present and to come, let us look up to God, and every man reform his own ways. Besides we are come here amongst a Christian people, full of piety and humanity: let us not bring that confusion of face upon ourselves, as to show our vices or unworthiness before them. Yet there is more. For they have by commandment (though in form of courtesy) cloistered us within these walls for three days: who knoweth whether it be not to take some taste of our manners and conditions? and if they find them bad, to banish us straightways; if good, to give us further time. For these men that they have given us for attendance may withal have an eye upon us. Therefore for God's love, and as we love the weal of our souls and bodies, let us so behave ourselves as we may be at peace with God, and may find grace in the eyes of this people.' Our company with one voice thanked me for my good admonition, and promised me to live soberly and civilly, and without giving any the least occasion of offence. So we spent our three days joyfully and without care, in expectation what would be done with us when they were expired. During which time, we had every hour joy of the amendment of our sick; who thought themselves cast into some divine pool of healing, they mended so kindly and so fast.

The morrow after our three days were past, there came to us a new man that we had not seen before, clothed in blue as the former was, save that his turban was white, with a small red cross on the top. He had also a tippet of fine linen. At his coming in, he did bend to us a little, and put his arms abroad. We of our parts saluted him in a very lowly and submissive manner; as looking that from him we should receive sentence of life or death. He desired to speak with some few of us: whereupon six of us only stayed, and the rest avoided the room. He said, 'I am by office governor of this House of Strangers, and by vocation I am a Christian priest; and therefore am come to you to offer you my service, both as strangers and chiefly as Christians. Some things I may tell you, which I think you will not be unwilling to hear. The state hath given you licence to stay on land for the space of six weeks: and let it not trouble you if your occasions ask further time, for the law in this point is not precise; and I do not doubt but myself shall be able to obtain for you such further time as may be convenient. Ye shall also understand, that the Strangers' House is at this time rich, and much aforehand; for it hath laid up revenue these thirty-seven years; for so long it is since any stranger arrived in this part: and therefore take ye no care; the state will defray you all the time you stay; neither shall you stay one day the less for that. As for any merchandise ye have brought, ye shall be well used, and have your return either in merchandise or in gold and silver: for to us it is all one. And if you have any other request to make, hide it not. For ye shall find we will not make your

so kindly: so naturally.
tippet: cape. *avoided*: left.
aforehand: provided for the future.*

countenance to fall by the answer ye shall receive. Only this I must tell you, that none of you must go above a *karan*' (that is with them a mile and a half) 'from the walls of the city, without especial leave.' We answered, after we had looked awhile one upon another admiring this gracious and parent-like usage; 'That we could not tell what to say: for we wanted words to express our thanks; and his noble free offers left us nothing to ask. It seemed to us that we had before us a picture of our salvation in heaven; for we that were awhile since in the jaws of death, were now brought into a place where we found nothing but consolations. For the commandment laid upon us, we would not fail to obey it, though it was impossible but our hearts should be inflamed to tread further upon this happy and holy ground.' We added; 'That our tongues should first cleave to the roofs of our mouths, ere we should forget either his reverend person or this whole nation in our prayers.' We also most humbly besought him to accept of us as his true servants, by as just a right as ever men on earth were bounden; laying and presenting both our persons and all we had at his feet. He said; 'He was a priest, and looked for a priest's reward: which was our brotherly love and the good of our souls and bodies.' So he went from us, not without tears of tenderness in his eyes; and left us also confused with joy and kindness, saying amongst ourselves, 'That we were come into a land of angels, which did appear to us daily and prevent us with comforts, which we thought not of, much less expected.'

The next day, about ten of the clock, the governor came to us again, and after salutations said familiarly, 'That he was come to visit us': and called for a chair, and sat him down: and we, being some ten of us, (the rest were of the meaner sort, or else gone abroad,) sat down with him. And when we were set, he began thus: 'We of this island of Bensalem,' (for so they call it in their language,) 'have this; that by means of our solitary situation, and of the laws of secrecy which we have for our travellers, and our rare admission of strangers, we know well most part of the habitable world, and are ourselves unknown. Therefore because he that knoweth least is fittest to ask questions, it is more reason, for the entertainment of the time, that ye ask me questions, than that I ask you.' We answered; 'That we humbly thanked him that he would give us leave so to do: and that we conceived by the taste we had already, that there was no worldly thing on earth more worthy to be known than the state of that happy land. But above all,' (we said,) 'since

admiring: wondering at.

that we were met from the several ends of the world, and hoped assuredly that we should meet one day in the kingdom of heaven, (for that we were both parts Christians,) we desired to know (in respect that land was so remote, and so divided by vast and unknown seas, from the land where our Saviour walked on earth,) who was the apostle of that nation, and how it was converted to the faith?' It apeared in his face that he took great contentment in this our question: he said, 'Ye knit my heart to you, by asking this question in the first place; for it sheweth that you 'first seek the kingdom of heaven'; and I shall gladly and briefly satisfy your demand.

'About twenty years after the ascension of our Saviour, it came to pass that there was seen by the people of Renfusa, (a city upon the eastern coast of our island,) within night, (the night was cloudy and calm,) as it might be some mile into the sea, a great pillar of light; not sharp, but in form of a column or cylinder, rising from the sea a great way up towards heaven: and on the top of it was seen a large cross of light, more bright and resplendent than the body of the pillar. Upon which so strange a spectacle, the people of the city gathered apace together upon the sands, to wonder; and so after put themselves into a number of small boats, to go nearer to this marvellous sight. But when the boats were come within about sixty yards of the pillar, they found themselves all bound, and could go no further; yet so as they might move to go about, but might not approach nearer: so as the boats stood all as in a theatre, beholding this light as an heavenly sign. It so fell out, that there was in one of the boats one of the wise men of the society of Salomon's House; which house or college (my good brethren) is the very eye of this kingdom; who having awhile attentively and devoutly viewed and contemplated this pillar and cross, fell down upon his face; and then raised himself upon his knees, and lifting up his hands to heaven, made his prayers in this manner:

'"Lord God of heaven and earth, thou hast vouchsafed of thy grace to those of our order, to know thy works of creation, and the secrets of them; and to discern (as far as appertaineth to the generations of men) between divine miracles, works of nature, works of art, and impostures and illusions of all sorts. I do here acknowledge and testify before this people, that the thing which we now see before

in respect: considering that.
first seek: Cf. Matthew 6:33, and Luke 12:31.
within night: by night.
very eye: the true centre of intelligence and light.

our eyes is thy Finger and a true Miracle; and forasmuch as we learn in our books that thou never workest miracles but to a divine and excellent end, (for the laws of nature are thine own laws, and thou exceedest them not but upon great cause,) we most humbly beseech thee to prosper this great sign, and to give us the interpretation and use of it in mercy; which thou dost in some part secretly promise by sending it unto us."[n]

'When he had made his prayer, he presently found the boat he was in moveable and unbound; whereas all the rest remained still fast; and taking that for an assurance of leave to approach, he caused the boat to be softly and with silence rowed towards the pillar. But ere he came near it, the pillar and cross of light brake up, and cast itself abroad, as it were, into a firmament of many stars; which also vanished soon after, and there was nothing left to be seen but a small ark or chest of cedar, dry, and not wet at all with water, though it swam. And in the fore-end of it, which was towards him, grew a small green branch of palm; and when the wise man had taken it with all reverence into his boat, it opened of itself, and there were found in it a Book and a Letter; both written in fine parchment, and wrapped in sindons of linen. The Book contained all the canonical books of the Old and New Testament, according as you have them, (for we know well what the Churches with you receive); and the Apocalypse itself, and some other books of the New Testament which were not at that time written, were nevertheless in the Book. And for the Letter, it was in these words:

' "I Bartholomew,[n] a servant of the Highest, and Apostle of Jesus Christ, was warned by an angel that appeared to me in a vision of glory, that I should commit this ark to the floods of the sea. Therefore I do testify and declare unto that people where God shall ordain this ark to come to land, that in the same day is come unto them salvation and peace and goodwill from the Father, and from the Lord Jesus."

'There was also in both these writings, as well the Book as the Letter, wrought a great miracle, conform to that of the Apostles in the original Gift of Tongues. For there being at this time in this land Hebrews, Persians, and Indians, besides the natives, every one read upon the Book and Letter, as if they had been written in his own language. And thus was this land saved from infidelity (as the remain of the old world was from water) by an ark, through the

sindons: pieces of fine linen.
conform to: corresponding to.
remain: remnant.

apostolical and miraculous evangelism of St. Bartholomew.' And here he paused, and a messenger came, and called him from us. So this was all that passed in that conference.

The next day, the governor came again to us immediately after dinner, and excused himself, saying, 'That the day before he was called from us somewhat abruptly, but now he would make us amends, and spend time with us, if we held his company and conference agreeable.' We answered, 'That we held it so agreeable and pleasing to us, as we forgot both dangers past and fears to come, for the time we heard him speak; and that we thought an hour spent with him, was worth years of our former life.' He bowed himself a little to us, and after we were set again, he said; 'Well, the questions are on your part.' One of our number said, after a little pause; 'That there was a matter we were no less desirous to know, than fearful to ask, lest we might presume too far. But encouraged by his rare humanity towards us, (that could scarce think ourselves strangers, being his vowed and professed servants,) we would take the hardiness to propound it: humbly beseeching him, if he thought it not fit to be answered, that he would pardon it, though he rejected it.' We said; 'We well observed those his words, which he formerly spake, that this happy island where we now stood was known to few, and yet knew most of the nations of the world; which we found to be true, considering they had the languages of Europe, and knew much of our state and business; and yet we in Europe (notwithstanding all the remote discoveries and navigations of this last age,) never heard any of the least inkling or glimpse of this island. This we found wonderful strange; for that all nations have inter-knowledge one of another either by voyage into foreign parts, or by strangers that come to them: and though the traveller into a foreign country doth commonly know more by the eye than he that stayeth at home can by relation of the traveller; yet both ways suffice to make a mutual knowledge, in some degree, on both parts. But for this island, we never heard tell of any ship of theirs that had been seen to arrive upon any shore of Europe; no, nor of either the East or West Indies; nor yet of any ship of any other part of the world that had made return from them. And yet the marvel rested not in this. For the situation of it (as his lordship said) in the secret conclave of such a vast sea might cause it. But then that they should have knowledge of the languages, books, affairs, of those that lie such a distance from them, it was a thing we could not tell what to make of; for that

inter-knowledge: mutual knowledge.*

it seemed to us a condition and propriety of divine powers and beings, to be hidden and unseen to others, and yet to have others open and as in a light to them.' At this speech the governor gave a gracious smile, and said; 'That we did well to ask pardon for this question we now asked; for that it imported as if we thought this land a land of magicians, that sent forth spirits of the air into all parts, to bring them news and intelligence of other countries.' It was answered by us all, in all possible humbleness, but yet with a countenance taking knowledge that we knew that he spake it but merrily, 'That we were apt enough to think there was somewhat supernatural in this island; but yet rather as angelical than magical. But to let his lordship know truly what it was that made us tender and doubtful to ask this question, it was not any such conceit, but because we remembered he had given a touch in his former speech, that this land had laws of secrecy touching strangers.' To this he said; 'You remember it aright; and therefore in that I shall say to you I must reserve some particulars, which it is not lawful for me to reveal; but there will be enough left to give you satisfaction.

'You shall understand (that which perhaps you will scarce think credible) that about three thousand years ago[n] or somewhat more, the navigation of the world, (specially for remote voyages,) was greater than at this day. Do not think with yourselves that I know not how much it is increased with you within these six-score years;[n] I know it well: and yet I say greater then than now; whether it was, that the example of the ark, that saved the remnant of men from the universal deluge, gave men confidence to adventure upon the waters; or what it was; but such is the truth. The Phoenicians, and especially the Tyrians, had great fleets. So had the Carthaginians, their colony, which is yet further west. Toward the east, the shipping of Egypt and of Palestina was likewise great. China also, and the great Atlantis (that you call America), which have now but junks and canoes, abounded then in tall ships. This island (as appeareth by faithful registers of those times) had then fifteen hundred strong ships, of great content. Of all this there is with you sparing memory, or none; but we have large knowledge thereof.

'At that time, this land was known and frequented by the ships and vessels of all the nations before named. And (as it cometh to pass) they had many times men of other countries, that were no sailors, that came with them; as Persians, Chaldeans, Arabians; so as almost

propriety: property.
taking knowledge: with a look of intelligence.
conceit: idea.

all nations of might and fame resorted hither; of whom we have some stirps and little tribes with us at this day. And for our own ships, they went sundry voyages, as well as to your Straits, which you call the Pillars of Hercules,[n] as to other parts in the Atlantic and Mediterrane Seas; as to Paguin (which is the same with Cambaline) and Quinzy, upon the Oriental Seas, as far as to the borders of the East Tartary.

'At the same time, and an age after, or more, the inhabitants of the great Atlantis did flourish. For though the narration and description which is made by a great man[n] with you, that the descendants of Neptune planted there; and of the magnificent temple, palace, city, and hill; and the manifold streams of goodly navigable rivers, (which, as so many chains, environed the same site and temple); and the several degrees of ascent whereby men did climb up to the same, as if it had been a *scala coeli*; be all poetical and fabulous: yet so much is true, that the said country of Atlantis, as well that of Peru, then called Coya, as that of Mexico, then named Tyrambel, were mighty and proud kingdoms in arms, shipping, and riches: so mighty, as at one time (or at least within the space of ten years) they both made two great expeditions; they of Tyrambel through the Atlantic to the Mediterrane Sea; and they of Coya through the South Sea upon this our island. And for the former of these, which was in to Europe, the same author amongst you (as it seemeth) had some relation from the Egyptian priest whom he citeth. For assuredly such a thing there was. But whether it were the ancient Athenians that had the glory of the repulse and resistance of those forces, I can say nothing: but certain it is, there never came back either ship or man from that voyage. Neither had the other voyage of those of Coya upon us had better fortune, if they had not met with enemies of greater clemency. For the king of this island (by name Altabin) a wise man and a great warrior, knowing well both his own strength and that of his enemies, handled the matter so, as he cut off their land-forces from their ships; and entoiled both their navy and their camp with a greater power than theirs, both by sea and land; and compelled them to render themselves without striking stroke: and after they were at his mercy, contenting himself only with their oath that they should no more bear arms against him, dismissed them all in safety. But the Divine Revenge overtook not long after those proud enterprises. For within less than

stirps: families.
scala coeli: a ladder to heaven, Genesis 28:12.
entoiled: trapped.

the space of one hundred years, the great Atlantis was utterly lost and destroyed: not by a great earthquake, as your man saith, (for that whole tract is little subject to earthquakes,) but by a particular deluge or inundation; those countries having, at this day, far greater rivers and far higher mountains to pour down waters, than any part of the old world. But it is true that the same inundation was not deep; not past forty foot, in most places, from the ground: so that although it destroyed man and beast generally, yet some few wild inhabitants of the wood escaped. Birds also were saved by flying to the high trees and woods. For as for men, although they had buildings in many places higher than the depth of the water, yet that inundation, though it were shallow, had a long continuance; whereby they of the vale that were not drowned, perished for want of food and other things necessary. So as marvel you not at the thin population of America, nor at the rudeness and ignorance of the people; for you must account your inhabitants of America as a young people; younger a thousand years, at the least, than the rest of the world; for that there was so much time between the universal flood and their particular inundation. For the poor remnant of human seed which remained in their mountains peopled the country again slowly, by little and little; and being simple and savage people, (not like Noah and his sons, which was the chief family of the earth,) they were not able to leave letters, arts, and civility to their posterity; and having likewise in their mountainous habitations been used (in respect of the extreme cold of those regions) to clothe themselves with the skins of tigers, bears, and great hairy goats, that they have in those parts; when after they came down into the valley, and found the intolerable heats which are there, and knew no means of lighter apparel, they were forced to begin the custom of going naked, which continueth at this day. Only they take great pride and delight in the feathers of birds, and this also they took from those their ancestors of the mountains, who were invited unto it by the infinite flights of birds that came up to the high grounds, while the waters stood below. So you see, by this main accident of time, we lost our traffic with the Americans, with whom of all others, in regard they lay nearest to us, we had most commerce. As for the other parts of the world, it is most manifest that in the ages following (whether it were in respect of wars, or by a natural revolution of time,) navigation did every where greatly decay; and specially far voyages (the rather by the

particular: partial.
civility: civilization.
in respect of: because of.

use of galleys, and such vessels as could hardly brook the ocean,) were altogether left and omitted. So then, that part of intercourse which could be from other nations to sail to us, you see how it hath long since ceased; except it were by some rare accident, as this of yours. But now of the cessation of that other part of intercourse, which might be by our sailing to other nations, I must yield you some other cause. For I cannot say (if I shall say truly,) but our shipping, for number, strength, mariners, pilots, and all things that appertain to navigation, is as great as ever: and therefore why we should sit at home, I shall now give you an account by itself: and it will draw nearer to give you satisfaction to your principal question.

'There reigned in this island, about nineteen hundred years ago, a King, whose memory of all others we most adore; not superstitiously, but as a divine instrument, though a mortal man; his name was Solamona;[n] and we esteem him as the lawgiver of our nation. This king had a *large heart*,[n] inscrutable for good; and was wholly bent to make his kingdom and people happy. He therefore, taking into consideration how sufficient and substantive this land was to maintain itself without any aid at all of the foreigner; being five thousand six hundred miles in circuit, and of rare fertility of soil in the greatest part thereof; and finding also the shipping of this country might be plentifully set on work, both by fishing and by transportations from port to port, and likewise by sailing into some small islands that are not far from us, and are under the crown and laws of this state; and recalling into his memory the happy and flourishing estate wherein this land then was, so as it might be a thousand ways altered to the worse, but scarce any one way to the better; thought nothing wanted to his noble and heroical intentions, but only (as far as human foresight might reach) to give perpetuity to that which was in his time so happily established. Therefore amongst his other fundamental laws of this kingdom, he did ordain the interdicts and prohibitions which we have touching entrance of strangers; which at that time (though it was after the calamity of America) was frequent; doubting novelties,[n] and commixture of manners. It is true, the like law against the admission of strangers without licence is an ancient law in the kingdom of China, and yet continued in use. But there it is a poor thing; and hath made them a curious, ignorant, fearful, foolish nation. But our lawgiver made his law of another temper. For first, he hath preserved all points of humanity, in taking

substantive: self-sufficing.

order and making provision for the relief of strangers distressed; whereof you have tasted.' At which speech (as reason was) we all rose up, and bowed ourselves. He went on. 'That king also, still desiring to join humanity and policy together; and thinking it against humanity to detain strangers here against their wills, and against policy that they should return and discover their knowledge of this estate, he took this course: he did ordain that of the strangers that should be permitted to land, as many (at all times) might depart as would; but as many as would stay should have very good conditions and means to live from the state. Wherein he saw so far, that now in so many ages since the prohibition, we have memory not of one ship that ever returned; and but of thirteen persons only, at several times, that chose to return in our bottoms. What those few that returned may have reported abroad I know not. But you must think, whatsoever they have said could be taken where they came but for a dream. Now for our travelling from hence into parts abroad, our Lawgiver thought fit altogether to restrain it. So is it not in China. For the Chineses sail where they will or can; which sheweth that their law of keeping out strangers is a law of pusillanimity and fear. But this restraint of ours hath one only exception, which is admirable; preserving the good which cometh by communicating with strangers, and avoiding the hurt; and I will now open it to you. And here I shall seem a little to digress, but you will by and by find it pertinent. Ye shall understand (my dear friends) that amongst the excellent acts of that king, one above all hath the pre-eminence. It was the erection and institution of an Order or Society which we call *Salomon's House*; the noblest foundation (as we think) that ever was upon the earth; and the lanthorn of this kingdom. It is dedicated to the study of the Works and Creatures of God. Some think it beareth the founder's name a little corrupted, as if it should be Solamona's House. But the records write it as it is spoken. So as I take it to be denominate of the King of the Hebrews, which is famous with you, and no stranger to us. For we have some parts of his works which with you are lost; namely, that Natural History which he wrote, of all plants, from the *cedar of Libanus* to the *moss*[n] *that groweth out of the wall*, and of all *things that have life and motion*. This maketh me think that our king, finding himself to symbolize in many things with that king of the Hebrews (which

taking order: taking measures.
as reason was: as was right.
bottoms: vessels. *denominate of*: named after.
symbolize . . . with: agree with, resemble.

lived many years before him), honoured him with the title of this foundation. And I am the rather induced to be of this opinion, for that I find in ancient records this Order or Society is sometimes called Salomon's House[n] and sometimes the College of the Six Days' Works; whereby I am satisfied that our excellent king had learned from the Hebrews that God had created the world and all that therein is within six days; and therefore he instituting that House for the finding out of the true nature of all things, (whereby God might have the more glory in the workmanship of them, and men the more fruit in the use of them,) did give it also that second name. But now to come to our present purpose. When the king had forbidden to all his people navigation into any part that was not under his crown, he made nevertheless this ordinance; That every twelve years there should be set forth out of this kingdom two ships, appointed to several voyages; That in either of these ships there should be a mission of three of the Fellows or Brethren of Salomon's House; whose errand was only to give us knowledge of the affairs and state of those countries to which they were designed, and especially of the sciences, arts, manufactures, and inventions of all the world; and withal to bring unto us books, instruments, and patterns in every kind; That the ships, after they had landed the brethren, should return; and that the brethren should stay abroad till the new mission. These ships are not otherwise fraught, than with store of victuals, and good quantity of treasure to remain with the brethren, for the buying of such things and rewarding of such persons as they should think fit. Now for me to tell you how the vulgar sort of mariners are contained from being discovered at land; and how they that must be put on shore for any time, colour themselves under the names of other nations; and to what places these voyages have been designed; and what places of *rendez-vous* are appointed for the new missions; and the like circumstances of the practique; I may not do it: neither is it much to your desire. But thus you see we maintain a trade, not for gold, silver, or jewels; nor for silks; nor for spices; nor any other commodity of matter; but only for God's first creature, which was *Light*: to have *light* (I say) of the growth of all parts of the world.'[n] And when he had said this, he was silent; and so were we all. For indeed we were all astonished to hear so strange things so probably told. And he, perceiving that we were willing to say somewhat but had it not ready, in great

designed: destined.
practique: practice.

courtesy took us off, and descended to ask us questions of our voyage and fortunes; and in the end concluded, that we might do well to think with ourselves what time of stay we would demand of the state; and bade us not to scant ourselves; for he would procure such time as we desired. Whereupon we all rose up, and presented ourselves to kiss the skirt of his tippet; but he would not suffer us; and so took his leave. But when it came once amongst our people that the state used to offer conditions to strangers that would stay, we had work enough to get any of our men to look to our ship, and to keep them from going presently to the governor to crave conditions. But with much ado we refrained them, till we might agree what course to take.

We took ourselves now for free men, seeing there was no danger of our utter perdition; and lived most joyfully, going abroad and seeing what was to be seen in the city and places adjacent within our tedder; and obtaining acquaintance with many of the city, not of the meanest quality; at whose hands we found such humanity, and such a freedom and desire to take strangers as it were into their bosom, as was enough to make us forget all that was dear to us in our own countries: and continually we met with many things right worthy of observation and relation; as indeed, if there be a mirror in the world worthy to hold men's eyes, it is that country. One day there were two of our company bidden to a Feast of the Family,[n] as they call it. A most natural, pious, and reverend custom it is, shewing that nation to be compounded of all goodness. This is the manner of it. It is granted to any man that shall live to see thirty persons descended of his body alive together, and all above three years old, to make this feast; which is done at the cost of the state. The Father of the Family, whom they call the *Tirsan*, two days before the feast, taketh to him three of such friends as he liketh to choose; and is assisted also by the governor of the city or place where the feast is celebrated; and all the persons of the family, of both sexes, are summoned to attend him. These two days the Tirsan sitteth in consultation concerning the good estate of the family. There, if there be any discord or suits between any of the family, they are compounded and appeased. There, if any of the family be distressed or decayed, order is taken for their relief and competent means to live. There, if any be subject to vice, or take ill courses, they are reproved and censured. So likewise direction is given touching marriages, and the courses of life which any of them should take, with divers other

took us off: relieved us. *presently*: at once. *tedder*: tether.

the like orders and advices. The governor assisteth, to the end to put in execution by his public authority the decrees and orders of the Tirsan, if they should be disobeyed; though that seldom needeth; such reverence and obedience they give to the order of nature. The Tirsan doth also then ever choose one man from amongst his sons, to live in house with him: who is called ever after the Son of the Vine. The reason will hereafter appear. On the feast-day, the Father or Tirsan cometh forth after divine service into a large room where the feast is celebrated; which room hath an half-pace at the upper end. Against the wall, in the middle of the half-pace, is a chair placed for him, with a table and carpet before it. Over the chair is a state, made round or oval, and it is of ivy; an ivy somewhat whiter than ours, like the leaf of a silver asp, but more shining; for it is green all winter. And the state is curiously wrought with silver and silk of divers colours, broiding or binding in the ivy; and is ever of the work of some of the daughters of the family; and veiled over at the top with a fine net of silk and silver. But the substance of it is true ivy; whereof, after it is taken down, the friends of the family are desirous to have some leaf or sprig to keep. The Tirsan cometh forth with all his generation or lineage, the males before him, and the females following him; and if there be a mother from whose body the whole lineage is descended, there is a traverse placed in a loft above on the right hand of the chair, with a privy door, and a carved window of glass, leaded with gold and blue; where she sitteth, but is not seen. When the Tirsan is come forth, he sitteth down in the chair; and all the lineage place themselves against the wall, both at his back and upon the return of the half-pace, in order of their years without difference of sex; and stand upon their feet. When he is set; the room being always full of company, but well kept and without disorder; after some pause there cometh in from the lower end of the room a *Taratan* (which is as much as an herald) and on either side of him two young lads; whereof one carrieth a scroll of their shining yellow parchment; and the other a cluster of grapes of gold, with a long foot or stalk. The herald and children are clothed with mantles of sea-water green satin; but the herald's mantle is streamed with gold, and hath a train. Then the herald with three curtesies, or rather inclinations, cometh up as far as the half-pace; and there first taketh into his hand the scroll. This scroll is the King's Charter, containing gift of revenew, and many

needeth: is necessary.
state: canopy.
traverse: screen or curtain.
half-pace: dais.
broiding: interlacing.
return: at the side of the dais.

privileges, exemptions, and points of honour, granted to the Father of the Family; and is ever styled and directed, *To such an one our well-beloved friend and creditor*: which is a title proper only to this case. For they say the king is debtor to no man, but for propagation of his subjects. The seal set to the king's charter is the king's image, imbossed or moulded in gold; and though such charters be expedited of course, and as of right, yet they are varied by discretion, according to the number and dignity of the family. This charter the herald readeth aloud; and while it is read, the father or Tirsan standeth up, supported by two of his sons, such as he chooseth. Then the herald mounteth the half-pace, and delivereth the charter into his hand: and with that there is an acclamation by all that are present in their language, which is thus much: *Happy are the people of Bensalem.* Then the herald taketh into his hand from the other child the cluster of grapes, which is of gold, both the stalk and the grapes. But the grapes are daintily enamelled; and if the males of the family be the greater number, the grapes are enamelled purple, with a little sun set on the top; if the females, then they are enamelled into a greenish yellow, with a crescent on the top. The grapes are in number as many as there are descendants of the family. This golden cluster the herald delivereth also to the Tirsan; who presently delivereth it over to that son that he had formerly chosen to be in house with him: who beareth it before his father as an ensign of honour when he goeth in public, ever after; and is thereupon called the Son of the Vine. After this ceremony ended, the father or Tirsan retireth; and after some time cometh forth again to dinner, where he sitteth alone under the state, as before; and none of his descendants sit with him, of what degree or dignity soever, except he hap to be of Salomon's House. He is served only by his own children, such as are male; who perform unto him all service of the table upon the knee; and the women only stand about him, leaning against the wall. The room below the half-pace hath tables on the sides for the guests that are bidden; who are served with great and comely order; and towards the end of dinner (which in the greatest feasts with them lasteth never above an hour and an half) there is an hymn sung, varied according to the invention of him that composeth it, (for they have excellest poesy,) but the subject of it is (always) the praises of Adam and Noah and Abraham; whereof the former two peopled the world, and the last was the Father of the Faithful: concluding ever with a thanksgiving for the nativity of our Saviour, in whose birth the births of all are only blessed. Dinner

of course . . .: issued as matters of course and of right.

being done, the Tirsan retireth again; and having withdrawn himself alone into a place where he maketh some private prayers, he cometh forth the third time, to give the blessing; with all his descendants, who stand about him as at the first. Then he calleth them forth by one and by one, by name, as he pleaseth, though seldom the order of age be inverted. The person that is called (the table being before removed) kneeleth down before the chair, and the father layeth his hand upon his head, or her head, and giveth the blessing in these words: 'Son of Bensalem, (or Daughter of Bensalem,) thy father saith it; the man by whom thou hast breath and life speaketh the word; The blessing of the everlasting Father, the Prince of Peace, and the Holy Dove be upon thee, and make the days of thy pilgrimage good and many.' This he saith to every of them; and that done, if there be any of his sons of eminent merit and virtue, (so they be not above two,) he calleth for them again; and saith, laying his arm over their shoulders, they standing; 'Sons, it is well ye are born, give God the praise, and persevere to the end.' And withal delivereth to either of them a jewel, made in the figure of an ear of wheat, which they ever after wear in the front of their turban or hat. This done, they fall to music and dances, and other recreations, after their manner, for the rest of the day. This is the full order of that feast.

By that time six or seven days were spent, I was fallen into strait acquaintance with a merchant of that city, whose name was Joabin. He was a Jew,[n] and circumcised: for they have some few stirps of Jews yet remaining among them, whom they leave to their own religion. Which they may the better do, because they are of a far differing disposition from the Jews in other parts. For whereas they hate the name of Christ, and have a secret inbred rancour against the people amongst whom they live: these (contrariwise) give unto our Saviour many high attributes, and love the nation of Bensalem extremely. Surely this man of whom I speak would ever acknowledge that Christ was born of a Virgin, and that he was more than a man; and he would tell how God made him ruler of the Seraphims which guard his throne; and they call him also the *Milken Way*,[n] and the *Eliah* of the *Messiah*; and many other high names; which though they be inferior to his divine Majesty, yet they are far from the language of other Jews. And for the country of Bensalem, this man would make no end of commending it: being desirous, by tradition among the Jews there, to have it believed that the people thereof were of the generations of Abraham, by another son, whom they

strait: close.

call Nachoran; and that Moses by a secret cabala ordained the laws of Bensalem which they now use; and that when the Messiah should come, and sit in his throne at Hierusalem, the king of Bensalem should sit at his feet, whereas other kings should keep a great distance. But yet setting aside these Jewish dreams, the man was a wise man, and learned, and of great policy, and excellently seen in the laws and customs of that nation. Amongst other discourses, one day I told him I was much affected with the relation I had from some of the company, of their custom in holding the Feast of the Family; for that (methought) I had never heard of a solemnity wherein nature did so much preside. And because propagation of families proceedeth from the nuptial copulation, I desired to know of him what laws and customs they had concerning marriage; and whether they kept marriage well; and whether they were tied to one wife? For that where population is so much affected, and such as with them it seemed to be, there is commonly permission of plurality of wives. To this he said, 'You have reason for to commend that excellent institution of the Feast of the Family. And indeed we have experience, that those families that are partakers of the blessing of that feast do flourish and prosper ever after in an extraordinary manner. But hear me now, and I will tell you what I know. You shall understand that there is not under the heavens so chaste a nation as this of Bensalem; nor so free from all pollution or foulness. It is the virgin of the world. I remember I have read in one of your European books, of an holy hermit amongst you that desired to see the Spirit of Fornication; and there appeared to him a little foul ugly Aethiop. But if he had desired to see the Spirit of Chastity of Bensalem, it would have appeared to him in the likeness of a fair beautiful Cherubin. For there is nothing amongst mortal men more fair and admirable, than the chaste minds of this people. Know therefore, that with them there are no stews, no dissolute houses, no courtesans, nor any thing of that kind. Nay they wonder (with detestation) at you in Europe, which permit such things. They say ye have put marriage out of office: for marriage is ordained a remedy for unlawful concupiscence; and natural concupiscence seemeth as a spur to marriage. But when men have at hand a remedy more agreeable to their corrupt will, marriage is almost expulsed. And therefore there are with you seen infinite men that marry not, but chuse rather a libertine

cabala: oral tradition handed down from Moses.
seen in: skilled in.
much affected: much desired.
expulsed: driven out.

and impure single life, than to be yoked in marriage; and many that do marry, marry late, when the prime and strength of their years is past. And when they do marry, what is marriage to them but a very bargain; wherein is sought alliance, or portion, or reputation, with some desire (almost indifferent) of issue; and not the faithful nuptial union of man and wife, that was first instituted. Neither is it possible that those that have cast away so basely so much of their strength, should greatly esteem children, (being of the same matter,) as chaste men do. So likewise during marriage, is the case much amended, as it ought to be if those things were tolerated only for necessity? No, but they remain still as a very affront to marriage. The haunting of those dissolute places, or resort to courtesans, are no more punished in married men than in bachelors. And the depraved custom of change, and the delight in meretricious embracements, (where sin is turned into art,) maketh marriage a dull thing, and a kind of imposition or tax. They hear you defend these things, as done to avoid greater evils; as advoutries, deflowering of virgins, unnatural lust, and the like. But they say this is a preposterous wisdom; and they call it *Lot's offer*, who to save his guests from abusing, offered his daughters: nay they say farther that there is little gained in this; for that the same vices and appetites do still remain and abound; unlawful lust being like a furnace, that if you stop the flames altogether, it will quench; but if you give it any vent, it will rage. As for masculine love, they have no touch of it; and yet there are not so faithful and inviolate friendships in the world again as are there; and to speak generally, (as I said before,) I have not read of any such chastity in any people as theirs. And their usual saying is, *"That whosoever is unchaste cannot reverence himself;"* and they say, *"That the reverence of a man's self is, next religion, the chiefest bridle of all vices."'* And when he had said this, the good Jew paused a little; whereupon I, far more willing to hear him speak on than to speak myself, yet thinking it decent that upon his pause of speech I should not be altogether silent, said only this; 'That I would say to him, as the widow of Sarepta said to Elias; that he was come to bring to memory our sins; and that I confess the righteousness of Bensalem was greater than the righteousness of Europe.' At which speech he bowed his head, and went on in this manner: 'They have also many wise and excellent laws touching marriage. They allow no polygamy. They have ordained that none do intermarry or contract, until a month be

advoutries: adulteries.
Lot's offer: See Genesis 19:8.
Elias: See I Kings 17:8.

passed from their first interview. Marriage without consent of parents they do not make void, but they mulct it in the inheritors: for the children of such marriages are not admitted to inherit above a third part of their parents' inheritance. I have read in a book of one of your men, of a Feigned Commonwealth,[n] where the married couple are permitted, before they contract, to see one another naked. This they dislike; for they think it a scorn to give a refusal after so familiar knowledge: but because of many hidden defects in men and women's bodies, they have a more civil way; for they have near every town a couple of pools, (which they call *Adam and Eve's pools,*) where it is permitted to one of the friends of the man, and another of the friends of the woman, to see them severally bathe naked.'

And as we were thus in conference, there came one that seemed to be a messenger, in a rich huke, that spake with the Jew: whereupon he turned to me and said; 'You will pardon me, for I am commanded away in haste.' The next morning he came to me again, joyful as it seemed, and said, 'There is word come to the governor of the city, that one of the Fathers of Salomon's House will be here this day seven-night: we have seen none of them this dozen years. His coming is in state; but the cause of his coming is secret. I will provide you and your fellows of a good standing to see his entry.' I thanked him, and told him, 'I was most glad of the news.' The day being come, he made his entry. He was a man of middle stature and age, comely of person, and had an aspect as if he pitied men. He was clothed in a robe of fine black cloth, with wide sleeves and a cape. His under garment was of excellent white linen down to the foot, girt with a girdle of the same; and a sindon or tippet of the same about his neck. He had gloves that were curious, and set with stone; and shoes of peach-coloured velvet. His neck was bare to the shoulders. His hat was like a helmet, or Spanish Montera; and his locks curled below it decently: they were of colour brown. His beard was cut round, and of the same colour with his hair, somewhat lighter. He was carried in a rich chariot without wheels, litter-wise; with two horses at either end, richly trapped in blue velvet embroidered; and two footmen on each side in the like attire. The chariot was all of cedar, gilt, and adorned with crystal; save that the fore-end had pannels of sapphires, set in borders of gold, and the hinder-end the like of emeralds of the Peru colour. There was also a sun of gold, radiant, upon the top, in the midst; and on the top before, a small

huke: a hooded cape.
sindon or tippet: a piece of linen or cape.
Montera: a Spanish hunting cap, with flaps over the ears.

cherub of gold, with wings displayed. The chariot was covered with cloth of gold tissued upon blue. He had before him fifty attendants, young men all, in white sattin loose coats to the mid-leg; and stockings of white silk; and shoes of blue velvet; and hats of blue velvet; with fine plumes of divers colours, set round like hat-bands. Next before the chariot went two men, bare-headed, in linen garments down to the foot, girt, and shoes of blue velvet; who carried the one a crosier, the other a pastoral staff like a sheep-hook; neither of them of metal, but the crosier of balm-wood, the pastoral staff of cedar. Horsemen he had none, neither before nor behind his chariot: as it seemeth, to avoid all tumult and trouble. Behind his chariot went all the officers and principals of the Companies of the City. He sat alone, upon cushions of a kind of excellent plush, blue; and under his foot curious carpets of silk of divers colours, like the Persian, but far finer. He held up his bare hand as he went, as blessing the people, but in silence. The street was wonderfully well kept: so that there was never any army had their men stand in better battle-array, than the people stood. The windows likewise were not crowded, but every one stood in them as if they had been placed. When the shew was past, the Jew said to me: 'I shall not be able to attend you as I would, in regard of some charge the city hath laid upon me, for the entertaining of this great person.' Three days after, the Jew came to me again, and said; 'Ye are happy men; for the Father of Salomon's House taketh knowledge of your being here, and commanded me to tell you that he will admit all your company to his presence, and have private conference with one of you that ye shall choose: and for this hath appointed the next day after to-morrow. And because he meaneth to give you his blessing, he hath appointed it in the forenoon.' We came at our day and hour, and I was chosen by my fellows for the private access. We found him in a fair chamber, richly hanged, and carpeted under foot, without any degrees to the state. He was set upon a low throne richly adorned, and a rich cloth of state over his head, of blue satin embroidered. He was alone, save that he had two pages of honour, on either hand one, finely attired in white. His under-garments were the like that we saw him wear in the chariot; but instead of his gown, he had on him a mantle with a cape, of the same fine black, fastened about him. When we came in, as we were taught, we bowed low at our first entrance; and when we were come near his chair, he stood up, holding forth his hand ungloved, and in posture of blessing; and we every one of us stooped

displayed: outspread. *tissued*: interwoven. *girt*: belted.
without any degrees: without any steps leading to the canopy.

down, and kissed the hem of his tippet. That done, the rest departed, and I remained. Then he warned the pages forth of the room, and caused me to sit down beside him, and spake to me thus in the Spanish tongue:

'God bless thee, my son; I will give thee the greatest jewel I have. For I will impart unto thee, for the love of God and men, a relation of the true state of Salomon's House. Son, to make you know the true state of Salomon's House, I will keep this order. First, I will set forth unto you the end of our foundation. Secondly, the preparations and instruments we have for our works. Thirdly, the several employments and functions whereto our fellows are assigned. And fourthly, the ordinances and rites which we observe.

'The End of our Foundation is the knowledge of Causes, and secret motions of things; and the enlarging of the bounds of Human Empire, to the effecting of all things possible.

'The Preparations and Instruments are these. We have large and deep caves[n] of several depths: the deepest are sunk six hundred fathom; and some of them are digged and made under great hills and mountains: so that if you reckon together the depth of the hill and the depth of the cave, they are (some of them) above three miles deep. For we find that the depth of a hill, and the depth of a cave from the flat, is the same thing; both remote alike from the sun and heaven's beams, and from the open air. These caves we call the Lower Region. And we use them for all coagulations, indurations, refrigerations, and conservations of bodies. We use them likewise for the imitation of natural mines; and the producing also of new artificial metals, by compositions and materials which we use, and lay there for many years. We use them also sometimes, (which may seem strange,) for curing of some diseases, and for prolongation of life[n] in some hermits that choose to live there, well accommodated of all things necessary; and indeed live very long; by whom also we learn many things.

'We have burials in several earths, where we put divers cements, as the Chineses do their porcellain. But we have them in greater variety, and some of them more fine. We have also great variety of composts, and soils, for the making of the earth fruitful.

'We have high towers; the highest about half a mile in height; and some of them likewise set upon high mountains; so that the vantage of the hill with the tower is in the highest of them three miles at least. And these places we call the Upper Region: accounting the air between the high places and the low, as a Middle Region. We use these towers, according to their several heights and situations, for insolation, refrigeration, conservation; and for the view of divers meteors; as winds, rain, snow, hail; and some of the fiery meteors also. And upon them, in some places, are dwellings of hermits, whom we visit sometimes, and instruct what to observe.

'We have great lakes both salt and fresh, whereof we have use for the fish and fowl. We use them also for burials of some natural bodies: for we find a difference in things buried in earth or in air below the earth, and things buried in water. We have also pools, of which some do strain fresh water out of salt; and others by art do turn fresh water into salt.[n] We have also some rocks in the midst of the sea, and some bays upon the shore, for some works wherein is required the air and vapour of the sea. We have likewise violent streams and cataracts, which serve us for many motions: and likewise engines[n] for multiplying and enforcing of winds, to set also on going divers motions.

'We have also a number of artificial wells and fountains, made in imitation of the natural sources and baths; as tincted upon vitriol, sulphur, steel, brass, lead, nitre, and other minerals. And again we have little wells for infusions of many things, where the waters take the virtue quicker and better than in vessels or basons. And amongst them we have a water which we call Water of Paradise, being, by that we do to it, made very sovereign for health, and prolongation of life.

'We have also great and spacious houses, where we imitate and demonstrate meteors; as snow, hail, rain, some artificial rains of bodies and not of water, thunders, lightnings; also generations of bodies in air; as frogs, flies, and divers others.

'We have also certain chambers, which we call Chambers of Health, where we qualify the air as we think good and proper for the cure of divers diseases, and preservation of health.

'We have also fair and large baths, of several mixtures, for the cure of diseases, and the restoring of man's body from arefaction: and others for the confirming of it in strength of sinews, vital parts, and the very juice and substance of the body.

'We have also large and various orchards and gardens, wherein we do not so much respect beauty, as variety of ground and soil, proper for divers trees and herbs: and some very spacious, where

meteors: meteorological phenomena.
tincted upon: tinctured with.
arefaction: drying up.

trees and berries are set whereof we make divers kinds of drinks, besides the vineyards. In these we practise likewise all conclusions of grafting and inoculating, as well of wild-trees as fruit-trees, which produceth many effects. And we make (by art) in the same orchards and gardens, trees and flowers to come earlier or later than their seasons; and to come up and bear more speedily than by their natural course they do. We make them also by art greater much than their nature; and their fruit greater and sweeter and of differing taste, smell, colour, and figure, from their nature. And many of them we so order, as they become of medicinal use.

'We have also means to make divers plants rise by mixtures of earths without seeds; and likewise to make divers new plants, differing from the vulgar; and to make one tree or plant turn into another.

'We have also parks and inclosures of all sorts of beasts and birds, which we use not only for view or rareness, but likewise for dissections and trials; that thereby we may take light what may be wrought upon the body of man. Wherein we find many strange effects; as continuing life in them, though divers parts, which you account vital, be perished and taken forth; resuscitating of some that seem dead in appearance; and the like. We try also all poisons and other medicines upon them, as well of chirurgery as physic. By art likewise, we make them greater or taller than their kind is; and contrariwise dwarf them, and stay their growth: we make them more fruitful and bearing than their kind is; and contrariwise barren and not generative. Also we make them differ in colour, shape, activity, many ways. We find means to make commixtures and copulations of different kinds; which have produced many new kinds, and them not barren, as the general opinion is. We make a number of kinds of serpents, worms, flies, fishes, of putrefaction; whereof some are advanced (in effect) to be perfect creatures, like beasts or birds; and have sexes, and do propagate. Neither do we this by chance, but we know beforehand of what matter and commixture what kind of those creatures will arise.

'We have also particular pools, where we make trials upon fishes, as we have said before of beasts and birds.

'We have also places for breed and generation of those kinds of worms and flies which are of special use; such as are with you your silk-worms and bees.

'I will not hold you long with recounting of our brew-houses, bake-houses, and kitchens, where are made divers drinks, breads, and

conclusions: experiments.
take light: gain understanding.

meats, rare and of special effects. Wines we have of grapes; and drinks of other juice of fruits, of grains, and of roots: and of mixtures with honey, sugar, manna, and fruits dried and decocted. Also of the tears or woundings of trees, and of the pulp of canes. And these drinks are of several ages, some to the age or last of forty years. We have drinks also brewed with several herbs, and roots, and spices; yea with several fleshes, and white meats; whereof some of the drinks are such, as they are in effect meat and drink both: so that divers, especially in age, do desire to live with them, with little or no meat or bread. And above all, we strive to have drinks of extreme thin parts, to insinuate into the body, and yet without all biting, sharpness, or fretting; insomuch as some of them put upon the back of your hand will, with a little stay, pass through to the palm, and yet taste mild to the mouth. We have also waters which we ripen in that fashion, as they become nourishing; so that they are indeed excellent drink; and many will use no other. Breads we have of several grains, roots, and kernels: yea and some of flesh and fish dried; with divers kinds of leavenings and seasonings: so that some do extremely move appetites; some do nourish so, as divers do live of them, without any other meat; who live very long. So for meats, we have some of them so beaten and made tender and mortified, yet without all corrupting, as a weak heat of the stomach will turn them into good chylus, as well as a strong heat would meat otherwise prepared. We have some meats also and breads and drinks, which taken by men enable them to fast long after; and some other, that used make the very flesh of men's bodies sensibly more hard and tough, and their strength far greater than otherwise it would be.

'We have dispensatories, or shops of medicines. Wherein you may easily think, if we have such variety of plants and living creatures more than you have in Europe, (for we know what you have,) the simples, drugs, and ingredients of medicines, must likewise be in so much the greater variety. We have them likewise of divers ages, and long fermentations. And for their preparations, we have not only all manner of exquisite distillations and separations, and especially by gentle heats and percolations through divers strainers, yea and substances; but also exact forms of composition, whereby they incorporate almost, as they were natural simples.

last: duration.
chylus: food ready to be absorbed into the blood.
simples: medicinal herbs.
exquisite: carefully prepared and so of excellent quality.
natural simples: medicines of a single constituent, as found in nature.

'We have also divers mechanical arts, which you have not; and stuffs made by them; as papers, linen, silks, tissues; dainty works of feathers of wonderful lustre; excellent dyes, and many others; and shops likewise, as well for such as are not brought into vulgar use amongst us as for those that are. For you must know that of the things before recited, many of them are grown into use throughout the kingdom; but yet if they did flow from our invention, we have of them also for patterns and principals.

'We have also furnaces of great diversities, and that keep great diversity of heats;[n] fierce and quick; strong and constant; soft and mild; blown, quiet; dry, moist; and the like. But above all, we have heats in imitation of the sun's and heavenly bodies' heats, that pass divers inequalities and (as it were) orbs, progresses, and returns, whereby we produce admirable effects. Besides, we have heats of dungs, and of bellies and maws of living creatures, and of their bloods and bodies; and of hays and herbs laid up moist; of lime unquenched; and such like. Instruments also which generate heat only by motion. And farther, places for strong insolations; and again, places under the earth, which by nature or art yield heat. These divers heats we use, as the nature of the operation which we intend requireth.

'We have also perspective-houses,[n] where we make demonstrations of all lights and radiations; and of all colours; and out of things uncoloured and transparent, we can represent unto you all several colours; not in rain-bows, as it is in gems and prisms, but of themselves single. We represent also all multiplications of light, which we carry to great distance, and make so sharp as to discern small points and lines; also all colorations of light: all delusions and deceits of the sight, in figures, magnitudes, motions, colours: all demonstrations of shadows. We find also divers means, yet unknown to you, of producing of light originally from divers bodies. We procure means of seeing objects afar off; as in the heaven and remote places; and represent things near as afar off, and things afar off as near; making feigned distances. We have also helps for the sight, far above spectacles and glasses in use. We have also glasses and means to see small and minute bodies perfectly and distinctly; as the shapes and colours of small flies and worms, grains and flaws in gems, which cannot otherwise be seen; observations in urine and blood, not otherwise to be seen. We make artificial rainbows, haloes, and circles about light. We represent also all manner of reflexions, refractions,

principals: originals.
insolations: exposing to the heat of the sun.

and multiplications of visual beams of objects.

'We have also precious stones of all kinds, many of them of great beauty, and to you unknown; crystals likewise; and glasses of divers kinds; and amongst them some of metals vitrificated, and other materials besides those of which you make glass. Also a number of fossils, and imperfect minerals, which you have not. Likewise loadstones of prodigious virtue; and other rare stones, both natural and artificial.

'We have also sound-houses,[n] where we practise and demonstrate all sounds, and their generation. We have harmonies which you have not, of quarter-sounds, and lesser slides of sounds. Divers instruments of music likewise to you unknown, some sweeter than any you have; together with bells and rings that are dainty and sweet. We represent small sounds as great and deep; likewise great sounds extenuate and sharp; we make divers tremblings and warblings of sounds, which in their original are entire. We represent and imitate all articulate sounds and letters, and the voices and notes of beasts and birds. We have certain helps which set to the ear do further the hearing greatly. We have also divers strange and artificial echoes, reflecting the voice many times, and as it were tossing it: and some that give back the voice louder than it came; some shriller, and some deeper; yea, some rendering the voice differing in the letters or articulate sound from that they receive. We have also means to convey sounds in trunks and pipes, in strange lines and distances.

'We have also perfume-houses; wherewith we join also practices of taste. We multiply smells, which may seem strange. We imitate smells, making all smells to breathe out of other mixtures than those that give them. We make divers imitations of taste likewise, so that they will deceive any man's taste. And in this house we contain also a confiture-house; where we make all sweet-meats, dry and moist, and divers pleasant wines, milks, broths, and sallets, far in greater variety than you have.

'We have also engine-houses, where are prepared engines and instruments for all sorts of motions. There we imitate and practise to make swifter motions than any you have, either out of your muskets or any engine that you have; and to make them and multiply them more easily, and with small force, by wheels and other

fossils: any rocks or minerals dug out of the earth.
slides: fine shades.
extenuate: made thin.
trunks: tubes.
sallets: salads.

means: and to make them stronger, and more violent than yours are; exceeding your greatest cannons and basilisks.[n] We represent also ordnance and instruments of war, and engines of all kinds: and likewise new mixtures and compositions of gun-powder, wildfires burning in water, and unquenchable. Also fire-works of all variety both for pleasure and use. We imitate also flights of birds; we have some degrees of flying in the air; we have ships and boats for going under water,[n] and brooking of seas; also swimming-girdles and supporters. We have divers curious clocks, and other like motions of return, and some perpetual motions. We imitate also motions of living creatures, by images of men, beasts, birds, fishes, and serpents. We have also a great number of other various motions, strange for equality, fineness, and subtilty.

'We have also a mathematical house,[n] where are represented all instruments, as well of geometry as astronomy, exquisitely made.

'We have also houses of deceits of the senses; where we represent all manner of feats of juggling, false apparitions, impostures, and illusions; and their fallacies. And surely you will easily believe that we that have so many things truly natural which induce admiration, could in a world of particulars deceive the senses, if we would disguise those things and labour to make them seem more miraculous. But we do hate all impostures and lies: insomuch as we have severely forbidden it to all our fellows, under pain of ignominy and fines, that they do not show any natural work or thing, adorned or swelling; but only pure as it is, and without all affectation of strangeness.

'These are (my son) the riches of Salomon's House.

'For the several employments and offices of our fellows; we have twelve that sail into foreign countries, under the names of other nations, (for our own we conceal;) who bring us the books, and abstracts, and patterns of experiments of all other parts. These we call Merchants of Light.

'We have three that collect the experiments which are in all books. These we call Depredators.

'We have three that collect the experiments of all mechanical arts; and also of liberal sciences; and also of practices which are not brought into arts. These we call Mystery-men.

'We have three that try new experiments, such as themselves think good. These we call Pioneers or Miners.

'We have three that draw the experiments of the former four into

motions of return: oscillations.

titles and tables, to give the better light for the drawing of observations and axioms out of them. These we call Compilers.

'We have three that bend themselves, looking into the experiments of their fellows, and cast about how to draw out of them things of use and practice for man's life, and knowledge as well for works as for plain demonstration of causes, means of natural divinations, and the easy and clear discovery of the virtues and parts of bodies. These we call Dowry-men or Benefactors.

'Then after divers meetings and consults of our whole number, to consider of the former labours and collections, we have three that take care, out of them. to direct new experiments, of a higher light, more penetrating into nature than the former. These we call Lamps.

'We have three others that do execute the experiments so directed, and report them. These we call Inoculators.

'Lastly, we have three that raise the former discoveries by experiments into greater observations, axioms, and aphorisms. These we call Interpreters of Nature.

'We have also, as you must think, novices and apprentices, that the succession of the former employed men do not fail; besides a great number of servants and attendants, men and women. And this we do also: we have consultations, which of the inventions and experiences which we have discovered shall be published, and which not: and take all an oath of secrecy, for the concealing of those which we think fit to keep secret: though some of those we do reveal sometimes to the state, and some not.

'For our ordinances and rites: we have two very long and fair galleries: in one of these we place patterns and samples of all manner of the more rare and excellent inventions: in the other we place the statua's of all principal inventors. There we have the statua of your Columbus, that discovered the West Indies: also the inventor of ships: your monk that was the inventor of ordnance and of gunpowder: the inventor of music: the inventor of letters: the inventor of printing: the inventor of observations of astronomy: the inventor of works in metal: the inventor of glass: the inventor of silk of the worm: the inventor of wine: the inventor of corn and bread: the inventor of sugars: and all these by more certain tradition than you have. Then have we divers inventors of our own, of excellent works; which since you have not seen, it were too long to make descriptions of them; and besides, in the right understanding of those descriptions you might easily err. For upon every invention of

bend: apply.

value, we erect a statua to the inventor, and give him a liberal and honourable reward. These statua's are some of brass; some of marble and touch-stone; some of cedar and other special woods gilt and adorned: some of iron; some of silver; some of gold.

'We have certain hymns and services, which we say daily, of laud and thanks to God for his marvellous works: and forms of prayers, imploring his aid and blessing for the illumination of our labours, and the turning of them into good and holy uses.

'Lastly, we have circuits or visits of divers principal cities of the kingdom; where, as it cometh to pass, we do publish such new profitable inventions as we think good. And we do also declare natural divinations of diseases, plagues, swarms of hurtful creatures, scarcity, tempests, earthquakes, great inundations, comets, temperature of the year, and divers other things; and we give counsel thereupon what the people shall do for the prevention and remedy of them.'

And when he had said this, he stood up; and I, as I had been taught, kneeled down; and he laid his right hand upon my head, and said: 'God bless thee, my son, and God bless this relation which I have made. I give thee leave to publish it for the good of other nations; for we here are in God's bosom, a land unknown.' And so he left me; having assigned a value of about two thousand ducats, for a bounty to me and my fellows. For they give great largesses where they come upon all occasions.

[THE REST WAS NOT PERFECTED.]

Magnalia Naturae,

PRAECIPUE QUOAD USUS HUMANOS.[n]

The prolongation of life.
The restitution of youth in some degree.
The retardation of age.
The curing of diseases counted incurable.
The mitigation of pain.
More easy and less loathsome purgings.
The increasing of strength and activity.
The increasing of ability to suffer torture or pain.
The altering of complexions, and fatness and leanness.
The altering of statures.
The altering of features.
The increasing and exalting of the intellectual parts.
Versions of bodies into other bodies.
Making of new species.
Transplanting of one species into another.
Instruments of destruction, as of war and poison.
Exhilaration of the spirits, and putting them in good disposition.
Force of the imagination, either upon another body, or upon the body itself.
Acceleration of time in maturations.
Acceleration of time in clarifications.
Acceleration of putrefaction.
Acceleration of decoction.
Acceleration of germination.
Making rich composts for the earth.
Impressions of the air, and raising of tempests.
Great alteration; as in induration, emollition, &c.
Turning crude and watry substances into oily and unctuous substances.
Drawing of new foods out of substances not now in use.
Making new threads for apparel; and new stuffs; such as paper, glass, &c.
Natural divinations.
Deceptions of the senses.
Greater pleasures of the senses.
Artificial minerals and cements.

Notes

p. 3 The Jewish law of sacrifice, as laid down in the books of Leviticus and Numbers, prescribed daily sacrifices. The freewill offering is mentioned in Leviticus 22:18, Numbers 15:3, Deuteronomy 12:17, and Psalms 119:108, but is not given much prominence.

p. 3 the ability to grasp or take in impressions, ideas, and knowledge, quantitatively conceived. James loved flattery, particularly of his intellectual powers. He had a quick mind and was generally regarded as very learned and eloquent. For Bacon's attitude to dedications, see below, I. iii. 9.

p. 4 i.e. inhibits free development of the subject-matter. James recommended his son, Prince Henry, to adopt a style that was plain, honest, natural, short, and sententious, avoiding both rustic crudeness and 'book language and pen and ink-horn terms'.

p. 5 Hermes Trismegistus, fabled to be an Egyptian priest, philosopher and king. To him were ascribed various Neoplatonic works written in the second or third centuries A.D. Bacon is referring to the introduction, by Ficinus, to Hermes' *Poemander* (1471).

p. 6 Aristotle and the Schoolmenuse 'First Cause' of God, the uncaused cause of everything. By 'second cause' Bacon means 'Nature', the means of growth and production whereby God operates in the world.

p. 6 Aquinas also says that the naming of the creatures in Paradise shows that Adam pursued natural knowledge. The temptation is described in Genesis 3:1 ff.

p. 7 Man is prevented from reaching an understanding of the highest law of Nature, not by any defect in the capacity of the human mind to understand, but by such hindrances as shortness of life, poor co-ordination of research, the imperfect transmitting of knowledge from age to age.

p. 8 Heraclitus, who flourished *c.* 500 B.C., is reported to have said that 'the dry soul is the wisest and best'. By a corruption in the Greek the sentence became 'the dry light (*lumen siccum*) is the wisest soul'. Heraclitus believed that the soul was a mixture of fire (the noble part) and water (the ignoble). The soul that has most fire he calls 'dry'. Bacon adopts this doctrine, arguing that knowledge can become 'steeped' or 'infused' (*madidum* or *maceratum*) with our emotions and desires. See *Novum Organum* i. 49.

p. 9 Referring to the myth of Icarus whose wings were cemented with wax. He flew too near the sun, which melted the wax, and Icarus fell into the sea.

p. 9 Bacon repeats these arguments in his essay 'Of Atheism'. An atheist, in Elizabethan usage, does not refer to one who denies the existence of God, but to one who identifies the creating principles of the universe with God. In 'Of Atheism' Bacon refers to Leucippus, Democritus, and Epicurus as examples of such atheists. Such philosophers do not believe in a God who transcends the creation, but accept that the universe is constructed of a fortuitous concourse of atoms, governed by necessity.

p. 9 Homer's reference (*Iliad* viii. 18 f.) to a golden cord that could be let down from heaven to earth, is the basis for the important concept of the chain of being. This postulates that God's creation has the qualities of plenitude, order, and unity, since every created thing is a link in the chain, which stretches from God down to inanimate objects. For a full discussion, see A. O. Lovejoy, *The Great Chain of Being* (Cambridge, Mass., 1936).

p. 10 Bacon is concerned to separate the study of God's word (the Bible, theology) from the study of God's works (the natural world, 'science'). The intermingling of the two was, he felt, largely the reason why scientific investigation had not progressed. See p. 42 and n. below.

p. 11 Alexander (356–323 B.C.); Aristotle was invited by Philip of Macedon to superintend the studies of the youthful Alexander.

p. 11 Plutarch, in his 'Life' of Agesilaus, refers to Epaminondas as 'famous at that time for his philosophy and learning, but he had not yet given proof of his capacity as a general'. Leading the armies of Thebes, he defeated the Spartans at Leuctra in 371 B.C.

p. 11 Xenophon was a friend and student of Socrates, of whose philosophy he gives an account in his *Memorabilia* and *Apology*. He was also a successful general and describes in his *Anabasis* how he led the retreat of 10,000 Greeks from Persia back to their homeland. See Bacon's account below, pp. 53–4.

p. 11 Montaigne, at the end of his essay 'Of Pedantisme', denies this.

p. 12 A person's 'complexion' was the particular way in which the 'humours' were intermixed in him, which explained his general character. The 'humours' (melancholy, phlegm, blood, and choler) are manufactured in the liver and carried round the body. The preponderance of one over the others, even slightly, gave a man his characteristic temperament.

p. 12 Seneca was Nero's tutor. During Nero's minority, Seneca and Burrus were responsible for government. Gordianus III was only thirteen when he succeeded his father in A.D. 238; he was well served by the learned Timesitheus or Misitheus, his Praetorian Prefect, whose daughter he later married. Alexander Severus succeeded at seventeen, in A.D. 222, and the government was in the hands of his mother and grandmother.

p. 14 Francesco Guicciardini's *Storia d'Italia* was printed in 1561. The only complete English translation is by A. P. Goddard (1753–6). Guicciardini was one of the principal advisers of Pope Clement VII, and partly responsible for the papal alliance with the French, which led to the sack of Rome by the Imperialists. His account of Clement VII (*History of Italy* xvi. 5) stresses the pope's irresolution.

p. 14 In his letter to Atticus (xvi. 7) Cicero describes how, after the murder of Julius Caesar, he set off for Greece but, being held up by contrary winds, decided to return to Rome after an interview with Brutus.

p. 14 Phocion (402–317 B.C.) was an Athenian general and statesman. Plutarch (in his 'Life' of Phocion) describes the inflexible honesty which led to his downfall. Although there was sufficient evidence that Nicanor was preparing to attack the Piraeus, Phocion allowed him to escape, and so to carry out his plan, because Nicanor had come to the Council trusting in Phocion's protection. For this, Phocion was impeached and later sentenced to death.

p. 14 In II. viii. 3 below Bacon recounts the fable of Ixion 'who designed to enjoy Juno, the goddess of power; and instead of her had copulation with a cloud, of which mixture were begotten centaurs and chimeras'.

p. 14 Perhaps Bacon is thinking of Cicero's comment on Cato, which is quoted below, I. iii. 5.

p. 15 Plutarch, in his 'Life' of Marcus Cato, 22, describes how Cato objected to the popularity in Rome of Greek philosophy and literature, and urged the Senate to hasten the departure of a Greek embassy which included Carneades and Diogenes. See I. ii. 1 above.

p. 16 The Committee of Thirty ruled Athens for eight months in 404–403 B.C. Socrates was tried and condemned, on charges of impiety and corrupting the young men of Athens, *after* this, in 399 B.C.

p. 18 *Annals*, iii. 76. 'They outshone them all, from the very fact that they were not to be seen.' Junia was the sister of Brutus and the wife of Cassius. At her funeral in A.D. 22 the images of these men were not carried in the procession since they were guilty of the murder of Caesar. It was Roman custom to carry the images of ancestors and relations of the deceased in funeral processions.

p. 19 Bacon is referring to the commentary on Joel 2:28 by Isaac Abrabanel (d. 1508), counsellor to Alphonso V of Portugal and Ferdinand of Spain. Cf. essay 'Of Youth and Age'.

p. 19 Montaigne begins his essay 'Of Pedantisme' with the same thought. Sir Philip Sidney recommends the introduction of the 'self-wise-seeming schoolmaster' in comedy, in his *Defence of Poesie* (1595), and has such a character in his masque *Lady of the May*. Shakespeare has Sir Hugh Evans and Holofernes. The pedant is found in Italian comedy of the period, and in English university plays. See K. M. Lea, *Italian Popular Comedy* (Oxford, 1934), ii. 393.

p. 19 After the foundation of the Society of Jesus in 1540, many colleges were established in Italy and Spain. The most famous were the Collegium Romanum and the Collegium Germanicum in Rome and the Collège de Clermont in France. The Jesuits offered free education in Greek, Hebrew, and Latin, and in rhetoric, logic, metaphysics, natural and moral philosophy, as well as theology—the full educational programme of the humanists. The colleges were not merely seminaries for intending members of the order.

p. 19 The negative seems to be superfluous here. Bacon means that he cannot find that any disgrace to learning can proceed from the manners or habits of learned men, inherent in them as men of learning. The punctuation suggests that we read, as Spedding suggests, 'not [I mean, from such manners as are] inherent to [i.e. in] them as men of learning.'

The Latin version reads—nullum occurrit dedecus literis ex litteratorum moribus, quatenus sunt literati, adhaerens.

p. 23 The reference is to the ugliness of Socrates. Plato, in the *Symposium*, 215b, says that Socrates' external form is 'like one of the sculptured Sileni' but within is 'temperance and wisdom'. What Bacon says here is a summary of the first paragraph of the prologue to Rabelais's *Gargantua and Pantagruel*.

p. 23 The story is in the essay 'On Salaried Posts in Great Houses', §§ 33, 34, which is about the humiliations to which a scholar is subjected in the family of a wealthy Roman. 'Cynic' is derived from the Greek for 'dog'.

p. 23 'Tout ces doctes esprits dont la voix flattereuse Change Hécube en Hélène, at Faustine en Lucrèce.' These are the opening lines of Du Bartas's 'Le Second Jour' in *La Creation du Monde ou Premiere Sepmaine*, 1578.

p. 25 Bacon suggests that the humanist revival of Greek and Latin learning was not unconnected with Luther's own interest in founding chairs of Greek and Hebrew at Wittenberg and in encouraging the study of the Bible and Saint Augustine. It was commonly said that those who became interested in Greek became interested in Lutheranism. Luther was 'a master of the biblical text, an authoritative church historian and a patristic scholar' (J. Atkinson, *Martin Luther and the Birth of Protestantism* (1968), p. 49). Like the humanists, Colet and Erasmus, Luther rejected scholastic theology and emphasized the authority of the Scriptures and the Fathers. These were the 'ancient authors' who, under Luther's influence, began generally to be read; their reading presupposed scholarly achievements in Greek, Hebrew, and Latin. See below, I. vi. 15.

p. 25 This whole paragraph is a succinct account of the 'revival of learning'. Bacon is thinking of the discovery of Greek texts—of Plato, Herodotus, Homer, Sophocles, Thucydides, Xenophon, Demosthenes, Plutarch, Lucian, etc.—and of Latin texts—Lucretius, Tacitus, Manilius, Plautus, Cicero's letters and orations. The consequence of these discoveries was a stress on grammar (the philological and textual study of classical works, the translation of Greek texts into Latin, and the editing of the Bible in Greek, Latin, and Hebrew versions), and on rhetoric (the study of elegance in literary expression, especially the imitation of newly discovered Latin models). These were the concerns of those scholars and university professors often called 'humanists' today. Such humanists as Erasmus and Colet had common ground with Luther in attacking the 'scholastics'. For Luther, the point of his attack on Aquinas, Duns Scotus, and William of Ockham was that the intense intellectualism and rationalism of their combination of Christian revelation and Aristotelian philosophy provided no satisfaction for the individual convinced of his sinfulness. He therefore turned to the Bible, and particularly to the epistles of St. Paul. Erasmus and Colet turned to the same works, but tended also to stress the barbarity of scholastic Latin and the unfruitfulness of the scholastic method—the posing of a question and the debating of it in syllogistic form. Bacon's aim of the restoration of scientific learning is clearly influenced by Erasmus's similar aim, to join battle with the schoolmen 'for the restoration, in its pristine brightness and dignity, of that old and true theology, which they have obscured by their subtleties'. But his complaint here is that the humanist delight in style for its own sake had produced an affectation not less detrimental to the pursuit of learning than the subtleties of the scholastics. On the substance of the whole paragraph, see P. O. Kristeller, *Renaissance Thought: the Classic, Scholastic and Humanist Strains* (1961), and J. A. Mazzeo, *Renaissance and Revolution* (1967), ch. 1. The authors in divinity who were revived include Eusebius, Basil, John Chrysostom, Gregory of Nazianus, etc., i.e. Greek Christian literature of the patristic and Byzantine period. These were translated into Latin. The works of the Latin Church Fathers—Ambrose, Lactantius, Jerome, and Augustine—had not ceased to be read in the Middle Ages, but were now read as 'Christian classics'.

p. 26 i.e. richness, fullness, copiousness of speech. Bacon is thinking of the sixteenth-century passion for imitating Cicero. Sidney adopts a similar position in his *Defence of Poetry* (1595) (ed. J. A. Van Dorsten, 1966, pp. 70–2) and Ben Jonson rearranges Bacon's comments in *Discoveries* (2116–24). Roger Ascham, in *The Scholemaster* (1570), ed. E. Arber (1897), p. 118 writes, 'Ye know not what hurt ye do to learning, that care not for words but for matter.' But he is arguing that the writers whose matter is of most importance to us are always those who are concerned with the eloquence of their style. Bacon is here following Erasmus

(*Ciceronianus*, 1528) and also Ramus and Gabriel Harvey, who, like Bacon, admire Cicero and imitate his use of symmetries and rhetorical figures, but condemn those fanatics who slavishly confine themselves to Ciceronian vocabulary, rhythms, and figures. With them, the stylistic purity of their Latin is more important than their matter.

p. 26 Jeronimo Osorio (1506–80), Bishop of Silves, the 'Portuguese Cicero' and Johann Sturm (1507–89), founder of the gymnasium at Strasburg, were friends of Roger Ascham (1515–68) Latin secretary to Edward VI, Mary, and Elizabeth. They all advocated and practised the imitation of Cicero in their Latin writings. Sturm produced school and college textbooks for this purpose. Ascham's *The Scholemaster* (1570) is the best English discussion of the topic. Nicholas Carr (1524–68) became Regius Professor of Greek in Oxford in 1547. He translated the orations of Demosthenes into Latin.

p. 26 Erasmus, *Colloquies*, 'Juvenis and Echo'. 'I have spent ten years reading Cicero.' The reply in Greek was 'oné' (in Latin, 'asine'), i.e. you ass.

p. 26 This sentence suggests that Bacon preferred the learning of the schoolmen to the stylistic affectation he is attacking. But he reserves his most brilliant writing for his condemnation of the schoolmen's 'vain matter' in I. iv. 5–7.

p. 26 The first letter, which was decorated, in a document issued by the king, or in an illuminated book (i.e. one illustrated with paintings).

p. 27 Bacon here distinguishes between two styles: (*a*) that of works intended for use in ordinary life and (*b*) that of works in which an author is extending the bounds of knowledge. In the former, rhetorical devices are acceptable. Plato, Bacon thought, supplied table-talk for men of culture and added grace and charm to everyday conversation (*The Masculine Birth of Time*). The distinction is elaborated in Book II (pp. 134–9).

p. 27 'Thou art no divinity.' Bacon noted this story in the *Promus*, fol. 16a. The scholiast on Theocritus, *Idylls* v. 21 attributes this story of Hercules to Cleander. See also Erasmus, *Adages* I. vii. 37.

p. 27 In the expanded version of the *Advancement*, published in Latin in 1623, Bacon here adds a paragraph attacking another kind of self-conscious style, used by Seneca and, to a less extent, by Tacitus and Pliny. This style succeeded the affected and verbose Ciceronianism and 'has been very pleasing unto the ears of our time'. It was characterized by pointed expressions and short periods, by conciseness, and, says Bacon, 'seems more witty and weighty than indeed it is'. He dismissed it as another form of stylistic affectation, 'no more than a jingle, or peculiar quaint affectation of words'. Since Bacon is concerned with the advancement of learning, he condemns the affectation of stylistic imitators of Cicero or Seneca. He uses both styles himself; see B. Vickers, *Francis Bacon and Renaissance Prose* (Cambridge, 1968), pp. 96–140.

p. 27 The most important theologians and philosophers of the twelfth and thirteenth centuries—Abelard, Albertus Magnus, Aquinas, Duns Scotus, and William of Ockham. Aquinas (*c*. 1225–74) adopted and adapted a number of Aristotelian theories in his *Summa theologiæ* and other works. In the next paragraph Bacon describes Aquinas's method of presenting his thought by means of questions, objections to his theory, and replies.

p. 29 i.e. they undertake to occupy large areas of knowledge, and are as intractable as animals who have been kept in the dark, and for the same reasons.

p. 30 The *Natural History* of the elder Pliny, a large compilation of scientific and pseudo-scientific facts.

p. 30 Jerome Cardan (1501–76), Italian physician, mathematician, and astrologer. In his *De subtilitate rerum* (1550) and *De rerum varietate* (1557) he collected experiments and inventions.

p. 30 Albertus Magnus (*c.* 1200–80), a scholastic philosopher, scientist, and theologian. He wrote commentaries on all Aristotle's works.

p. 30 Avicenna (980–1037), the Persian philosopher and physician, whose encyclopedic works on medicine, and the natural sciences were used in Europe from the twelfth century. Averroes (1126–98) whose commentaries on Aristotle were translated into Latin and used throughout Europe.

p. 30 Bacon refers first to Aristotle's collection *Historia Animalium* and then to a spurious work attributed in his day to Aristotle, *De Mirabilibus Auscultationibus*. Cf. II. i. 3–4 below.

p. 31 Bacon often uses ideas drawn from these 'sciences' in circumstances in which he considered it permissible to use the existing body of 'knowledge', e.g. he will write 'the contemplative planet carrieth me away wholly'. Further, he borrows much alchemical and magical lore in his compilation of experiments, the *Sylva Sylvarum*. He uses much of the terminology of alchemists and magicians, and their basic doctrine that man is the servant and interpreter of nature. But he objects to their narrow, haphazard experimentalism, the barrenness and pretentiousness of their claims. See P. Rossi, *Francis Bacon; From Magic to Science* (1968), ch. 1, and B. Farrington, *The Philosophy of Francis Bacon* (1964), ch. 10. The 'natural magicians' included Cornelius Agrippa, Paracelsus, Della Porta, Cardan, Dr. John Dee. They 'contemplated the virtues of all natural and celestial things' and 'made known the hidden and secret powers of nature'. By understanding nature, they hoped to produce wonderful effects. Dr. Faustus represents the natural magician, in his combination of charlatanry and a longing to achieve the ultimate understanding of the nature of things. See *Novum Organum* i. 85.

p. 31 Bacon is the first to use this word in a sense related to its original meaning —the turning of a stream into another channel. The phrase 'the derivations and prosecutions to these ends' means 'the subsidiary channels leading to these ends and the modes in which they have been followed'.

p. 31 Perhaps Bacon is thinking of the use of chemicals in medicine (which was seen by Paracelsus as the main aim of the alchemist), of the general usefulness of alchemical processes such as distillation, and of the development of apparatus and furnaces required by alchemists.

p. 31 Cf. *Novum Organum*: i. 74, 85, 108–14, 119, 120, 129. Printing by movable types began in Europe in the mid-fifteenth century; gunpowder was used in firearms from the fourteenth century; the combination of the wind-rose with the magnetic needle was developed in the fourteenth century. These are the three main inventions of the late Middle Ages, often pointed to in the sixteenth century as profoundly influencing civilization. Bacon's point is that technology develops, though slowly, while pure science deteriorates with the passage of time. Technologists co-operate to improve the invention; so-called scientists, instead of following the method of investigation of Aristotle, do no more than write commentaries on his works.

p. 32 Morbid, unhealthy; a term taken over by Bacon from the pathology based on the doctrine of the humours.

p. 33 'What we call antiquity is the youth of the world.' Cf. *Novum Organum* i. 84. The idea that 'the present time is the real antiquity' is to be found also in Vives, Bruno, Gilbert, and Galileo. The effectiveness of this paradox, in Bacon's argument, depends upon the inference that the authority accorded to 'antiquity' (i.e. the classical ages) should be transferred to the new 'antiquity' (i.e. the present). In II Esdras 14:10 we read 'the world has lost its youth, and time is growing old'. See W. von Leyden, 'Antiquity and Authority', *Journal of the History of Ideas* xix (1958), 473–92.

p. 33 A law passed in the time of Augustus, by which a person of sixty was not allowed to marry anyone under fifty.

p. 33 Bacon is here speaking generally of the greatest obstacle to the advancement of learning—despair of the possibility of progress. He uses as illustrations the exploits of Alexander and Columbus (eastward and westward from the Mediterranean). He saw his own intellectual venture (the regeneration of the sciences) in terms of the more recent geographical discoveries, using the Pillars of Hercules (the straits of Gibraltar) as the emblem of the boundary between existing knowledge and the new world of learning. In *Novum Organum* i. 92, 97 he applies to himself the quotation from Livy and the references to Alexander and Columbus.

p. 33 Aubrey tells the story of Hobbes, at the age of 40, casually reading Euclid's 47th proposition. 'By God,' says he, 'this is impossible.' By reading all the previous propositions 'he was demonstratively convinced of that truth. This made him in love with geometry.'

p. 33 A fiction of law by which two times (or other things) are identified, and for legal purposes regarded as one and the same (*OED*).

p. 33 i.e. after a variety of them have been propounded and examined.

p. 34 The arts were the branches of learning—grammar, rhetoric, dialectic, arithmetic, geometry, astronomy, and music, i.e. the liberal arts. The mechanical arts included navigation, agriculture, hunting, medicine, armaments (including architecture). By 'methods' Bacon means the presentation of knowledge in systematic form, using either the liberal arts and natural, moral, and rational philosophies as the organizing principle, or the six days of Creation. Works so arranged were encyclopedic in their contents. See V. K. Whitaker, *Francis Bacon's Intellectual Milieu* (Los Angeles, 1962). Book II of the *Advancement* itself is of course arranged in terms of faculty psychology—memory, imagination, and reason. By 'aphorisms' Bacon means the brief expression of ideas and observations, without linkage, as in the 1597 edition of the *Essays*. This important topic is discussed below, II. xvii.

p. 34 Defined below, II. v. 2 as 'primitive or summary philosophy' and II. vii. 3 as 'the common principles and axioms which are promiscuous and indifferent to several sciences'. In *Novum Organum* i. 79–80 the term used for this concept is 'philosophia naturalis'.

p. 34 i.e. devotees of the intellect or understanding. In philosophy the term refers to those who hold that knowledge is wholly or mainly derived from the action of the intellect, i.e. pure reason. Bacon is the first to use the word (*OED*). See *Novum Organum* i. 83, where Bacon refutes the accepted belief that 'the dignity of the human mind is impaired' by close observation of material things. He condemned Plato for diverting men's minds from observation to contemplation.

p. 34 Bacon frequently makes these observations, e.g. in *Novum Organum* i. 54, 63, 64, 96. He objects to Plato's doctrine that knowledge must be knowledge of Forms, which were distinct from matter; therefore his philosophy is ultimately concerned with objects that may be contemplated but never observed. His speculations are theological. To Aristotle Bacon objects that his ultimate concern is with categories, with logical and metaphysical concepts of Quality, Relation, Matter, Substance, Form, and Privation, with the technical forms and terminology of the syllogism.

In Proclus Bacon refers to the Neoplatonists and 'the Pythagorical and mystical arithmetic, which began to be recovered from Proclus' (*De Augmentis* iii. 6). Proclus (A.D. 410–85) wrote an introduction to Ptolemy's *Almagest* and a commentary on Book I of Euclid's *Elements*. His works were translated in the Renaissance by Pico della Mirandola and Ficino.

p. 34 The alchemists, concerned with the transmutation of metals, experimented with furnaces, but also constructed an esoteric philosophy concerned with the regeneration of the human soul and available only to chosen adepts.

p. 34 William Gilbert (1540–1603), physician to Elizabeth I and James I, wrote the first treatise on the magnet (1600), a work of thorough and ingenious experiments. But he went on, particularly in the posthumously published *De mundo nostro, sublunari Philosophia nova* (1651) to suggest that magnetism is animate, that the earth is a magnet, that magnetism explains the action of the tides, winds, etc. Gilbert was half scientist, half natural-magician.

p. 35 Referring to the story of the Choice of Hercules, told by Xenophon, *Memorabilia* II. i.

p. 35 See II. xvii below for a full discussion of the communication of knowledge.

p. 36 Cf. *Novum Organum* i. 81, 124. There he stresses the end of the sciences as the endowing of human life with new discoveries and new powers. He concludes, 'Truth and utility are here the very same thing; and works themselves are of greater values as pledges of truth than as contributing to the comforts of life.'

p. 36 Macrobius, in his commentary on Cicero's *Dream of Scipio* I. xii, describes Saturn and Jupiter in these terms. In astrological belief, the planets were thought to have special influences, and when in conjunction (i.e. both in the same sign of the zodiac) their influence on men and actions would be combined.

p. 36 Ovid, *Metamorphoses* x. 667. 'She leaves the path and picks up the rolling gold.' Atalanta would only marry the man who beat her in a race. The successful suitor dropped three golden apples in the course of the race, and Atalanta was delayed by picking them up.

p. 37 This is based on Augustine, *The City of God* xi. 29.

p. 37 Bacon seems to suggest that matter and form can—or did for a time at the Creation—exist independently. He is probably thinking of Aquinas's metaphysical theory that every material thing is composed of a substantial form (i.e. the determining principle that makes a thing what it is) and 'first matter' (i.e. an indeterminate potential element which has no definite form of its own and no definite characteristics). All visible matter is already informed and possesses determined characteristics. The 'first matter' is potentiality—it can be actualized by successive forms. In this sense, God could have made 'matter' on the first day, and actualized its potentialities for form afterwards. See F. C. Copleston, *Aquinas* (1963), pp. 85 f., 136 f., and Bacon's account of Cupid in *The Wisdom of the Ancients*, where he asserts that first matter had some form (cf. Lemmi, p. 57).

p. 38 A reference to *The Celestial Hierarchies*, a work which, until the sixteenth century, was attributed to Dionysius who was converted by St. Paul (Acts 17:34). Bacon summarizes his account of the angelic hierarchies, as it was accepted by the Church.

p. 38 Bacon frequently uses the image of light when speaking of wisdom or knowledge; cf. his essay on Truth. It was commonly accepted (by St. Augustine, for example, in xi. 19 and 32 of *The City of God*) that the angels were created when God said 'Let there be light.'

p. 38 See note to p. 6 above. On p. 6 he had explained the cause of the Fall in terms similar to those used here.

p. 38 See Genesis 4:2. Bacon is interpreting the story (as Augustine does in *The City of God* xv. 1), maintaining that while it is literally true it is also capable of allegorical interpretation. See II. xxv. 17 below. Discussion of the active versus the contemplative life is traditional and has its origin in Aristotle's *Ethics* i. 5; x. 7, 8. See below II. xx. 8.

p. 39 Genesis 4:21–2. Jubal, inventor of the harp and pipe, Tubal-cain, master of all copper-smiths and blacksmiths. The story of the confusion of tongues is in Genesis 11.

p. 39 The Egyptian learning ascribed to Moses was generally accepted as that contained in the writings supposedly by Hermes Trismegistus. These were considered to be of profound antiquity, but are in fact compilations of the second and third centuries A.D. They were shown to be late works by Isaac Casaubon in 1614. Moses is 'God's first pen' because he was supposed to have written the Pentateuch.

p. 39 The Pentateuch (the first five books of the Bible) contains the various sacrificial, social, and priestly codes to be observed by the Jews. Their observance is the 'badge' or distinguishing mark of the chosen people. But Bacon refers also to the tradition of interpreting the Old Testament stories and figures as prefigurations of the New Testament. In this tradition, Moses prefigures Christ in many ways. Bacon refers to the way in which Moses as establisher of the Old Law prefigures Christ who established the New Law. On this tradition see B. Smalley, *The Study of the Bible in the Middle Ages* (Oxford, 1941), and H. A. Wolfson, *The Philosophy of the Church Fathers* (Cambridge, Mass., 1956). There are also the Rabbinical interpretations of the Jews themselves. These non-literal interpretations of the Biblical text, called *midrash*, were common from the second century B.C. Through the works of Philo of Alexandria, a contemporary of Christ, they influenced early Christian exegesis of the Bible, beginning with Origen (A.D. 185–254). A second wave of Jewish influence began in the twelfth century, with the work of Andrew of St. Victor (see Smalley, op. cit., pp. 121 f.). Bacon discusses Biblical interpretation below, II. xxv. 17. See p. 19 n. above.

p. 39 Though Bacon here finds scientific learning in the book of Job, he does not generally advocate using the Bible as a source of such knowledge. Later on (II. xxv. 16) he mentions the school of Paracelsus as guilty of using the Bible in this way, and in *Novum Organum* i. 65 refers scornfully to those who try to found a system of natural philosophy on the first chapter of Genesis or the book of Job. He is thinking of the influence of the 13th-century Spanish *Zohar* (largely a commentary on Genesis) on the Cabalistic and Hermetic work of Pico, whose *Heptaplus* (1489) is also a commentary on the first chapter of Genesis.

p. 40 His petition (I Kings 3:5 f.) was to be able to discern between good and bad. God's gift is described in I Kings 4:29–34.

p. 40 In I Kings 4 : 33 there is a reference to this work by Solomon. The Authorized Version has 'hyssop', meaning any lowly plant, instead of 'moss'. The New English Bible has 'marjoram'. By 'rudiment' he means a half-way form or stage between two species; see *Novum Organum* ii. 30.

p. 41 The humour here accentuates the harmlessness and the goodness of scientific investigation, which Bacon is recommending to the King, who could appoint learned men and provide adequate finance to join God in this 'game'.

p. 41 Julian 'the Apostate', Emperor, A.D. 361–3. His edict forbad Christians to institute public schools of rhetoric and literature, but did not forbid their attending such schools. See Ammianus Marcellinus XXII. x. 7, and Augustine, *Confessions*, viii. 5.

p. 41 Pope Gregory the Great (590–604), who sent Augustine to convert the English in 597. He is supposed to have tried to extinguish all heathen antiquities and literature. Cf. Machiavelli, *Discourses* ii. 5. The reference to the invasions is to that of the Tartars in the fourth and of the Saracens in the seventh centuries.

p. 42 See notes to pp. 19, 25 above.

p. 42 The concept of God's two books, one the Bible, the other the natural world, is traditional. St. Paul (Romans 1:18–20) says that men can know God's invisible attributes, his power and deity, in the things he has made. Bacon believes that the study of the book of works will not undermine the book of God's word, but he knows that many divines have feared otherwise, see *Novum Organum* i. 89. In the following century the study of science as the wisdom of God in Creation became a commonplace. See W. P. Jones, *The Rhetoric of Science* (1967).

p. 43 As a reward for ridding the earth of so many tyrants and monsters, Jupiter had Hercules carried to heaven from the pyre on which his body was being cremated. Plutarch, in his 'Life' of Theseus, describes the sacrifices made to Theseus in Athens, because he was a protector of the distressed. Minos, mythical king of Crete, was a great law-giver and was rewarded with the office of judge in Hades. Romulus, the founder of Rome was, according to Plutarch, accorded the status of a martial god by the Romans.

p. 43 Ceres, goddess of corn and harvests, taught men the arts of ploughing, sowing, reaping, and the care of fruit trees. Bacchus taught men the use of the vine, the cultivation of the earth, and the manner of making honey. Mercurius or Hermes, messenger of the gods, invented the lyre, which he gave to Apollo; he was worshipped in many forms by the Greeks and Romans, not least because he was the god of eloquence. Apollo was the god of all the fine arts, being regarded as the inventor of medicine, music, poetry, and eloquence.

p. 43 This is the interpretation of the myth given in Bacon's *Wisdom of the Ancients* (1609) also. See Philostratus the Younger, *Imagines* 6 for an account of Orpheus.

p. 44 Gibbon also selects this period as the happiest for the human race. The six are Nerva, Trajan, Hadrian, Antoninus, Lucius Verus, and Marcus Aurelius, whose reigns covered almost a century, A.D. 96–180. With Commodus began the decline. Marcus Aurelius was the author of a famous set of stoic meditations.

p. 45 In the *De Augmentis* Bacon abridged the following section—it was indeed the wrong material for an intricate work; it holds up the argument by too much example.

p. 45 A Latin version of Homer, *Iliad* i. 42. 'Phoebus, with thy darts revenge our tears.' Nerva adopted the young Trajan shortly before his death. His actual message was 'may the Danubians expiate my tears under the strokes of thy darts'.

p. 45 Emperor A.D. 98–117. Tacitus and the younger Pliny were his friends, though he was himself a soldier. His public library, the Ulpian, was the most important library in Rome.

p. 45 See note to p. 41 above. Gregory was moved by a story of Trajan's love of justice to pray to God that Trajan's soul be received in heaven. The story is told in Paulus Diaconus, *Life of Gregory* 27, in the *Legenda Aurea*, etc., and is referred to in Dante's *Purgatorio* x. 73.

p. 46 The emperor Hadrian, A.D. 117–38, a man of learning, a skilled wrlter and orator, a musician and painter, learned in mathematics and sciences, and an expert in military discipline. See Dio's *Roman History* lxix. 3 f. It was not, however, Hadrian, but Alexander Severus (emperor, A.D. 222–35) who placed figures of Christ, Apollonius, Abraham, Orpheus, and others in one shrine. Apollonius of Tyana, almost contemporary with Christ, was a neo-Pythagorean philosopher who pretended to miraculous powers.

p. 47 Marcus Aurelius and Lucius Verus were emperors jointly until the death of Verus in A.D. 169. Marcus died in A.D. 180. There was much persecution of Christians during his reign.

p. 48 Bacon spoke of Elizabeth in similar terms in his letter to the Lord Chancellor, concerning the history of Britain (*Works*, x. 249–52). His 'Praise of Elizabeth' in the 1592 *Conference of Pleasure* (*Works* viii. 126–43) and the posthumously published 'Felicities of Queen Elizabeth' (*Works*, vi. 291 f.) show Bacon's fine historical sense.

p. 49 See p. 11 above.

p. 49 Only fragments of his writings survive. Bacon omits to mention that Alexander kept him in chains for seven months, after which time he died.

p. 50 The story is in Plutarch's *Apophthegms* but Bacon seems to be following Erasmus's version, in his *Apophthegms* iv. 17.

p. 51 In Plutarch's 'Life' of Alexander, 15, it is Perdiccas who asks the question. The reply is not 'hope' but 'that which I hope for'. Bacon sees himself as one resolving upon a great enterprise, in his plans for the regeneration of learning, and in *Novum Organum* i, 92 writes, 'I am now therefore to speak touching Hope.'

p. 51 Plutarch, in his 'Life' of Caesar, 11, tells how Crassus stood surety to Caesar's creditors for 830 talents, before he was allowed to take the Praetorship in Spain.

p. 51 This story Bacon heard in France in 1576 when Henry, third Duke of Guise (1550–88), was head of the Catholic League.

p. 51 Bacon omitted this sentence in the *De Augmentis*. Virgil was the subject of lengthy commentaries, by Donatus, Servius, and Fulgentius. The latter, a sixth-century work, interprets the *Aeneid* as a philosophical allegory of human life and progress. Macrobius' *Saturnalia* is also a commentary on Virgil, who was regarded as an authority on grammar and rhetoric (two basic elements in medieval and Renaissance learning), a prophet of Christianity (in the fourth Eclogue), and learned in agriculture and general sciences (in the *Georgics*). See D. C. Allen, *Mysteriously Meant* (Baltimore, 1970), ch. 6.

p. 52 This work does not survive, but fragments are quoted by Cicero and Aulus Gellius. In *De Augmentis* vi. 1, Bacon says that it was probably not a philosophical grammar, but a series of precepts of correct usage, distinguished from vulgarity and affectation.

p. 52 An image from painting, which seems to mean that words were made to accord with what reason suggested as being real.

p. 52 Suetonius, *Lives of the Caesars*, 'Julius Caesar' 40. In 46 B.C. the calendar was three months in advance of the real time. By decree, Caesar added 67 days to the year, and reorganized the calendar to fit in with the sun's course. The year was fixed at 365·25 days, and a leap-year every fourth year was instituted. Caesar was advised by the Alexandrian astronomer Sosigenes.

p. 52 Plutarch and Suetonius, in their lives of Caesar, and Cicero, in his *Letters to Atticus* (xii. 40, 41; xiii. 50) mention a work by Caesar in answer to Cicero's panegyric on Cato. It is quoted by Aulus Gellius, in his *Attic Nights* IV. xvi. 8.

p. 52 These are mentioned by Cicero and, as 'dicta collectanea', by Suetonius. They were suppressed by Augustus. Bacon himself collected a number of such anecdotes, embodying pointed sayings, in 1625.

p. 52 The three anecdotes are from Suetonius, *Life of Caesar* 70, 79, and Plutarch's 'Life' of Caesar, 35.

p. 53 In Xenophon's *Anabasis* II. i. 12. Modern editors give the speech to Theopompus, not to Xenophon. The Greeks had gone with Cyrus against his brother Artaxerxes in 401 B.C. Cyrus was killed, and Xeonophon led the Greek army home through enemy territory. Jason the Thessalonian planned his expedition just before his murder in 370 B.C. Agesilaus the Spartan invaded Persia in 396–394 B.C. and was accompanied by Xenophon. Alexander began his Persian expedition in 334 B.C.

p. 55 Pope expresses this

> A little learning is a dang'rous thing;
> Drink deep, or taste not the Pierian spring.
> (*Essay on Criticism*, ll. 215–16)

p. 55 Virgil, *Georgics* ii. 490–3.

Happy the man who, studying nature's laws,
Through known effects can trace the secret cause,
His mind possessing in a fearless state,
Fearless of Fortune, and resigned to Fate.
(Dryden)

p. 56 Goodness (*bonitas*) is the product of Truth (*veritas*), as evil is the product of error. The image is from setting a seal on wax. This is a Platonic belief, that moral goodness depends on intellectual attainment, 'virtue is knowledge'. Cf. II. xii. 1.

p. 59 Bacon is thinking of the Aristotelians. Aristotle, in the *De Anima*, seems to imply that there is one immortal intellect which functions in individual men while they are alive; but the individual intellect is, as it were, reabsorbed into the immortal intellect when a person dies. Aquinas argued against this belief, which was expounded by followers of Averroes as representing Aristotle's opinion. Pietro Pomponazzi, one of the school of Paduan Averroists, argued in his *On the Immortality of the Soul* (1516) that according to Aristotle's *Ethics* the ultimate end of man is contemplation, and that the satisfactory fulfilment of this end requires immortality.

p. 59 Tacitus, *Annals* xiv. 9. 'Let him kill his mother, provided that he becomes emperor.' This was said by Agrippina, when the astrologers foretold that her son Nero would become emperor and kill his mother. Nero ordered her assassination in A.D. 59. Agrippina committed many crimes herself for the sake of power.

p. 59 Cicero, *De Oratore* i. 44. 'He preferred his old wife to immortality.' Ulysses left Calypso, who promised him immortality (*Odyssey* v. 135 f.) for Penelope.

p. 60 Cf. Bacon's essay 'Of Parents and Children'.

p. 60 The name given by the ancients to the two rocks, Calpe (Gibraltar) and Abyla (Jebel el Mina) on either side of the Straits of Gibraltar. They were supposed to mark the western limits of Hercules' wanderings, and of geographical knowledge in that direction. They are used by Bacon metaphorically, to denote the limit of any investigation whatever.

p. 62 i.e. the professions of divinity, medicine, and law. Bacon regrets the absence, in universities, of provision for the study of arts and sciences other than those useful in the three professions.

p. 62 See *Coriolanus* I. i. 99 f. The story is told by Livy and by Plutarch.

p. 62 Pertaining to the professions, or pertaining to professors, and so to the learning already established in the Universities. *OED* gives this as the first usage in this sense. There were public lectures on anatomy (by the reader of the Company of Barbers and Surgeons) and on surgery (at the Royal College of Physicians) in London. And in 1598 Gresham College was founded, with professors of Rhetoric, Divinity, Music, Physics, Geometry, Astronomy, and Law.

p. 63 In 1611 Bacon suggested that some of Thomas Sutton's bequest be used to increase university salaries, instead of to found the Charterhouse (*Works* xi, 252–4).

p. 63 Minerva was the Roman goddess of wisdom, a personification of the thinking power. The Muses were the inspiring goddesses of song, poetry, the arts, and sciences. Vulcan was the Roman god of fire and furnaces. Fire was the most important agent of the alchemists. See J. Read, *Prelude to Chemistry* (1936, repr. 1961), pp. 143 f.

p. 64 i.e. medicine. Botanical gardens (for growing *Simples*, plants used in medicine) were established in many Italian cities during the sixteenth century. The one at Padua was established in 1546, when Andreas Vesalius, the founder of modern anatomical studies, was professor there. There was a famous one at Leiden (established in 1587). The one in Oxford was founded in 1621. John Gerarde's, in London, was founded in 1595.

p. 64 A mythical figure to whom the Greeks attributed developments in sculpture and architecture. He is reputed to have made wings for himself and his son Icarus.

p. 64 The story is told in Pliny, *Natural History* viii. 17. By 'arts of nature' Bacon means experimental sciences, as opposed to 'history of nature', the collecting of facts about the creation. In the *De Augmentis* the compiler of a natural history is likened to a wanderer through the woods, the 'travailer in arts of nature' to one who makes his way through a labyrinth (*Works* iii. 325).

p. 64 A neat phrasing of the principle used by James I in the opening pages of *A Counterblaste to Tobacco* (1604).

p. 64 The first two years of the course of study at Elizabethan Oxford and Cambridge were spent on logic and rhetoric. Bacon went to Trinity College when he was twelve and studied rhetoric (Cicero and Demosthenes), history as a source of analogies and examples in writing and speaking (Caesar, Livy, Sallust, and Xenophon), poetry (Homer) and moral philosophy (Aristotle's *Ethics* and Plato's *Laws*).

p. 65 Terms used in Cicero, *De Oratore* iii. 26 and *Orator* 24.

p. 65 Bacon is probably thinking of religious orders, with their local and general officers (provincials and generals).

p. 66 i.e. communities, societies. The reference is to trade guilds, in which members of the same trade formed a common fund and joined in certain spiritual exercises, with their own patron saint.

p. 66 In James 1:17, God is called 'the father of lights'.

p. 66 One of the reasons for the scarcity is that people have believed there was plenty.

p. 66 See Exodus 4 and 7. The rod that turned into a serpent and swallowed the others was Moses' rod, though Aaron cast it down; it is usually referred to as Aaron's rod. James I thought of it as Moses', see *Daemonologie* (1597; ed. G. B. Harrison, 1924), pp. 22–3.

p. 67 Though Bacon seems here to be subdividing human learning according to the concepts of the psychology of faculties, he is using the terms very loosely. Later, in ch. 1 of the *Description of the Intellectual Globe* (*c.* 1612, in *Works* iii. 727; v. 503) he expands his account and refers to the 'three faculties of the understanding' (*intellectus*) instead of the 'three parts'. In 1623, in Book ii, ch. 1

of the *De Augmentis* he refers to the 'three faculties of the rational soul' (*anima rationalis*). This last is closer to the doctrine of the soul and its faculties, which was elaborated by Albertus Magnus from the ideas of Aristotle and Avicenna. But in this doctrine, the Rational Soul exercised two faculties—*intellectus* or understanding, by which self-evident truths are perceived, and *ratio* or reason, the process of proceeding step by step to prove a truth. In addition there was the Sensitive Soul, whose faculties were Memory, Estimation (i.e. instinct), Fantasy (i.e. the compounding and dividing of mental images), Imagination (i.e. the retention of things perceived) and Common Sense (i.e. the basic faculty that 'turns mere sensation into coherent consciousness of myself as subject in a world of objects'). (See C. S. Lewis, *The Discarded Image* (Cambridge, 1967), pp. 152 ff.) Bacon has simplified this, and his doctrine may be compared with Spenser's in *The Faerie Queene* II. ix. 49–58 where Phantasy (Bacon's 'Imagination'), Memory, and Judgment (Bacon's 'Reason') are described in the House of Alma. Later in the *Advancement* (II. ix. 1) Bacon insists that dividing up knowledge is for convenience of handling only, and in II. xii. 1 he discusses the faculties. For a thorough account, see K. R. Wallace, *Francis Bacon on the Nature of Man* (Chicago, 1967). In *De Augmentis* ii. 13 (the account of Pan) Bacon refers to A. Donius, *De Natura Hominis* (1581), Book 2, ch. 21 as saying that the ancients fixed their eyes too eagerly on the memory, imagination, and reason, and neglected the Thinking Faculty, 'which holds the first place. For he who remembers or recollects, thinks; he who imagines, thinks; he who reasons, thinks.'

p. 68 When Bacon was writing, there were in existence histories of the Church and its affairs, and histories of particular countries and states. But there was no chronological survey of the development of human learning.

p. 68 Like a statue of Polyphemus, the Cyclops, who had only one eye, which was put out by Odysseus (*Odyssey* ix).

p. 68 i.e. an accurate history of learning. In *De Augmentis* ii. 4 Bacon gives a more detailed account of what such a work should be. For the important influence of this notion on Bacon's works, see B. Farrington, *The Philosophy of Francis Bacon* (1964), esp. chs. 8 and 9, and P. Rossi, *Francis Bacon, From Magic to Science*, pp. 44 f.

p. 68 St. Ambrose (340?–397), bishop of Milan. His works include commentaries on the Old Testament, ethical and theological treatises. St. Augustine (354–430), bishop of Hippo, the greatest thinker in Christian antiquity, a voluminous writer, and a profound influence on the Reformation. The Protestants were interested in Church history, because they felt they were restoring the original doctrine and practices. Bacon has in mind de Bèze's *Ecclesiastical History* (1580), and *The Magdeburg Centuries* by Flacius Illyricus *et al.* (Basle, 1560–74).

p. 68 Bacon is asking for a description of those natural phenomena that appear to be unusual, to supplement straightforward natural histories. In *Novum Organum* ii. 28, 29, he distinguishes between 'singular instances' and 'deviating instances'. The former have some quality of strangeness when considered in relation to other objects of a similar kind, e.g. 'the sun and moon amongst the heavenly bodies; the magnet amongst minerals; quicksilver amongst metals; the elephant amongst quadrupeds'. The latter are sports of nature, unusual variants of individual plants, animals, etc.

p. 69 Cf. pp. 100 and 208 below. The systematic refutation of popular errors in natural history was undertaken by Sir Thomas Browne in *Pseudodoxia Epidemica* (1646), a work expanded in editions up to the sixth in 1672. He deals with such topics as 'that an elephant hath no joints', 'that a salamander lives in the fire', 'that snails have no eyes'. In Book I, ch. 8 Browne surveys the authors, referred to here by Bacon, who have collected fabulous information, including Herodotus, Ctesias, Dioscorides, Pliny, Albertus Magnus, Cardan, and della Porta. See L. B. Wright, *Middle-Class Culture in Elizabethan England* (Chapel Hill, 1935), ch. 15, 'The Strange World of Science'.

p. 69 Browne discusses the survival of errors, as a result of the neglect of scientific examination, the respect paid to the authority of ancient writers, and the habit of using the supposed facts in figurative speech, in Book I, chs. 5–9. Poets keep alive the story of the Phoenix, the notion that a man carrying fern-seed will be invisible, that toads have a jewel in the head, etc. John Lyly, in his *Euphues* (1579), delighted in similes drawn from natural history, of the kind 'that the bull being tied to the fig tree looseth his strength'. They were derived from Pliny's *Natural History* and Bartholomaeus Anglicus's *De Proprietatibus Rerum*. Cf. II. xxv. 16 below.

p. 69 The treatise *De Mirabilibus Auscultationibus* is not now attributed to Aristotle. A translation will be found in the Loeb Classical Library.

p. 69 Bacon uses 'axiom' to denote any general principle arrived at during the stages of induction, from simple generalizations arrived at by experience to the highest generalizations about nature. A partial axiom is a generalization that is based on an examination of only part of the relevant data.

p. 69 Bacon (cf. *Novum Organum* ii. 29) believed that by understanding the freaks of nature we might discover ways of inducing changes in natural phenomena.

p. 69 Bacon himself records one of his dreams in his *Sylva Sylvarum* 986. Cardan was particularly interested in supernatural events and dreams.

p. 69 James VI, in his *Daemonologie* (1597), an investigation into possession by the devil and witchcraft. Cf. *Novum Organum* i. 120.

p. 70 In this and the next paragraph Bacon stresses one of his most important ideas—the necessity for the philosopher to learn from the technical and mechanical arts of mining and metallurgy, botany, hydrostatics, agriculture, etc. From these he would learn the patient observation and accumulation of facts about the natural world, and learn from the practical processes of common mechanical practices what power man can obtain over nature. For the mechanical arts had already transformed nature; Bacon in 1594 suggested a museum of inventions, 'whatsoever the hand of man by exquisite art or engine hath made rare in stuffs form or motion' (*Works*, viii. 335). Modern commentators stress Bacon's debt to George Agricola (1494–1555) and Bernard Palissy (1510–98), whose lectures in Paris on agriculture, mineralogy, and geology he may have attended. See Rossi, op. cit., ch. 1. The 'secrets, rarities and special subtleties' are what attracted natural magicians such as Cardan and Porta.

p. 70 Thales, in Plato, *Theaetetus* 174.

p. 71 Cf. *Novum Organum* i. 81, and note to p. 36 above.

p. 71 Virgil, *Georgics* iv. 387 f. and Ovid, *Metamorphoses* viii. 728 f., tell the story of Proteus, who could only be constrained to prophesy when he was prevented from changing shape by being held firm.

p. 71 For discussion of Bacon's theories of history, see Vickers, *Essential Articles*, pp. 211–50. For Elizabethan historians and interest in history, see L. B. Wright, op. cit., ch. 9, 'The Use of History'; T. D. Kendrick, *British Antiquity*, 1950; W. K. Ferguson, *The Renaissance in Historical Thought* (Cambridge, Mass., 1948).

p. 72 Perhaps Bacon is here thinking of such a work as William Camden's *Remaines concerning Britaine* (1605).

p. 72 Bacon is probably thinking of L. Annaeus Florus, who wrote an abridgement of Roman history, and of Sextus Aurelius Victor, who wrote a work *De Caesaribus*, or of contemporary abridgements of English histories. Ascham, *The Scholemaster* (ed. Arber), p. 110, objects to epitomes, and see II. xvii. 12 below.

p. 73 Thucydides' *History of the Peloponnesian War*, from 432 to 411 B.C.; Xenophon's account of the Persian expedition, the *Anabasis*, in 401 B.C. (see n. to p. 11 above); and Sallust's account of Catiline's conspiracy, the *Catilina*, in 63 B.C., all works written by contemporary observers and participants.

p. 73 A continuous narrative of Greek and Roman history, incorporating the text of Greek and Roman historians, would run from Theseus to Philopoemen, and from Romulus to Justinian. Plutarch begins his *Lives* with Theseus and Romulus. Philopoemen, one of the few great men that Greece produced in her decline, was sometimes called 'the last of the Greeks'. He died in 183 B.C. 'The last of the Romans', according to Tacitus, was Cassius. Justinian died in A.D. 565.

p. 74 i.e. George Buchanan (1506–82) who published his *Rerum Scoticarum Historia* at Edinburgh in 1582. Buchanan had been James's tutor during the period when he was writing his *Historia*. James later (1584) banned the book, because Buchanan attacked his mother, Mary Queen of Scots, for her complicity in the murder of her husband, and because he did not believe in the divine right of kings.

p. 74 i.e. a history of England from 1485 to 1603. Bacon himself wrote a *History of Henry VII* (1622) and left a fragment of a history of Henry VIII.

p. 75 Ariosto, *Orlando Furioso*, end of Book 34 and beginning of Book 35. The 'ancient fiction' on which this is based is that of the three fates, Clotho, Lachesis, and Atropos.

p. 76 i.e. a heraldry whereby the ranks of books as well as of persons may be distinguished. 'Annals' would contain records of more important matters, 'journals' of the less important.

p. 76 Plutarch, *Table-Talk* i. 6 and 'Life' of Alexander, 23.

p. 76 Bacon is probably thinking of such a work as Machiavelli's *Discourses on the First Ten Books of Titus Livius*. See below, II. xxiii. 8.

p. 77 The mingling of geography and history ('chorography') had been recommended by Erasmus. Camden's *Britannia* (1586), Thomas Danett's *Description of the Low Countreys* (1593), George Abbot's *Brief Description of the Whole World* (1599) are examples. See P. Burke, *The Renaissance Sense of the Past* (1969).

p. 77 The first voyage round the world was in 1519–22. This was a Spanish expedition, led by the Portuguese Magellan, who died on the voyage. Drake's voyage was in 1577–9.

p. 78 Daniel 12:4. Many shall run to and fro and knowledge shall be increased. Cf. *Novum Organum* i. 93. The frontispiece of the *Novum Organum* (1620) shows a ship in full sail, passing between two pillars with the motto 'multi pertransibunt, et augebitur scientia'.

p. 78 See II. xxiv. 1 below. In *Cogitata et Visa* (trs. in Farrington, op. cit., p. 95) Bacon identifies three periods of learning, the Greek, Roman, and Modern periods, and in *Novum Organum* i. 78 he says that each period lasted 200 years. This would be (1) the pre-Socratics, from Thales to Plato, (2) from Cicero to Marcus Aurelius, (3) from the beginning of printing to Bacon's time. But in the *Redargutio Philosophiarum* (trs. Farrington, op. cit., p. 106) Bacon dismisses the Romans and names Aristotle, Plato, Hippocrates, Galen, Euclid, and Ptolemy as the only sources of scientific knowledge.

p. 79 Cf. I Corinthians 2:14. The 'natural' man is man without the benefit of divine revelation.

p. 79 Coverdale, in his translation of Habbakuk 2:2 has 'that who so commeth by, may read it', and this is how Bacon takes the passage. The Authorized Version has the correct translation, 'that he may run (i.e. hasten to carry on the tidings) that readeth'.

p. 80 Bacon here uses language which suggests that he accepts faculty psychology, and the belief that reason, imagination, and memory were seated in particular ventricles of the brain. The memory was supposed to be at the back of the head. (For this doctrine, see Burton, *Anatomy of Melancholy*, Part I, Section 1, Member 2, Subsection 4.) But later on, at ix. 3, he suggests that this is a topic requiring investigation, and he strengthens his rejection in the *De Augmentis* iv. 1 *ad fin.*

p. 80 Bacon's account of poetry should be read in conjunction with Sidney's *Defence of Poetry* (ed. J. A. Van Dorsten, Oxford, 1966). Both accept the Aristotelian view that poetry is imitation and that the poet is a poet by virtue of the fable he invents, not by virtue of the verse he uses. But where Bacon thinks of poetry as 'one of the principal portions of learning', Sidney thinks of it as the supreme learning. Both accept (cf. Sidney, ed. cit., pp. 25, 28) that poetry embodies a glimpse of an ideal order, 'bridging the gap between the sinful state and the lost paradise'.

p. 80 For Bacon on the imagination, see below, II. xii. 1 (where poesy 'is rather a pleasure or play of the imagination, than a work or duty thereof'), and ch. 1 of the *Description of the Intellectual Globe* (*Works* v. 503–4), and the discussion in J. L. Harrison, 'Bacon's View of Rhetoric, Poetry and the Imagination', *Huntington Library Quarterly*, 20 (1957), 107–25 (*Essential Articles*, pp. 253–71).

p. 80 Cf. II. xvi. 5 below. Aristotle, in the *Poetics*, ch. 1, stresses that imitation makes the poet, not the metre. In the *De Augmentis* II. 13, where he expands his account of poetry, Bacon consigns metre to rhetoric, and refers Satire, Elegy, Epigram, and Ode to 'philosophy and arts of speech'.

p. 80 Bacon confines himself to invented stories, i.e. narrative poetry, including epic, drama, and allegory.

p. 80 Bacon is not looking for verisimilitude; he accepts that poetry delights by offering more variety, greater goodness, more heroic greatness, than normal experience supplies. By depicting 'poetic justice' the poet symbolically represents the ideal world for which man longs.

p. 81 Cf. Sidney, ed. cit., p. 25, where it is said of the poet that 'with the force of a divine breath he brings things forth surpassing [nature's] doings', an argument that, since the Fall, 'our erected wit [i.e. intellect] maketh us know what perfection is'. But Sidney does not adopt Plato's view of the poet as 'god-inspired' (*Ion*, 534b, c). Poetry is a stimulant to good works, not a soporific or a deceiver, for Bacon. But when he is writing of natural history or of ethics he naturally prefers the mind to be 'buckled and bowed' to the nature of things. He does not therefore expect men to live only in the grey world of reason; poetry satisfies a natural desire of the human mind, which Bacon shares and recognizes.

p. 81 Symbolical, figurative. *OED* gives this as the first usage. Bacon's interest in allegorical poetry developed. In 1609 he published an analysis of 31 classical myths (*De Sapientia Veterum*), and included three of his analyses in *De Augmentis* ii. 13. Here he means that parabolical poetry is 'typical history, by which ideas that are the objects of the intellect are represented in forms that are objects of the sense'.

p. 81 The fables attributed to Aesop (*fl.* 570 B.C.) are now known to be medieval works. The Seven Wise Men of Greece were Solon, Thales, Pittacus, Bias, Chilon, Cleobulus, and Periander (or Myso). See Plato, *Protagoras* 343. Their sayings were such as 'know thyself', 'nothing in extremes', which could be expanded and interpreted. Hieroglyphics were popular in the sixteenth century, Erasmus praising them in his *Adagies* as a universal language. They derived from the *Hieroglyphica* of Horapollo Niliacus, and were thought to have been instituted by Hermes Trismegistus himself. The symbols were interpreted as conveying religious precepts. See J. Seznec, *The Survival of the Pagan Gods* (New York, 1961), pp. 99 f. See below, II. xvi. 3; II. xvii. 10.

p. 81 Referring to the tradition of interpreting Biblical texts in four senses—literal, allegorical, moral, and anagogical. See n. to p. 39 above and II. xxv. 17 below.

p. 82 Virgil, *Aeneid* iv. 178–80. 'Her did mother Earth, enraged against the gods, bring forth last (it is said) of her offspring, sister to Coeus and Enceladus.' Cf. Bacon's essay 'Of Seditions and Troubles' and *De Sapientia*, ch. 9.

p. 82 As goddess of Wisdom Pallas fits the interpretation. But according to Homer (*Iliad* i. 396–404) it was Thetis, and Hesiod (*Theogony* 617 f.) says Gaia. Briareus was a famous giant with a hundred arms and fifty heads.

p. 82 Homer, *Iliad* xi. 832, Plutarch, *De Musica* xl. 4, Machiavelli, *The Prince*, ch. 18.

p. 82 A Stoic philosopher, born 280 B.C. Cicero, *On the Nature of the Gods* i. 15 mocks him for finding Stoic doctrines in the stories of the gods.

p. 82 Moral meanings were found in Homer by Heraclitus or Heraclides (4th century B.C.), in *Homeric Allegories* and Phornutus or Cornutus (1st century A.D.), in *Commentary on the Nature of the Gods*, and by the Neoplatonists, the school of Plotinus, Porphyry, and Iamblichus. See Seznec, op. cit., ch. 3, and Rabelais, Prologue to *Gargantua*. In *De Sapientia* Bacon accepts that the myths had inner meanings.

p. 82 It should be *second* part.

p. 83 Cf. II. xxii. 4, 6 below.

p. 83 Bacon is thinking of the third faculty in the normal faculty psychology, which used the terms memory, fantasy (imagination), and judgement.

p. 83 This is really a new start, subdividing knowledge according to its *sources* —revelation, or 'the notions of the mind and the reports of the senses'—instead of according to the faculties of the soul. Divinity he discusses in II. xxv. below.

p. 83 Philosophy is subdivided into divine (natural theology), natural and human, according to the *object* of knowledge.

p. 83 Bacon pauses, because all knowledge is interrelated, to define a branch of learning which correlates different disciplines and which discovers the underlying principles common to all. See II. vii. 3 below, and *Novum Organum* i. 79, 80, 107.

p. 83 A 'rhapsody' is a confused mixture. This suggests Aristotle's *Metaphysics* and scholastic developments of its doctrines. See F. C. Copleston, *Aquinas* (1963), ch. 2.

p. 84 Logic for Aristotle was 'a study not of words but of the thought of which words are signs', and it was from Aristotle that the study of logic derived. But Bacon goes on to list 'quantity, similitude, diversity', etc., which suggests Plato's 'common properties' (likeness and unlikeness, being and not-being, identity and difference, odd and even, unity and number) in *Theaetetus* 185, rather than Aristotle's list in the *Categories*.

p. 84 i.e. partaking of the nature of two or more different classes. In *De Augmentis* (iii. 1) and in *Novum Organum* ii. 30 the examples are moss (halfway between putrescence and a plant), flying fish (coming between birds and fish), bats, seals. Their study, Bacon thought, would help (1) in indicating the composition and structure of things, (2) in suggesting the cause for the number and quality of the ordinary species in existence, and (3) in prompting the understanding in the direction of what might be possible.

p. 84 Aristotle uses 'axiom' in the *Posterior Analytics* as equivalent to Euclid's 'common notions' and Bacon clearly has this work in mind here, since his first example is used by Aristotle. See D. Ross, *Aristotle* (1966), pp. 41 f.

p. 84 Euclid, *Elements*, Book I, Axiom 4. 'If equals be added to unequals, the result will be unequal.' Bacon's discussion of this is from Aristotle, *Ethics* v. 3, 4 (and see Sir Thomas Elyot, *The Governor* (1531), Book III. 2). Commutative justice (or corrective) takes no account of the persons involved, but only of the transaction, i.e. it assumes that the individuals began as equals, that by some act inequality has been introduced, and justice will therefore, by a penalty imposed on one, restore equality. Distributive justice is concerned with the determination of rights, i.e. with giving to everyone according to his deserts.

p. 84 Euclid, *Elements*, Book I, Axiom 1. 'Things that are equal to the same are equal to each other.' The syllogism unites propositions agreeing in a middle term, e.g. in the syllogism every Y is X, Z is Y, therefore Z is X', Z and X are both equal to Y, and therefore equal each other. See *Novum Organum* ii. 27, and R. Whately, *Elements of Logic* (1826), Book 2, ch. 3. Modern text books on logic devote little space to the syllogism, but see P. Alexander, *An Introduction to Logic* (1971).

p. 85 To their first principle. See Machiavelli, *Discourses on Livy* iii. 1, where it is argued that the taking of Rome by the Gauls was necessary to restore the Romans to their observation of rules of justice and religion. The 'Persian magic' of Zoroaster is mentioned in Plato, *Alcibiades I* 121e. See *Works* x. 89 f. The *Chaldean Oracles* attributed to Zoroaster were important to the Platonists of the 16th century, Pico, Patrizi, Bruno, etc.

p. 85 Cf. *Novum Organum* ii. 27 and *Sylva Sylvarum* ii. 113. A trope in rhetoric is a figure of speech; in Gregorian music it is a short cadence at the end of a melody (*OED*). See Cicero, *De Oratore* ii. 63, §255 and Quintilian, *Institutio Oratorio* VI. iii. 24.

p. 86 See Macrobius, *Commentary on the Dream of Scipio* ii. 12, and St. Augustine, *City of God* vii. 23.

p. 87 The schoolmen elaborated the theory of angels and of fallen spirits, using texts from scripture and Dionysius the Areopagite. See C. S. Lewis, *The Discarded Image* (1967), pp. 70 f., 122 f., and p. 38 above.

p. 88 See note to p. 63 above.

p. 88 Natural Magic stems from Ficino in the late fifteenth century. It involved belief in the power of the magus to affect the 'occult powers' of things by talismans, incantations, numbers, drawing down the virtues of the upper intellectual and celestial worlds by manipulating the animals, plants, metals, stones, etc. in the elemental world. See D. P. Walker, *Spiritual and Demonic Magic from Ficino to Campanella* (1958), and F. A. Yates, *Giordano Bruno and the Hermetic Tradition* (1964), ch. 4 (Ficino), ch. 7 (Agrippa), ch. 8 (magic and science). Bacon repudiates the magician's methods and assumptions, but wants to keep the name 'natural magic', because this is the area of 'applied' science. See below, II. viii. 3.

p. 88 The common form of 'metaphysics' at this time. It referred to that branch of speculative inquiry which treats of the first principles of things, including such concepts as being, substance, essence, time, space, cause, identity. Bacon will keep the word, but change its reference.

p. 89 Cf. *The Refutation of Philosophies* (in Farrington, op. cit., p. 113) and II. viii. 5 below. Bacon digresses to attack Aristotle and to avow that he is linking antiquity and the new learning.

p. 90 In a letter of June 1622 (*Works* xiv. 377) Bacon wrote, 'Be not troubled about the Metaphysics. When true Physics have been discovered, there will be no Metaphysics. Beyond the true Physics is divinity only.' But here, and in *De Augmentis* iii. 4, 5 and *Novum Organum* ii. 9 he assigns to Metaphysics (as he conceives it) the formal and final causes, to Physics the material and efficient causes. This is based on Aristotle's distinction of four kinds of causes for the existence of a thing:

(1) the material cause, the substance or matter composing it;
(2) the efficient cause, the force employed in shaping it;
(3) the formal cause, the pattern, type or design, according to which it is shaped; and
(4) the final cause, the end, motive or purpose of the work.

But since for Bacon matter is not an unknowable substance, but an actual existing 'thing', his notion of 'causes' is very different from Aristotle's. See F. M. Anderson, *The Philosophy of Francis Bacon* (Chicago, 1948), pp. 157 f.

p. 91 This is the ultimate goal for Bacon. In *Novum Organum* ii, he investigates the Form of Heat, which is the Law of Heat. The Forms are relatively few in number, including Light, Weight, Density, Cold, and they refer to the ultimate knowledge of matter. They are not abstractions such as Plato conceived. See *Works* i. 28 f. The reference to sounds and words is to Plato's *Philebus* 17.

p. 93 The Latin has 'those men who are inflated by their own knowledge and a daring insolence'.

p. 93 Though this is not the doctrine of Parmenides, it is expounded by Plato in the *Philebus* 16.

p. 94 i.e. the end, motive or purpose of a thing. The pursuit of final causes had been part of physics, instead of metaphysics, and therefore it is 'misplaced'. The examples given are explanations in terms of a thing's function. Galen's *Of the Use of Parts*, much used in the sixteenth century, is a treatise on final causes in animal physiology. See also Plato, *Timaeus* 44 f.; Aristotle, *Physics* ii. 8; Xenophon, *Memorabilia* i. 4. Cf. *Novum Organum* i. 48, 65; ii. 2.

p. 94 Sucking-fish, believed to have the power of staying the course of a ship.

p. 94 The contemporary of Socrates. See Diogenes Laertius ix. 44, 45. Bacon has in mind also Leucippus, Epicurus, and Lucretius. Whereas Aristotle, by seeking final causes in nature, tends to identify Nature with God, Democritus does not seek final causes in Nature, which he investigates as made up of particles of matter in motion coming together fortuitously. See Anderson, op. cit., pp. 118 f.; and Bertrand Russell, *History of Western Philosophy* (1946), ch. 9. Bacon will reserve for metaphysics the question, 'what purpose does this serve' (the answer to which leads one towards God), and to physics the question, 'what earlier circumstances caused this event' (the answer to which will lead one towards Nature). See Bacon's account of Pan in *The Wisdom of the Ancients* and in *De Augmentis* ii. 13.

p. 95 Quantity, as a general proposition, is referred to primary philosophy; mathematics, which deals with quantity relatively, to metaphysics. On the theories of number of Democritus and Pythagoras, see Russell, *History of Western Philosophy*, chs. 3, 9; G. de Santillana, *Origins of Scientific Thought* (New York, 1961), chs. 4, 9.

p. 97 The study of the occult forces of nature by natural means, by Agrippa, della Porta, Cardan, John Dee, etc. See Rossi, ch. 1; M. Boas, *The Scientific Renaissance, 1540–1630* (1970), ch. 6; and notes to pp. 31, 88 above.

p. 98 Bacon thought that all things were made of particles of matter and 'spirits', 'real, subtle and invisible bodies'. See Lemmi, *Classical Deities*, pp. 74 f.; K. R. Wallace, *Bacon on the Nature of Man*, ch. 2. On the making of gold and the prolongation of life, see Bacon's experiments in *Sylva Sylvarum* 327, 738.

p. 98 This inventory is referred to in the *Valerius Terminus* (*Works* iii. 234).

p. 99 Pope Alexander VI, referring to the expedition of the French under Charles VIII in 1494.

p. 99 A Roman legal formula, by which the judge declared his inability to decide on the guilt or innocence of the accused.

p. 99 A collection of mathematical, optical, musical, physiological and medical problems, attributed to Aristotle.

p. 100 See note to p. 69 above, and II. xxv. 16 below.

p. 100 Bacon would like a collection of the surviving views on these Presocratic philosophers, whose works are now fragmentary. They have now been collected in H. Diels and W. Kranz, *Die Fragmente der Vorsokratiker*, 10th ed., 3 vols. (Berlin, 1960–1). The texts are translated in K. Freeman, *The Pre-Socratic Philosophers* (Oxford, 1946), and in G. S. Kirk and J. E. Raven, *The Presocratic Philosophers* (Cambridge, 1957). There is a survey in E. Hussey, *The Presocratics* (1972).

p. 101 The Copernican theory, that the earth rotated and revolved around the sun, was not provable, and as a theory was not more acceptable than the Ptolemaic. Tycho Brahe, a great practical astronomer, did not accept it. Bacon wrote more fully on astronomy in *Description of the Intellectual Globe* and *Thema Coeli* (see *Works* iii. 715 f., v. 503 f.).

p. 101 The *Placita Philosophorum* attributed to Plutarch. See Farrington, pp. 84 f. For an account of the sources Bacon has in mind, see J. Burnet, *Early Greek Philosophy* (1945), pp. 31–8.

p. 101 In his *Annals*; Suetonius in his *Lives of the Caesars*.

p. 101 Theophrastus of Hohenheim (1493–1541) a voluminous writer on alchemy and medicine. Peter Severinus, a Danish physician (1542–1602) compiled a harmony of Paracelsus's ideas in *Idea Medicinae Philosophicae* (Basle, 1571). See Farrington, pp. 66, 71.

p. 101 Bernadino Telesius (1508–88), author of *De Rerum Natura* (Naples, 1565, 1570, 1586). Bacon thought him 'the best of the novelists' and refers to him frequently (see *Works* i. 50 f.; iii. 720 f.; v. 654–5). P. O. Kristeller, *Eight Philosophers of the Italian Renaissance* (Stanford, 1964), has a study and bibliography. His philosophy is 'pastoral' because placid and superficial speculation, though Bacon found it suggestive.

p. 101 Antonio Donius, author of *De natura hominis* (Lyon, 1581), is commended for his theory that the sensitive soul has physical movement, in Bacon's account of Pan in *Wisdom of the Ancients*.

p. 101 Girolamo Fracastorio (1478–1553), Italian humanist, poet, physician, and astronomer, who wrote on contagious diseases, on astronomy, and on sympathy and antipathy in natural magic. He wrote a poem *Syphilis* (Verona, 1530), from which the disease was named. He was the first to speak of the magnetic pole (1530) and to understand the organic nature of fossils.

p. 101 William Gilbert (1544–1603), Elizabeth's physician, author of *De Magnete* (1600), based on careful experiments. For Xenophanes (sixth century B.C.), see Burnet, *Early Greek Philosophy*, pp. 112 f. But in the Latin it is Philolaus, for whom see Burnet, pp. 277 f., 297 f.

p. 102 'Know thyself', said to have been written over the temple of Apollo at Delphi.

p. 102 Cf. Seneca, *Epistles to Lucilius* lxxxix. 2.

p. 102 In *De Oratore* iii. 16.

p. 102 See note to p. 101 above.

p. 102 Among the humanists the dignity of man was a favourite theme; there are treatises by Gianozzo Manetti and Pico for example. But the medieval theme of the miseries of man, *de contemptu mundi*, was continued and developed also. George Gascoigne translated Innocent III's treatise (*c.* 1200) on the subject in his *The Drum of Dooms Day* (1576). See also Montaigne's 'Apology for Raymond Sebond' in his *Essays*, and T. Spencer, *Shakespeare and the Nature of Man* (Cambridge, 1943), ch. 2.

p. 103 The *Physiognomonica* attributed to Aristotle deals briefly with the connection between bodily and mental characteristics, of the kind, 'Drawn back lips are the mark of the acid temper'. Hippocrates discusses dreams and health in *Regimen IV* (see the Loeb edition of Hippocrates, iv. 421 f.).

p. 103 James I, *Basilicon Doron* (Edinburgh, 1599), p. 135. Physiognomy in its magical contexts was not neglected by della Porta and Robert Fludd. See B. L. Joseph, *Elizabethan Acting* (Oxford, 1951), pp. 100 f. Cf. 'There is no art To find the mind's construction in the face', *Macbeth*, I. iv. 7, but Thomas Hill's *The Contemplation of Mankind* (1571) supplied such information.

p. 104 The four 'humours' were thought to be engendered of food and drink in the liver. The balance of these in the body could therefore be affected by diet, blood-letting, vomits, purgatives, and medicines. The humours influenced the mind; physicians therefore prescribed physical remedies for mental states. A good account of medical practices is in Burton, *The Anatomy of Melancholy*, Part 2, Sections 4 and 5. The passions could also affect the 'humours', since the passions, seated in the heart, modified the temperature of the 'spirits' whose task it was to maintain the humours in equilibrium by preserving the correct heat. See L. B. Campbell, *Shakespeare's Tragic Heroes*, chs. 5, 6, 7. In *Sylva Sylvarum* x. 974 Bacon recommends the roasted brains of hares, hens, or deer, taken in wine, to strengthen the memory.

p. 104 Pythagoras (*fl.* 532 B.C.) established such precepts as, 'to abstain from beans', 'not to eat from a whole loaf', 'not to walk on highways'. The Manichees, followers of Mani (A.D. 216–76) forbad the eating of meat. Muslims do not drink wine or eat pork, and observe a month's fast in Ramadan. The Jewish laws are in Leviticus 3:17; 7:22–7; 11.

p. 104 The imagination creates and responds to images and was conceived by Bacon to be an activity of spirit. The effect of imagination could therefore be on the body of the person imagining. He considers such effects in *Sylva Sylvarum* x. 902, 945, but is unable to suggest experiments. Cf. K. R. Wallace, *Bacon on the Nature of Man*, pp. 88 f., and Montaigne's essay 'Of the force of imagination'. For the effect of the passions on the body see *Sylva Sylvarum* viii. 713–22 (e.g. anger causes paleness, trembling).

p. 105 Said by Socrates of a work by Heraclitus, which Euripides lent him, in Diogenes Laertius, *Lives of Eminent Philosophers* ii. 22; ix. 12.

p. 105 In the *Timaeus* 69–71. It was commonly believed that the imagination resided in the front of the brain, the reason in the middle, and memory at the rear.

p. 105 The idea that the body of man reproduced in little the macrocosm or universe (the head being the heavens, the blood the rivers, his breath the air) was developed from Plato's *Timaeus* 43 f. by the Neo-platonists and transmitted by Boethius and Bernard of Tours. Paracelsus believed in the detailed correspondences and the possibility of mastering control of them. See *Sylva Sylvarum*, Introduction to Century X. Cf. Tillyard, *Elizabethan World Picture*, and J. Seznec, *Survival of the Pagan Gods*, pp. 49, 64 f.

p. 106 Ovid, *Metamorphoses* i. 518, 521.

p. 107 Virgil, *Aeneid* vii. 772–3, 'Himself [Jupiter] with thunder hurled down to the Stygian floods Apollo's offspring [Aesculapius], the inventor of medicine and art.' And Virgil, *Aeneid* vii. 11, '[Circe's land] where with incessant song the wealthy daughter of the Sun makes her inaccessible groves resound'. Aesculapius, god of medicine, and Circe, famous for her knowledge of magic and venomous herbs.

p. 108 The father of medicine, *fl.* before 400 B.C. His case histories are in his *De Epidemiis*.

p. 109 The anatomy of Galen was based on the dissection of monkeys. Andreas Vesalius, professor at Padua (1514–64), published detailed illustrations of human anatomy in *De humani corporis fabrica* (Basle, 1543). The humours *were* thought to have particular 'nestlings', choler in the gall, melancholy in the spleen, phlegm in the kidneys.

p. 109 The cause on which a disease depends so immediately, that it continues so long as the disease remains; translating Latin *causa continens*. This is the first English usage.

p. 109 A Roman physician, *fl.* A.D. 14–27. His *De Re Medica* was rediscovered by Guarino in 1426.

p. 110 In 81 B.C. Sulla, as dictator of Rome, ordered the execution of thousands of his enemies; see Plutarch's 'Life' of Sulla. The Triumvirs, Octavius, Antony, and Lepidus, drew up a list of 2,300 proscribed persons, including Cicero, after the murder of Caesar.

p. 110 For Augustus, see Suetonius, *Life of Augustus* 99; for Antoninus, his life by Capitolinus, 12.

p. 110 Latin version of a Greek epigram in Diogenes Laertius, x. 15.

p. 111 Theriaca, a nostrum invented by Andromachus, Nero's physician. It was a medical compound of many ingredients used as an antidote to poisons, venomous bites, and malignant diseases. *Mithridatum*: a similar compound used as a universal antidote; so called from Mithridates VII, King of Pontus, who is said to have fortified his body against poison by drinking antidotes. *Diascordium*: a medicine made of the dried leaves of herbs, including water-germander, invented by Fracastorius.

p. 111 They are made up for convenience, and not with reference to a particular disease.

p. 111 Mineral drugs were advocated by Paracelsus, in opposition to drugs of vegetable origin, which were accepted in the Galenic tradition. Paracelsus wrote on mineral waters; the many treatises on this topic were gathered in *Collectio de Balneis* (Venice, 1533), Cf. p. 240 below.

p. 113 i.e. juggling and conjuring. Cf. p. 245 below.

p. 113 Bacon is the only user of this word (*OED*). He means 'of the nature of an addition from without, not essentially inherent'. Discussion of the nature of the soul is found in Aristotle, Plato, Plotinus, Augustine, and in most theological writers. In the sixteenth century there were important discussions of the immortality of the soul by Ficinó, Pomponazzi, and Telesius, from whose *De Rerum Natura* (Naples, 1565) Bacon probably derived his notion of the two souls in man (see *Works* i. 49 f.). The immortality of the soul was proclaimed as Catholic dogma at the Lateran Council in 1513. For Aquinas's views, see F. C. Copleston, op. cit., ch. 4. See Sir John Davies's poem *Nosce Teipsum* (1599) for an account of the soul. By 'a producat' Bacon means an act of creation such as in the fifth day of Creation (Genesis 1:20, 24), 'Let the earth bring forth.'

p. 114 The Greeks and Romans believed in the omens listed here by Bacon. Chaldean astrology was the 'science' of understanding the influence of the planets on men, introduced by the Babylonians to the Greeks and Romans, and flourishing throughout the Middle Ages and Renaissance. Conjunctions (the apparent proximity of two planets, in the same sign of the Zodiac), aspects (the relative positions of planets), and eclipses could be predicted by astronomical calculations.

p. 115 The imagination reproduces and creates images. Bacon is here interested in the force of images, and discusses the topic at length in *Sylva Sylvarum*, Century X. Agrippa and Paracelsus believed in the power of images, i.e. talismans, to drawn down supernatural powers (see F. A. Yates, *Giordano Bruno*, ch. 7). On fascination as the power of spirits exercised through the eye, see Craig, *Enchanted Glass*, pp. 43 f. For Bacon's interest, see Wallace, *Bacon on the Nature of Man*, pp. 88 f.

p. 116 Here Bacon begins a lengthy discussion of topics which interest him profoundly, (*a*) invention, judgement, memory, and tradition (communication), i.e. logic, (*b*) the nature of good and the culture of the mind, i.e. ethics. The human powers, or faculties, of reason and understanding are employed to acquire true knowledge; the powers of will, appetite, and affection are employed to produce good actions. To arrive at true knowledge and good actions men use the powers of imagination and memory also. Affections are the passions of joy, sorrow, fear, anger, envy, compassion, shame, love, hope. The nearest equivalent to 'appetite' is desire guided by pleasure and pain, impelling man to action. The will is the power that impels to action, but it exercises the power of choosing to act or not to act. See K. R. Wallace, *Francis Bacon on the Nature of Man*. The imagination is a sort of messenger and ambassador.

p. 116 This is very much an adaptation of Aristotle, *Politics* i. 5.

p. 116 See II. xviii. 2 on the office of rhetoric. When men's minds are wrought on by rhetorical artifices, the imagination is roused till it triumphs over reason, and as it were does violence partly by blinding and partly by exciting it.

p. 116 i.e. into Memory, Imagination and Reason at II. i. 1. Bacon is here concerned not with the product of imagination, poetry, but with the imagination as itself a subject of study, and this can be studied partly as a branch of the doctrine of the Soul (II. xi. 3) and partly under the heading of Rhetoric.

p. 118 Bacon means 'finding out', and his major concern is for 'finding out' new facts. The term is normally used in rhetoric, where it applies to the collecting of the material relevant to the discourse, none of which need be 'new'. Bacon desires a logic of invention, a method whereby discoveries can be made. Existing logic is only useful as an instrument for handling the known. Bacon's new logic is set out in the *Novum Organum.*

p. 118 Aristotle, *Prior Analytics* i. 30. 'To believe every artist in his own art'. Celsus, in the Preface to his *De Re Medica* (written in the first century A.D., rediscovered in 1426 by Guarino) is in fact attributing the statement in the text to empirics, to whom he is opposed.

p. 118 In the *De Augmentis* (v. 2) it is 'Plato more than once intimates'; the reference to *Theaetetus* being clearly wrong.

p. 119 Virgil, *Aeneid* xii. 412–15:

> A branch of healing dittany she brought,
> Which in the Cretan fields with care she sought.
> Rough is the stem, which woolly leaves surround,
> The leaves with flowers, the flowers with purple crown'd;
> Well known to wounded goats; a sure relief,
> To draw the pointed steel, and ease the grief.
>
> (Dryden)

Plutarch, in *Whether Land or Sea Animals are Cleverer*, ch. 20, lists man's indebtedness to animal cleverness.

p. 119 *Aeneid* viii. 698–9. 'Her [Cleopatra's] monstrous gods of every form, and barking Anubis, opposed to Neptune, Venus and Minerva, are wielding their weapons.'

p. 119 See Polydore Virgil's *De Rerum Inventoribus* (Venice, 1499, expanded in an edition at Basle, 1521). Bacon uses Acosta's *Natural History of the Indies* (iii. 2) for the story of the use of sticks to make fire, Pliny for the ibis, Polydore for the invention of gunpowder. Aristotle (*Ethics* vi. 4) quotes Agathon, 'Art is in love with luck, and luck with art.' See Pope, *Essay on Man* iii. 169 f.

p. 119 Persius, Prologue to the *Satires*, 1. 8. 'Who taught the parrot to say "how d'ye do"?' The raven and the ant are in Pliny, *Natural History* x, 43 and xi. 30.

p. 120 Induction is the process of reasoning whereby we discover the general laws governing the numberless events and facts in the natural world. The process begins with observation of individual instances. If all possible instances are observed, to which the conclusion can refer, it said to be a perfect induction. In the natural world, inductions are imperfect, because it is impossible to examine all cases. In *Novum Organum* i. 105 Bacon praises Plato for his use of the method of induction for the purpose of discussing definitions and ideas. (See F. H. Anderson, *The Philosophy of Francis Bacon* (Chicago, 1948), ch. 11.) Socrates proceeds to define a concept by analysing a group of instances, and thereby abstracts the element common to them all. This is not the method of induction which Bacon proposes in the *Novum Organum* as likely to lead to the discovery of the Forms in Nature, since it is imperfect induction.

p. 120 The *lictores* attended Roman magistrates to push the crowd aside (i.e. to act as *whifflers*) and the *viatores* summoned persons before him (i.e. acted as *sergeants*).

p. 121 See Matthew 18:3. In *Valerius Terminus* (*Works* iii. 224) Bacon says that to enter the kingdom of knowledge man must become as a child, i.e. humbly learn to read the book of nature.

p. 121 Bacon is using the technical language of syllogistic logic—middle proposition (or term), deduced, reduction. He means that in natural sciences you would not deduce truths from the axioms incidentally arrived at by inductions, because the syllogism only relates words to each other, and the words do not necessarily correspond to facts. See Bacon on the Idols of the Market-Place, in *Novum Organum* i. 43, 59, 70, and also II. xvi. 2, 3 below.

p. 121 Cf. Aristotle, *De Interpretatione* I. i. 2.

p. 121 The school of Pyrrho, the Sceptics, was earlier than the New Academy, since Pyrrho was contemporary with Alexander the Great. The Sceptic argued that it is impossible to know whether we know anything or not. The New Academy flourished in the third and second centuries B.C. under Arcesilaus and Carneades, and maintained the doctrine of *acatalepsia*, that nothing can be known with certainty. Socrates affected ignorance, not to induce scepticism, but to cleanse the mind of preconceived opinions, to prepare it for the impartial quest for truth. Cf. *Novum Organum* i. 37, 67, 126.

p. 121 Cf. Cicero, *Academic Questions* ii. 5, §15, 'Under pretence of ignorance he affected knowledge.' Cicero belonged to the New Academy, having as his teachers Philo of Larissa and Antiochus of Ascalon.

p. 121 In the *De Augmentis* (*Works* i. 622) the reading is 'simpliciter et integre', i.e. in simplicity and integrity.

p. 122 In *Novum Organum* ii. 39 f. he mentions microscopes, telescopes, measuring rods, and astrolabes.

p. 122 Cf. *Novum Organum* i. 61, 122 where he suggests that the application of his method would be mechanical and lead to certainty.

p. 122 i.e. learned experience. In *De Augmentis* v. 2 he explains that this is a kind of wisdom, a proceeding from experiment to experiment not haphazardly but with some direction. He calls it 'the hunt of Pan', thinking of the fable that it was Pan who, by 'a happy accident as he was hunting', discovered the hiding place of Proserpine (*Wisdom of the Ancients*, s.v. Pan). See *Novum Organum* i. 70, 100–10.

p. 122 i.e. the interpretation of nature. This is his Novum Organum, proceeding from experiments to axioms, which then lead to new experiments. Cf. *Novum Organum* i. 26 f.

p. 122 For a clear account of 'invention' in rhetoric and logic, see Sister Miriam Joseph, *Rhetoric in Shakespeare's Time* (New York, 1962), pp. 22–30, 308–53 ('Topics of Invention').

p. 123 i.e. topics. In rhetoric and logic, works listing 'topics' (e.g. definition, genus and species, contraries, similarities) gave guidance on finding matter and argument for the writer and orator.

p. 123 A technical term in jurisprudence signifying the recourse to general principles of justice to correct or supplement the provisions of the law.

p. 124 'Every act of discovery advances the art of discovery.' This is, in effect, the last sentence of Book I of the *Novum Organum*. Bacon wrote a list of particular topics (i.e. specific things to look for) respecting Light and Luminous Matter (*Works* ii. 317 f., v. 409 f.), and Heavy and Light (*De Augmentis* v. 3).

p. 124 i.e. middle term. See n. to. p. 84 above; R. Whately's *Elements of Logic*, or W. S. Jevons's *Elementary Lessons in Logic*, will explain the details of the syllogism referred to in the following paragraphs. In a 'probation ostensive' you prove either the same conclusion or one which implies it. *Per incommodum* is 'by absurdity', i.e. you prove not directly that the original conclusion is true, but that it cannot be false. The basic doctrine is laid down in Aristotle's *Prior Analytics*.

p. 125 i.e. a logical refutation of a proposition that has been syllogistically defended. The doctrine's true use is to answer sophistical arguments, the corrupt use to invent sophistical objections. Aristotle handles the topic in his *Sophistici Elenchi*. Bacon uses the word to mean 'cautions'.

p. 126 A reference to Aristotelian and scholastic discussions of substance and attributes, or of essence. In logic and metaphysics such discussions lead to attempts to define such concepts as 'act', 'existence', etc. Bacon is concerned with cautions against ambiguity of words.

p. 126 'Predicaments' is the Latin equivalent of the word. Aristotle, in his *Categories*, deals with the main types of meaning of the words and phrases that can be combined to make sentences. Words 'mean' one of the following things—substance, quantity, quality, relation, place, date, posture, possession, action, passivity.

p. 127 In *De Augmentis* v. 4 this section is enlarged to include the doctrine of the idols or false appearances which hinder man's pursuit of truth. See *Novum Organum* i. 39–68 for this important doctrine.

p. 127 Bacon is thinking of what he later called the Idols of the Tribe, inherent in men. Cf. *Novum Organum* i. 39–40.

p. 127 Girolamo Fracastoro, in *Homocentrics* (1538), was one of those who proposed homocentric spheres, to explain planetary movement, and to get rid of epicycles and eccentrics. Eccentrics, in Ptolemaic astronomy, explained the movement of the planets around the earth, and their varying brightness.

p. 127 It should be *monadica*, meaning unique. Cf. *Novum Organum* i. 45, ii. 28.

p. 127 Empedocles recognized four elements, earth, air, fire, and water. Each was assigned an orbit, that of fire being the outermost, just below the moon and completely transparent. See C. S. Lewis, *The Discarded Image*, ch. 5. The new Copernican cosmology 'quite put out' the element of fire.

p. 127 The measure of all things. This was the doctrine of Protagoras, see Plato, *Theaetetus* 152, Aristotle, *Metaphysics* X. i. 20.

p. 127 A sect which flourished in the fourth and as late as the tenth centuries. The monastic sect of Audaeans, in Mesopotamia in the fourth century, maintained that the expression, 'God created man in his own image', meant that God was shaped like a man. They interpreted Biblical references to the eye, ear, arm, or hand of God literally. For the Epicureans see Cicero, *On the Nature of the Gods* ii. 17.

p. 128 Roman magistrates in charge of buildings, who on festal occasions adorned them with lights and statues. See Cicero, *On the Nature of the Gods* i. 9.

p. 128 i.e. Again, let us . . . He turns now to the Idols of the Cave, originating in the constitution, habits, etc., of the individual. Plato's myth is in Book vii of the *Republic*. See I. v. 6 above.

p. 128 The Idols of the Market-Place.

p. 129 i.e. from things better known. Demonstration by congruity is by the use of analogy. The reference is to Aristotle, *Prior Analytics* ii. 5.

p. 129 See II. xvi. 2 below where he calls it 'the organ of tradition'.

p. 129 Elizabethan students collected ideas and facts from their reading, and entered them under suitable headings in commonplace books, which formed a store-house of useful material for writing and oratory. Bacon's own *Colours of Good and Evil*, Antitheta, and *Promus of Formularies* are highly original commonplace collections. Volumes of commonplaces were published, making the original wide reading unnecessary. Bacon urges the independent compilation of such works. See K. R. Wallace, *Bacon on Communication and Rhetoric*, pp. 63 f.

p. 130 Cf. *De Augmentis* v. 5 for a full discussion, and *Novum Organum* i. 127, ii. 26. F. A. Yates, *The Art of Memory* (1966), surveys the topic from classical to Renaissance times. The art was part of rhetoric, to enable an orator to memorize his speech, but Bacon suggests that it could have practical use for the scientist. What one wished to remember was reduced to images, which were then 'placed', in imagination, on (usually) parts of a building. The mind then visited the parts in the required order, drawing from each place the images 'placed' on them. Bacon calls the 'places' *prenotions* and the images *emblems*.

p. 131 Bacon learned this from Joseph de Acosto's *Moral History of the East and West Indies* (1604), Book 6, ch. 5. The Chinese written characters are not such as he describes them. But Bacon's interest in 'real characters', a language in which signs would be really in contact with the things they expressed, stimulated interest in the following century. See John Wilkins, *An Essay towards a Real Character and a Philosophical Language* (1668).

p. 131 See n. to p. 81 above.

p. 132 The idea that there are names that are by nature the right ones is the subject of Plato's *Cratylus*. Leibniz suggested that the language of Adam, when he named the creatures, must have been close to reality. See D. P. Walker, 'Leibniz and Language', *Journal of the Warburg and Courtauld Institutes*, 35 (1972), 294–307. In *Novum Organum* i. 43, 59, 60 Bacon discusses words.

p. 133 This implied condemnation of the attempts to write English verse in classical metres was strengthened in the *De Augmentis* (vi. 1). Ascham had recommended the use of classical metres, Sidney and Spenser had been attracted to the notion. Daniel's *Defence of Rhyme* (1603) against Thomas Campion was perhaps in Bacon's mind here.

p. 133 The account is expanded in *De Augmentis* vi. 1. See *Works* i. 658, 841–4.

p. 134 The controversy, which raged in Cambridge when Bacon was an undergraduate, was that relating to Ramus's reform of dialectic and rhetoric. See W. S. Howell, *Logic and Rhetoric in England, 1500–1700* (Princeton, 1956), ch. 4, and N. W. Gilbert, *Renaissance Concepts of Method* (New York, 1960). It was a controversy between those 'concerned with the efficiency of teaching and with the application of the useful arts to life', and those 'concerned with methods of strict demonstration and proof' (Gilbert, p. 222). The most important work to come from the controversy was Descartes's *Discours de la méthode* (1637).

p. 134 '"Method" is a branch of Logic or Rhetoric which teaches how to arrange thoughts and topics for investigation, exposition or literary composition' *OED*). For Bacon, as for Ramus, it is a branch of Logic.

p. 134 The magistral method is dogmatic or doctrinal. It is used when a teacher wishes to convey information which is to be believed, and not questioned. Examples would be the *Sentences* of Peter Lombard and the *Loci communes theologici* of Melanchthon, collections of the main points in theology, arranged in order to give a 'system' of doctrine (Gilbert, pp. 108 f.). The method of probation is that used when the matter delivered is to be questioned and examined. The magistral method is for use when existing knowledge is to be applied; the probational method is for use when the bounds of knowledge are to be extended; the former for the ordinary audience, the latter for the learned audience. See Rossi, pp. 175–7.

p. 135 i.e. to be spun continuously, without a break.

p. 135 i.e. secret (or esoteric) and open or exoteric. In the latter case, special effort was made to be clear and orderly, but Bacon speaks only of the former here. The alchemists and the 'magicians', such as Agrippa and Cardan, urged that their esoteric knowledge should be given only to the initiated, and they presented it in an obscure manner.

p. 135 The aphorism is usually a brief, pithy statement or observation; it is authoritative and profound. As such it was common in works of medicine, politics, and Roman law. Bacon planned a collection of 300 legal maxims (see *Works* vii. 309 ff., and *De Augmentis* viii. 3). He saw the aphorism as a way of presenting truth, in a way that invited further thought. The lack of argumentative or systematic link with other aphorisms prevented the thinking from becoming a rigid system. The presentation of thought in 'methods' or system, by careful organizing of the discourse (which is what Bacon does in the *Advancement*) he considered less useful for the presentation of new knowledge. See B. Vickers, *Francis Bacon and Renaissance Prose*, ch. 3.

p. 136 Actual things are not arranged in any pattern, therefore observations on them are best left separate and unconnected. Knowledge carefully arranged suggests that it is complete and inhibits further inquiry.

p. 136 See I. iv. 6 above. This is the scholastic method. Bacon dislikes disputatious learning. He goes on with an image of a 'piece' or stronghold. A general should not leave a strongly held enemy fort in his rear, while he goes ahead to subdue lightly held places. So a teacher will not use the question and answer method for anything but important prejudices held by a learner.

p. 136 The single method of Ramus (Pierre de la Ramée) the French humanist philosopher (1515–72) who provoked the hostility of the Aristotelians by his attempts to reform the science of logic. For Ramus 'method' was arrangement, whereby subject-matter is arranged in a descending order of generality, the most general being placed first, the least general last. See Gilbert, op. cit., ch. 5. On *the particular topics of invention* see II. xiii. 10 above.

p. 137 The *De Augmentis* reads here 'another diversity of method to be used with judgement', which is what we would expect.

p. 137 Probably a reference to Aristotle, *Ethics* vi. 3 where Aristotle discusses 'knowing' and refers, perhaps, to Socrates' use of similes when discussing the same topic in Plato's *Theaetetus* 197. 'Democritus' is therefore an error. Bacon himself uses figurative language when doctrines quite new are to be imparted.

p. 137 *Analysis* and *systasis* (or synthesis) are traditional terms in logic; *cryptic* = the secret method of communicating knowledge.

p. 138 These Greek terms are defined by Aristotle in *Posterior Analytics* i. 4 when writing of the principles of demonstration. These principles were elevated by Ramus into three fundamental 'laws', of Truth, of Justice and of Wisdom. Bacon lists first what was Ramus's Law of Wisdom, that to be universal a proposition must be true of its subject *qua ipsum*. 'The predicate must belong to the subject not only necessarily but in virtue of the specific nature of the subject, not in virtue of the generic character which it shares with other species. For thus alone will the subject contain nothing irrelevant to the predicate' (see D. Ross, *Aristotle*, 1966, p. 46). Next Bacon lists Ramus's Law of Truth, that the predicate must be affirmable in all cases and at all times, of the subject of a proposition. The Law of Justice is referred to in the next paragraph. This states that whatever is so connected with the essence of a thing as to be involved in its definition is said to belong to it *per se* (*Καθαυτό*). Thus a line belongs *per se* to the notion of a triangle, because the definition of a triangle involves the conception of a line. See *Works* iii. 203–5; Rossi, op. cit., pp. 193 f.

In the *De Augmentis* Bacon refers not to 'the canker of epitomes' (the passion for which was attributed to the influence of Ramus, see n. to p. 72 above) but to Ramus's 'peculiar method and dichotomies'.

p. 138 Failing to advance, not making progress. *incurring*, running into themselves. One of the dangers of the Law of Wisdom is of arguing in a circle, of taking the conclusion itself as one of the premises, or of using in a definition a word that is a synonym of the word defined.

p. 138 This is the Law of Justice.

p. 138 Abraham Ortelius, *Theatrum orbis terrarum* (1570), the first great atlas of the world.

p. 138 Raymond Lully (1235–1316). Lully's art was an art of invention. By symbolizing all the elements of nature, and the categories, and by establishing laws governing all the possible combinations of the symbols, he proposed to answer all questions. It is mocked by Swift in ch. 5 of the *Voyage to Laputa*. But it attracted great attention in the Renaissance, Pico, Agrippa, and Bruno adopting the Art. Descartes and Leibniz were also influenced by his quest for a universal key to knowledge. See F. A. Yates, *The Art of Memory*, ch. 8.

p. 138 A system intended as an aid to learning, in which words or terms are grouped according to types or classes (*OED*). Bacon adopts the word, from the Latin; Alessandro Citolini's *Tipocosmia* was printed at Venice in 1561.

p. 139 See K. R. Wallace, *Francis Bacon on Communication and Rhetoric*, for a full discussion of this section, and Sister Miriam Joseph, *Rhetoric in Shakespeare's Time*, for general information about Elizabethan rhetoric.

p. 139 The analysis is based on faculty psychology and sees the function of rhetoric not as persuading to belief but as influencing action. It is a part of rational philosophy, because rhetoric uses the imagination on behalf of reason, making argumentative discourse vivid. For Bacon, rhetoric is to the imagination what Logic is to the understanding. His theory is adopted largely from Plato in the *Phaedrus*. Control of Will is fought for by Reason and Passion, each of which strives to enlist the aid of Imagination. Rhetoric is the diplomat whose task is to enlist the aid of Imagination on the side of Reason, so that man may lead a rational life. The whole paragraph is the most obviously balanced piece of writing in Bacon's works.

p. 140 In *Phaedrus* 250. Pope, *Essays on Man* ii. 217–18 has the opposite,

> Vice is a monster of so frightful mien,
> As, to be hated, needs but to be seen.

Bacon does not see rhetoric as being limited to popular discourse. All men, learned and unlearned, tend towards unreason, prejudice, and emotion, and cannot be won by mere proofs.

p. 140 In Cicero, *De Finibus* iv. 18–19.

p. 140 A technical term in rhetoric, meaning a kind of exordium to a speech designed artfully to win over the hearer.

p. 141 A common image, borrowed from Zeno by Cicero and Quintilian, to explain the preoccupation of Logic with the discourse of philosophy, and Rhetoric with the discourses of orators. The complete theory of communication, it is implied, requires both. See W. S. Howell, *Logic and Rhetoric in England, 1500–1700* (1956), pp. 364–75.

p. 142 i.e. popular opinions about good and evil. Each statement, such as he gives in his example, is apparently true, but not really so. A man should therefore have ready a prepared refutation of such statements. Rhetoric must second reason, so the bad reasoning implied in these accepted sophisms must be shown. The suggestion for this comes from Aristotle, *Rhetoric* i. 6, 7, and partly from the *Topics* i. 12 f. and the *Sophistici Elenchi*. Bacon published a collection of these *Colours* with his *Essays* in 1597. They are useful in deliberative discourse, i.e. in persuading or dissuading.

p. 142 *Sophism*: 'What men praise, is good; what men condemn, is bad.' *Refutation*: 'The merchant praises what he wants to sell' (Horace, *Epistles* II. ii. 11). *Malum est*: Proverbs 20:14, 'A bad buy, a bad buy, says the purchaser, but he goes off to brag about it.'

p. 142 In *De Augmentis* vi. 3 Bacon gives 47 examples, many of which formed the basis for essays. They are on such topics as a speaker may expect to handle frequently. In brief form are collected points for and against, on topics such as Beauty, Life, Health, Pride, and not only on legal topics. They are useful in demonstrative discourse (praising or dispraising) and judicial (proving or disproving a case) as well as in deliberative discourse.

p. 143 *For the words of the law*: 'Interpretation which departs from the letter is not interpretation but divination. When the letter is departed from, the Judge becomes a Lawgiver.' *For the intention of the law*: 'The sense according to which each word is to be interpreted must be collected from all the words together.'

p. 143 Bacon's orator has always equipped himself beforehand with colours, antitheses, and formulae, i.e. the sentences and phrases that are needed for different kinds of opening, conclusions, links, transitions, repartees etc. One of his MS. collections of formulae has survived, the *Promus of Formularies and Elegancies* (*c.* 1594), ed. Mrs. Henry Pott (1883). His example here is how to conclude a passage in a persuasive argument.

p. 143 Acts 9:25 'He was let down in a basket' (*sportam*), emended to 'by the gate' (*portam*). Although this is a possible error, the Vulgate has *in sporta*. The editing of good Latin texts was a major activity of the humanists.

p. 144 i.e. pedagogical. In the *De Augmentis* vi. 4 he adds, 'consult the schools of the Jesuits, for nothing better has been put into practice'.

p. 145 *Annals* i. 16–22. In *De Augmentis* he recommends acting as useful in education, and notes that the Jesuits use it, but not that it was common in English universities and schools.

p. 146 i.e. passions and will, the subject of Ethics. Bacon wishes to build on the basis of observation of 'ordinary and common matters', of how men behave.

p. 146 Cf. Aristotle, *Ethics* ii. 1; x. 9.

p. 147 Discussions of happiness and the highest good by Plato, Aristotle, Cicero, etc., were a kind of theology to the heathens.

p. 148 These topics are discussed in Aristotle, *Ethics* i. 8; x. 6; iii. 4.

p. 149 Moses in Exodus 32:32 and St. Paul in Romans 9:3. Cf. essay 'Of Goodness and Goodness of Nature'.

p. 149 In Aristotle, *Ethics* x, 7, 8 who decides for the contemplative. The topic was frequently debated in the Renaissance.

p. 149 Cicero, *Tusculan Questions* v. 3, though it was Leo, not Hiero.

p. 149 Augustine, *City of God* xiv. 9, speaks of St. Paul as a spectacle in the theatre of this world.

p. 150 The controversies are those between two parties, (*a*) that of Zeno (founder of the Stoic ethic, who died 263 B.C.) and of Socrates, as represented in Plato's dialogues. These and their Stoic and Platonic followers represented happiness as resulting from virtue exercised in society. And (*b*) the sects listed in the rest of the paragraph. The *Cyrenaics* were founded by Aristippus of Cyrene (*fl.* 366 B.C.), who, like Epicurus later, taught that pleasure is the aim of the happy life. Lucretius, *De Rerum Natura*, is the main source of our information about Epicureanism.

p. 150 The Golden Age of Saturn, when it was eternal spring, and men were free from fears, was succeeded by the Age of Silver under Jupiter. See Ovid, *Metamorphoses* i. 89 ff.

p. 150 Herillus was a Stoic of Carthage, *fl.* 264 B.C. He distinguished only between what we do willingly and what unwillingly.

p. 150 i.e. the extreme German sect. The parallel with Herillus is tenuous, being based only on the Anabaptists' stress on individual faith and withdrawal from contact with all outsiders.

p. 150 The *Enchiridion* of Epictetus, first century A.D.

p. 150 Fernandez Gonsalvo of Cordova (d. 1515). See Guicciardini, *History of Italy* vi. 2.

p. 151 See the story in Aulus Gellius, *Attic Nights* ix. 2.

p. 151 This should be Aristippus, see Diogenes Laertius, *Lives of the Philosophers* ii. 75. *in praecipitio*: even on the steepest precipice.

p. 151 The officer who dispenses stores and the one who collects them.

p. 152 Seneca, *Naturales Quaestiones* II. lix. 37. 'We think it a great matter to be a little longer in dying.' Proverbs 27:1. 'Do not flatter yourself about tomorrow, for you never know what a day will bring forth.'

p. 152 First used by Bacon, meaning giant-like. Cf. note to p. 82 above. Thinking of the rebellion of the giants against the gods, he implies 'those who trouble the world on a gigantic scale'. For Sulla, Dictator of Rome in 81 B.C., see Plutarch's 'Life' of him, to the last sentence of which Bacon refers.

p. 153 Virgil, *Aeneid* vi. 730. 'They have a fiery vigour and a heavenly origin.' Bacon means that there are nobler natures in existence, to which the inferior aspire—in man, to the perfection of the angel.

p. 154 Plutarch's 'Life' of Solon, 7. ' "Enjoy not, that you may not desire; desire not, that you may not fear." This is the precaution of cowardice and pusillanimity.'

p. 154 Cf. Plato, *Phaedo* 64.

p. 156 In Cicero, *De Oratore* ii. 18.

p. 156 James's *Basilicon Doron*, a book of instructions for his son, Prince Henry, was printed in 1599, and, in large numbers, in 1603. The critique of it is full of strangeness—*sick of dizziness*, i.e. confused and disordered; *of convulsions*, i.e. contorted by cramming in irrelevances.

p. 156 James's *The True Law of Free Monarchies*, 1603.

p. 157 Cicero's oration in 46 B.C. thanking Caesar for pardoning Marcellus.

p. 157 'The *little king* of serpents, commonly called the cockatrice' (Sir Thomas Browne, *Pseudodoxia Epidemica*, Book 3, ch. 7, where there is a full account of the basilisk).

p. 157 *The Prince* (1513) and the *Discourses on Livy* (1512–17) appealed to Bacon as accounts of the actual behaviour of men, necessary to be understood by those who aim at public good. 'The pages of Tacitus breathe a livelier and truer observation of morals and institutions' than the admired ethical works of Plato and Aristotle (*Masculine Birth of Time*, in Farrington, pp. 71–2). The substance of this paragraph is found in the *Meditationes Sacrae*, 'Of the Innocency of the Dove and the Wisdom of the Serpent', printed with the *Essays* in 1597. This kind of knowledge is deficient, he adds in *De Augmentis* vii. 2, and calls it 'Serious Satire' or 'The Treatise of the Inner Nature of Things'.

p. 158 Virgil, *Aeneid* vi. 822. 'It was an unhappy deed, whatever posterity might say of it.' Livy ii. 5 tells the story of Brutus' execution of his sons, because they conspired to restore the Tarquins.

p. 159 Aristotle, *Magna Moralia* i. 1. 'It is necessary then to speak of virtue, both what it is and whence it proceeds. It would be almost useless to know what virtue is, but be ignorant of the ways and means of acquiring it. Therefore we must inquire not only to what kind virtue belongs, but also how it may be obtained. For we wish to be acquainted with the thing itself, and to gain possession of it; and we shall not fully succeed in this, unless we know both the whence and the how.'

p. 160 The elaborate section on the Culture of the Mind is parallel to the *Essays* in part. See R. S. Crane, 'The Relation of Bacon's *Essays* to his Program for the Advancement of Learning', in *Essential Articles*, pp. 272 f.

p. 161 The study of the influences of the heavenly bodies on events, plants, minerals and human psychology. The dispositions listed are influenced by the planets in this order—Saturn, Jupiter, Mars, the Sun, Venus, Mercury, and the Moon.

p. 162 Accounts of the conclaves at which Popes were elected were circulated. See Leopold Ranke, *History of the Popes* (trs. E. Foster, 1891), Appendix, sections 5, 6 (iii. 258 f.).

p. 162 Perception deriving from the senses. *dry*, impassive. *formal*, precise, stiff. *real*, sincere, straightforward. *humorous*, fanciful or moody (i.e. dependent on some 'humour'). *certain*, trustworthy, confident. *huomo*, i.e. the impulsive man and the cautious man.

p. 162 i.e. rebuke them sharply. These are St. Paul's words, when he quotes (Titus 1:12–13) an account of the Cretans from Epimenides, 'the Cretans are always liars, vicious brutes and lazy gluttons'.

p. 162 Sallust, *History of the war against Jugurtha* 113. 'For the most part the desires of kings are as changeable as they are strong, and often contradictory.'

p. 162 Pindar, *Olympian Odes* i. 55. 'Who are not able to digest great prosperity.'

p. 163 *Rhetoric* ii. 12–17.

p. 163 i.e. politicians in democratic states. The reference is to a fragment of Solon's. On the affections as winds, cf. Pope, 'Reason the Card, but Passion is the gale', *Essay on Man* ii. 108.

p. 163 The *Nicomachean Ethics*, *Eudemian Ethics*, and *Magna Moralia*. See *Rhetoric* ii. 1–11, for discussion of Anger, Fear, Pity, Indignation, Envy. The discussion of pleasure and pain is in [*Nicomachean*] *Ethics* ii. 3 f.

p. 163 Referring not to Seneca, whose Stoic *Epistles to Lucilius* he often quotes, but to the Greek school of Zeno.

p. 164 Plutarch and Seneca wrote on anger; Plutarch, Seneca, and Servius Sulpicius on consolation, Plutarch on bashfulness, Sir Thomas More and Cardan on 'comfort'.

p. 164 Rewards and punishments.

p. 164 A list of topics, on some of which Bacon wrote Essays. He selects 'custom and habit' (cf. Essays 'Of Nature in Men' and 'Of Custom and Education') and 'books and studies' (cf. essay 'Of Studies') as examples.

p. 166 The wine of devils. St. Augustine, in his *Confessions* i. 16. 26, quotes from Terence, and says that he blames not the words, which are 'precious vessels' but 'that wine of error which is drunk to us in them by intoxicated teachers'. St. Jerome says that 'the songs of the poets are the food of devils'. Agrippa in *The Vanity and Uncertainty of Arts and Sciences*, ch. 4 quotes both. Cf. essay 'Of Truth'.

p. 166 In *Ethics* i. 3, it is 'political science is not a proper study for the young'. But for Aristotle ethics is a branch of political science. Cf. Shakespeare, *Troilus and Cressida* II. ii. 165–7.

p. 172 In the courtesy books of the period, della Casa's *Galateo* (trs. 1576) and Stefano Guazzo's *La Civile Conversatione* (trs. 1581, 1586) being frequently published and imitated. See L. B. Wright, *Middle-Class Culture* (1935), ch. 5.

p. 177 The symbol in the Renaissance of the union of supreme insight and supreme power. Some Biblical texts offer a deeper insight into God's purpose than others.

p. 178 Machiavelli's *Discourses on Livy* and *The Prince*. Bacon clearly enjoys expounding Solomon's aphorisms. But he recommends basing knowledge of this kind on historical facts. He would rather build up an axiom from facts than illustrate an axiom with them.

p. 178 Cicero's *Letters to Atticus*, and his letters to his friends, to his brother, to Brutus, are full of detailed autobiographical and historical facts of the exciting period of Rome under Pompey, Caesar, and Antony.

p. 180 In II. xxiii. 48 below he quotes Revelation 4:6 in support of his notion that this globe seems dark to us but to God it is as crystal. He seems to have transferred this and made it a basic law for all inquiries of truth—that all matters can be apprehended and meditated in axioms.

p. 181 The god of raillery and censure. He complained that Vulcan had not provided man with a window in the breast. See Lucian, *Herotimus* 20.

p. 185 i.e. men should account their abilities less than they are and their disabilities as greater than they are.

p. 185 Caesar Borgia, son of Pope Alexander VI, who was made Duke of the Valentinois. See Guicciardini, *History of Italy* vi. 3.

p. 186 Caesar's friends, Antony, Hirtius, Pansa, Dolabella, Pollio, were active men of affairs but of no great character or reputation.

p. 197 The first usage of this word, meaning inability to hold one's tongue. Sisyphus was punished because he betrayed the designs of the gods, Tantalus divulged the secrets of Zeus. For the way in which Bacon thought that 'soul' moved the body, see K. R. Wallace, *Francis Bacon on the Nature of Man.*

p. 200 Sarah (Genesis 18:12) laughed when God told Abraham that she would bear a son in her old age, but Abraham believed, 'and it was imputed to him for righteousness' (Romans 4:22).

p. 201 Bacon distinguishes sharply between truth of religion and truth of science. See B. Willey, *The Seventeenth Century Background* (1949), ch. 2.

p. 202 i.e. admits and rejects the use of reason and disputation, but according to just limitations. By 'the old law' he means that of the covenant of the Old Testament. Its law of sacrifice was not unreasonable. Heathen, i.e. Greek and Roman religion, was lacking in doctrine; there were the gods and their attendant ritual, and the varied beliefs expressed by Plato, Aristotle, the Stoics, the Epicureans, the Sceptics, etc.

p. 202 In revealing His mysteries to us, God uses 'the language of accommodation', that is, speaks in a way that we can understand, and to understand we must use our reason. But in this we use reason only to understand, not to dispute. We can also use our reason in a secondary degree to derive doctrine from the revealed facts. This is much like playing a game of chess or administering laws; in these there are basic, unquestionable *données*, and our reason is exercised within the limitations imposed by them.

p. 203 A mixed metaphor. The opiate refers to the curious speculations and the bridle to the fury of controversies.

p. 205 Bacon advocates toleration of religious sects in several works. See W. K. Jordan, *Development of Religious Toleration in England, 1603–40* (1936), pp. 457–72.

p. 205 The 'cistern' method seems to be that of collecting from scripture all statements relating to doctrine and reducing them to a series of positions as, for example in Aquinas's *Summa Theologiae* and *Summa contra Gentiles.*

p. 206 Peter Lombard, the Paris theologian of the twelfth century whose *Four Books of Opinions* (i.e. of the opinions of the Fathers on the Trinity, Creation and Sin, Incarnation and the Virtues, and the Sacraments and the Last Things) were known as *The Sentences.* The standard theological textbook until *c.* 1600, it was lectured on by Aquinas, Duns Scotus, William of Ockham, and Luther. Its contents are largely from St. Augustine.

p. 206 Tribonianus was one of the sixteen commissioned by Justinian (*c.* A.D. 530) to compile a complete body of law. The *Digest* collected what was valuable in preceding jurists. Civil law is Roman law. On the 'modern' glosses on the civil law, see B. Smalley, *Study of the Bible in the Middle Ages*, pp. 36 f., and M. P. Gilmore, *Humanists and Jurists* (Cambridge, Mass. 1963), ch. 3.

p. 206 Bacon's dislike of knowledge presented in 'methods' extends to divinity; cf. p. 34 above.

p. 207 In the medieval practice of interpreting texts of scripture in a fourfold manner, the anagogical sense extracted was that spiritual sense that applied to the life to come.

p. 207 Cf. note to p. 39 above. Paracelsus (1493–1541) the alchemist and his followers are often charged with founding science on texts from Genesis (e.g. 'What generates is an element' Genesis 1:11, 21, 24). Richard Bostocke stresses the Biblical character of Paracelsism and its opposition to 'heathen' science in his *The Difference between the ancient physic, first taught by the godly forefathers, and the later physic proceeding from Idolaters* (1585).

p. 208 Cf. Luke 24:5, and *Novum Organum* i. 65. In scripture, everything which concerns the passing interests of the body is called dead; the only living knowledge having regard to the eternal interest of the soul (Devey).

p. 209 The interpretation of scripture in these three ways was first defined by Origen and was practised until the seventeenth century. See B. Willey, *The Seventeenth Century Background*, ch. 4.

p. 209 See M. MacLure, *The Paul's Cross Sermons, 1534–1642* (Toronto, 1958) and J. W. Blench, *Preaching in England . . . 1540–1600* (Oxford, 1964).

p. 209 This paragraph is a summary of St. Augustine's comments in *The City of God* ii.

p. 210 Cf. Plato, *Protagoras* 348 d. The phrase 'by thought, word and deed' occurs in the general confession said at Communion in the Anglican Church.

p. 211 In James's *Daemonologie* (1597), iii, 6.

p. 214 The world discovered to the west of Europe was *Novus Orbis*, the New World. Parts of that world were called Hispania Nova, Nova Francia, Nova Scotia, New England, hence Bacon's *New Atlantis*. Plato's Atlantis was sometimes identified with America; Bacon places his in the Pacific (the South Sea), roughly in the area where the Solomon Islands had been discovered in 1568.

p. 214 The *New Atlantis* was published by William Rawley, Bacon's chaplain, in 1627, at the end of the volume of natural history, the *Sylva Sylvarum*, which Bacon compiled when he left this fable unfinished. Between April and October 1623 Bacon had hopes of becoming Provost of Eton College. Had he been successful he would probably have wanted to set up a research institution there, on the lines of the College of the Six Day's Works. See Introduction, p. x n. 16 above.

p. 223 The Father of the House of Solomon is both scientist and priest, and he alone is allowed to approach the pillar and find the ark. His prayer is that of a man whose only knowledge is that of God's works, and, as such, an important part of Bacon's argument that such knowledge does not produce atheism. In the fable Bacon presents a people whose scientific knowledge was not impeded by Greek philosophy, and the Christian Revelation, when it comes, is easily absorbed into their scientific culture.

p. 223 The apostle Bartholomew was supposed to have travelled to India. Bacon stresses the miraculous nature of this revelation to the people of New Atlantis. It is the coming of God's Word to add to their knowledge of God's Works, and includes the 'gift of tongues', a second Pentecost.

p. 225 According to the then accepted chronology of world history, this would be *c.* 1400 B.C., at the time, in Biblical history, of the Exodus.

p. 225 This takes us back to 1500. The dates of the earliest voyages are Diaz, 1488; Columbus, 1492; Vasco da Gama, 1498; Vespucci, 1497–1504; Magellan, 1519–22.

p. 226 The straits of Gibraltar. *Paguin* or *Cambaline* is Peking or Cambalu. *Quinzy* is Hangchowfoo.

p. 226 Plato, in the *Timaeus* and *Critias*, describes Atlantis and its sudden disappearance. Bacon enjoys elaborating his explanation for the isolation of New Atlantis in terms of conjectures involving anthropology and history.

p. 228 The lawgiver is necessary to enable Bacon to complete his fable by a description of ideal laws, a subject dear to him who had passed his life in the study and practice of law.

p. 228 Cf. I Kings 4:29 and Proverbs 25:3.

p. 228 i.e. fearing innovations. As a lawyer and politician, Bacon deprecated innovation.

p. 229 See I Kings 4:33. 'Moss' is Bacon's word; the Vulgate and translations have 'hyssop'.

p. 230 The foundation of such an institution was Bacon's hope. The Royal Society, founded in 1662, in part fulfilled his dream. Note that the Lawgiver and the founder of the House of Salomona, are one. *Six Days' Works*: The arrangement of human knowledge according to the pattern set in the Biblical account of the six days of creation was common in medieval and Renaissance encyclopedists. See V. K. Whitaker, *Francis Bacon's Intellectual Milieu* (Clark Memorial Library, 1962).

p. 230 i.e. knowledge, in whatever parts of the world it is to be found.

p. 231 Bacon enjoys inventing ceremonial. He is here replying to Plato's argument in the *Republic* that the family unit should be abolished. For the Christian, the family is the natural unit. See also *Works* i. 798 n. on the Pragmatic Sanction promulgated in Spain in 1622 giving special privileges to those who had six children (cf. Bacon's essay 'Of the true greatness of Kingdoms and Estates'.)

p. 234 Bacon emphasizes the Hebraic nature of Bensalem. The traditional European view of Jews, which Bacon goes on to give, is that found in Marlowe's *Jew of Malta* and Shakespeare's *Merchant of Venice* (see ed. J. R. Brown, 1955, pp. xxxi, xxvii). The narrator, as a stranger, naturally makes contact in Bensalem with the alien, though tolerated, Jew.

p. 234 To the Jews the Milky Way was a river flowing through the heavens proceeding from the throne of God. *Eliah*: The Jews expected Elijah to return to herald the Messiah (Malachi 4:5). Christ thought of John the Baptist as the Elijah (Mark 9:2–13).

p. 237 In More's *Utopia*, Book 2, and in Plato, *Laws* 771 e ff.

p. 239 Deep caves were thought useful for preserving vegetable matter, which decayed in air, for refrigeration, and for producing metals (because these were originally created underground). Bacon believed, with Telesius, that 'heat and cold are Nature's two hands, whereby she chiefly worketh' (see *Sylva Sylvarum* 68–75; *Novum Organum* ii. 50; 'De Calore et Frigore' in *Works* iii. 644–52).

p. 239 See *Works* v. 283 (The 'History of Life and Death') for Bacon's belief that exclusion of the air delayed the desiccation of bodies and, by preventing perspiration, retained the 'spirits'. He recommends life in caves or on mountains.

p. 240 This is the first experiment in the *Sylva Sylvarum.*

p. 240 For Bacon's experiments with windmills, see *Works* v. 185–7.

p. 243 Heat and the controlled variation of heat was regarded as vital for all transmutations, and so of particular importance in alchemy. See John Read, *Prelude to Chemistry* (1936), pp. 143 f. *Dry* heat is ordinary fire, *moist* heat that found in horse-dung (see *Works* iv. 240). Bacon's conjectures on the nature of fire in the firmament are in his *Thema Coeli* (*Works* v. 547 ff.). Here he suggests that it would be possible to make heats corresponding to the various heats in the moon, comets, planets, and sun, and reproduce the movements of such heats in circles, straight lines, and bends, similar to the movements of the fiery bodies in the heavens. *Novum Organum*, Book ii, is devoted to Bacon's investigation of the Form of Heat.

p. 243 See *Works* v. 409 ff. for Bacon's tract on 'Light and Luminous Matter', and *Novum Organum* ii. 39 for his account of microscopes and telescopes.

p. 244 The material of Bacon's *Historia Soni et Auditus* (*Works* iii. 653 ff.) was taken over into the *Sylva Sylvarum.*

p. 245 A kind of cannon. Bacon chose Motion as the first subject to which his new method was to be applied. See *Works* iii. 621 f., v. 433 f.; *Novum Organum* ii. 36, 48.

p. 245 Cornelis Drebbel (1572–1633) exhibited a submarine in the Thames in 1620. See R. L. Colie in *Huntington Library Quarterly*, 18 (1955), 245–60. Drebbel's laboratory at Eltham was equipped for experiments in refrigeration, the production of artificial rain, baking and brewing, dyeing, various kinds of heat, etc., and Drebbel was well known for his lenses, magic lantern, instruments of motion and music, fountains, etc. Leibniz reinvented the submarine (letter to H. J. Friedrich, October 1671).

p. 245 Bacon was no mathematician, and, unlike Galileo, did not believe that the basic structure of reality was mathematical. Descartes, who was a great mathematician, was often visited by learned men who asked to see his instruments. John Aubrey comments, 'in those days mathematical learning lay much in the knowledge of instruments'.

p. 249 The wonderful works of Nature, chiefly such as benefit mankind. The phrase is often used by Paracelsus. Bacon wrote (*Works* x. 87) in 1603 that 'the knowledge that we now possess will not teach a man even what to *wish*'. This is his list of 'wishes', printed at the end of the *New Atlantis* in 1625; they relate to the experiments in the *Sylva Sylvarum*, but indicate some of the things the Fathers of Solomon would have discovered. Many of the ideas are to be found in Agrippa and della Porta.

Index